MECHANISMS AND GENETICS OF NEURODEVELOPMENTAL COGNITIVE DISORDERS

MECHANISMS AND GENETICS OF NEURODEVELOPMENTAL COGNITIVE DISORDERS

Moyra Smith

Academic Press is an imprint of Elsevier
125 London Wall, London EC2Y 5AS, United Kingdom
525 B Street, Suite 1650, San Diego, CA 92101, United States
50 Hampshire Street, 5th Floor, Cambridge, MA 02139, United States
The Boulevard, Langford Lane, Kidlington, Oxford OX5 1GB, United Kingdom

Library of Congress Cataloging-in-Publication Data
A catalog record for this book is available from the Library of Congress

British Library Cataloguing-in-Publication Data
A catalogue record for this book is available from the British Library

ISBN: 978-0-12-821913-3

For information on all Academic Press publications visit our website at https://www.elsevier.com/books-and-journals

Publisher: Nikki Levy
Acquisitions Editor: Natalie Farra
Editorial Project Manager: Ruby Smith
Production Project Manager: Omer Mukthar
Cover Designer: Mark Rogers

Typeset by TNQ Technologies

Dedication

This work is dedicated to Dr Simon Prinsloo. Thank you, Simon, for standing behind me at critical times many years ago and for your encouragement in recent years.

Contents

5. Gene expression, regulation, and epigenetics in brain

6. Neuroimmunology

7. Neurodevelopmental, neurocognitive, and behavioral disorders

8. Epilepsy and movement disorders

9. Health and well-being

10. Brain and mind

Acknowledgments

I wish to express deep gratitude to those who have taught and inspired me over the years. They include teachers, colleagues, students, and patients. I also wish to express gratitude to researchers and authors who make new knowledge available and to those who support their activities.

I express sincere thanks to those at Elsevier (ELSA), especially to Ruby Smith and Robin James Sulit for their expert assistance with this publication.

CHAPTER

1

Brain, early development cortices, architecture, cell types, connectivity, networks

1. Early brain development

Stiles and Jernigan (2010) noted that a major step in brain development includes development of neural tube by day 27 of embryonic life (E27), and the neural tube is lined with the neural plate from which neuroprogenitor cells originate. The hollow neural tube eventually forms the ventricles of the brain. Neuroprogenitor cells in the frontal area of the neural tube (the rostral region) give rise to the brain. Neuroprogenitor cells in the caudal region of the neural tune give rise to the hindbrain and spinal cord. The hindbrain eventually develops into the medulla, pons, and cerebellum.

Stiles and Jernigan documented the following neural tube–derived regions identified by day 49 of embryonic life in humans (E49) (Table 1.1):

- Diencephalon mesencephalon
  - Cephalic flexure
    - Metencephalon
    - Myelencephalon
    - Cervical flexure
    - Spinal cord

The embryonic period extends from conception to the eighth week of gestation. Stiles and Jernigan noted that by the end of the embryonic period, rudimentary central and peripheral structures were in place.

Neural tube defects in humans have been associated with folate deficiency (Fig. 1.1).

Primary progenitor cells of neurons, migrations, and radial glial cells

Penisson et al. (2019) noted that studies of stages and processes in early neurodevelopment are important given that developmental malformations in the cortex are associated with cognitive impairment, epilepsy, and other disorders.

Primary progenitor cells of neurons are derived from the epithelial cells of the neural tube. Neuroepithelial cells can undergo symmetric and asymmetric divisions. Neuroepithelial cells also give rise to polarized radial glial cells that extend processes toward the ventricular and pial brain surfaces. Radial glial cells can undergo divisions to give rise to different cells including basal radial glial cells with cell bodies in the inner and outer subventricular can divide to give rise to immature neurons or intermediate progenitors.

Mechanisms and Genetics of Neurodevelopmental Cognitive Disorders
https://doi.org/10.1016/B978-0-12-821913-3.00013-5

TABLE 1.1 Regions visible by 49 days postconception.

Regions
Telencephalic vesicle
Diencephalon mesencephalon
Cephalic flexure
Metencephalon
Myelencephalon
Cervical flexure
Spinal Cord

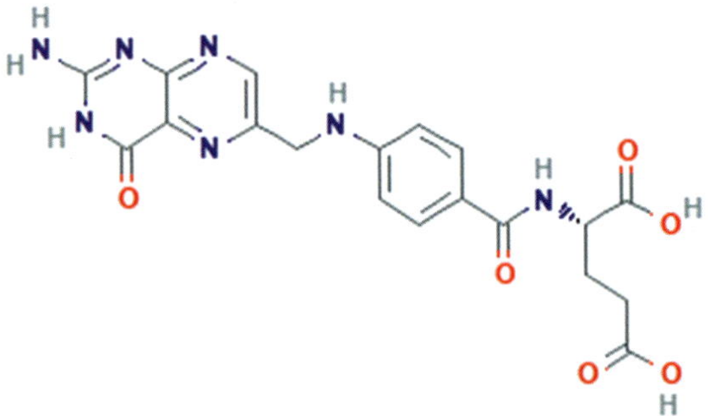

FIGURE 1.1 **Structure of folic acid.** Folic acid deficiency can lead to neural tube defects in humans. *From https://pubchem.ncbi.nlm.nih.gov/compound/Folic-acid, (Li et al., 2018); (Penisson et al., 2019).*

Penisson et al. noted that radial glial cells are key to brain development in primates and species defined as gyrencephalic, where brains have folds, gyri. These features distinguish primates from smooth brained lissencephalic species. In gyrencephalic species, bodies of radial glial cells are in subventricular regions. Some radial glial cells located apically have processes that project into the ventricles.

In recent decades, detailed transcriptomic studies have been carried out on cells that constitute the brain from embryonic life on. Basal glial cells were reported to undergo multiple cell division. Penisson et al. documented expression of key signaling pathways in radial glial cells. These included FGF (fibroblast growth factor), MAPK (mitogen-activated kinase signaling), SHH (sonic hedgehog), PTEN (phosphatase tensin), AKT (serine threonine kinase), and PDGF (platelet-derived growth factor).

Division of neuroepithelial cells therefore leads to generation of radial glial cells and ultimately leads to formation of precursor cells of the cortex within several regions, the ventricular zone, the subventricular zone, subplate, cortical plate, and marginal zone. The basal glial cells are highly neurogenic, and this neurogenesis is key to production of gyri.

Brain development in the fetal period

The fetal period is defined as the second trimester beginning at approximately day 63. Stiles and Jernigan noted that key changes during this period involved transformation of the smooth brain (lissencephaly) to a brain with folds and grooves (gyri and sulci).

Neuronal migration

Neurons migrate from the ventricular region, and migration occurs initially using radial glial cells as guides. After the first migration, neurons migrate in stages to their final positions where they then develop dendrites and axon.

Earliest-born neurons ultimately occur in the deeper cortical layers; later-born neurons are more superficial and closer to the pia. Cajal—Retzius cell are more superficial and are close to the pia. Cajal—Retzius cells in the marginal zone are thought to play important roles in organizing the cortical layers. Cortical layers in the early brain from exterior to interior include the following:

- Pial surface
- Marginal zone
- Cortical plate
- Intermediate zone
- Outer subventricular zone
- Inner subventricular zone
- Ventricular zone
- Ventricles

Genomic, epigenomic, and transcriptome studies of the developing brain

Unique resources have been developed for studies of the brain. The BrainSpan consortium and the Allen Institute (www.brainspan.org) have constructed an atlas of the developing human brain. Cell diversity in different brain regions has been defined. The developmental transcriptome has been defined through RNA sequencing.

Li et al. (2018) carried out a genomic, epigenomic, and transcriptome study at the tissue and cellular levels on a series of samples derived from 1230 human brains at different life stages, embryonic, fetal, infancy, childhood, adolescence, and adult, from 5 weeks postconception to 64 years. Genomic studies were carried out and revealed no genomic abnormalities. Transcriptomic analyses using only cellular mRNA were carried out on the earliest embryonic samples, 5 and 6 weeks postconception. Both cellular and tissue mRNA analyses were carried out on 607 brains. Anatomical regions analyzed included neocortex, cerebellar cortex, hippocampus, mesodorsal thalamus nucleus, amygdala, and striatum. These areas were selected since they were considered regions primarily involved in cognition and behavior.

Epigenomic studies included analyses of cytosine methylation using methylation bead chips, chromatin immunoprecipitation associated with DNA sequencing, and histone analyses including histone methylation analysis and histone acetylation analysis.

Regulation analyses included identification of CTCF-binding sites, DNA analyses of promoters, enhancers, and insulators.

The transcriptomic analyses revealed differences in levels of expression with clear differences in expression between embryonic and midfetal stages and in stages from late infancy onward. Importantly the authors determined that there were major alterations of expression with a transition beginning just before birth. In the period starting in late fetal life and extending beyond that, they observed decreased differences in levels of gene expression. They attributed this to the fact that in the regions analyzed, there was an increase in transcription from mature neurons and increased expression of genes involved in dendrite formation and synapse development.

Methylation was shown to differ in different regions in postnatal period. Relationships were also established between methylation signatures and enhancer activity. Enhancers active in the fetal period were reported to undergo methylation in the postnatal stages.

Li et al. identified enhancers of expression of the following genes that were also noted to have identified with defects in specific neurodevelopmental disorders.

- MEF2C transcription enhancer factor 2; MEF2C defects occur in nonsyndromic intellectual disability.
- SATB2 SATB homeobox 2, DNA-binding protein involved in transcription regulation chromatin remodeling.
- TCF4 transcription factor 4, defective in Pitt–Hopkins syndrome, intellectual disability, dysmorphology.
- TSHZ3 zinc finger homeobox domain regulates neocortical organization and circuitry.
- Sex differential expression was documented in 783 genes (Li et al., 2018).

Key gene products in specific neuronal cell types

The BrainSpan project documented key genes that were used to distinguish the different neuronal cell types that occurred in different brain regions and the products produced by key genes. These are listed in the following.

- GAD1 glutamate decarboxylase 1
- ADARB2 member of the double-stranded RNA adenosine deaminase family of RNA-editing enzymes
- LAMP5 lysosomal associated membrane protein family member 5
- VIP vasoactive intestinal peptide
- SST somatostatin binds to high-affinity G protein—coupled somatostatin receptors
- SLC17A7 solute carrier family 17 member 7 specifically expressed in neuron-rich regions of the brain
- CUX2 cut like homeobox 2, has DNA binding domain
- RORB RAR-related orphan receptor B, involved in organogenesis and differentiation
- PVALB parvalbumin high-affinity calcium ion-binding protein

Zeng and Sanes (2017) emphasized the great numbers and diversity of neurons and synapses. They also noted that it was important to note that neurons form groups.

In establishing their cell classification, Zeng and Sanes set out to define three characteristics, morphology, physiology, and function. Morphology included cell shape, connections, and branching patterns. They illustrated 5 types of neurons based on shape and patterns of connection. Interestingly the 5 types of cells could also be distinguished with specific protein and enzyme markers.

Cell type and specific protein or enzyme markers

- Sparse neurogliaform cells, HTR3A+ 5-hydroxytryptamine receptor
- 3A bipolar cells, VIP + vasoactive intestinal peptide
- Martinotti cells, SST + somatostatin
- Basket cells, PVALB + parvalbumin
- Thick tufted cells, RBP + retinol binding protein 4

Molecular signatures of the cells were defined based on transcription and single cell mRNA sequencing and intensity of expression of the following:

- SNAP25 synaptosome—associated protein 25
- GAD1 glutamate decarboxylase 1
- VIP vasoactive intestinal peptide
- SST somatostatin
- PVALB parvalbumin
- SLC17A7 solute carrier family 17 member 7

Martinotti cells are described as small multipolar cells with short branch dendrites. They are sometimes described as being somatostatin and calbindin positive. Bipolar cells are cells with two extensions one axon and one dendrite, which are reported to be predominant in the sensory system. Pyramidal cells have a triangular cell body, thick axon, and thick dendrite; they are characteristics of gray matter. Granular cells have very small cell bodies and branch out into dendritic arbors. Cajal—Retzius cells also known as horizontal cells have a tangential arrangement.

In the brain, four types of glial cells occur: astrocytes, oligodendrocytes, microglia, and oligodendrocyte precursor cells.

Zeng and Sanes (2017) emphasized that even within a particular cell type, there was heterogeneity in the level of expression, in part related to developmental stage. They noted that further information on morphology is being obtained from enhanced light microscopic imaging and electron microscopy.

As an example of a specific study, they provided details on classification of cerebral cortex neurons. In a broad classification, they separated glutamatergic excitatory neurons from GABAergic inhibitory neurons. They defined classes of neurons based on their projections: the 5 classes were defined as projections into layers.

Subclasses of inhibitory neurons were separated on the bases of production of the following gene products:

- PVALB + parvalbumin
- SST + somatostatin
- VIP + vasoactive intestinal protein
- HTR3+ VIP-5-hydroxytryptamine receptor 3A + vasoactive intestinal protein

2. Neuronal cell polarity microtubules centrosome and cytoskeleton

Kuijpers and Hoogenraad (2011) noted that the centrosome is the main cytoskeleton organizing center and the site of microtubule nucleation. They emphasized the importance of the centrosome that exists at the base of the microtubule array and noted that the centrosome is also involved in cell division. They noted that the centrosome is composed of cylindrical centrioles surrounded by pericentriolar material and that the microtubule growing ends are embedded in pericentriolar material. They emphasized that defects in centrosomal proteins predisposed to neurological deficits.

Cilia

Guo et al. (2015) noted that in the brain, progenitor cells and neurons have primary cilia and that defects in cilia functions are known to lead to brain abnormalities referred to as ciliopathies. They defined cilia as microtubule structures involved in integrating signal transduction. Cilia were reported to play roles in neuronal migration and in determining organization within the cortical layers. Ciliary proteins were also found to play roles in postmigratory differentiation of neurons including projection of axons and dendrites.

Cilia also have functions on many other tissues and organs.

Van Dam et al. (2019) reported establishment of Cilia Carta a compendium of ciliary genes. They noted that it was likely that many cilia genes had not yet been identified.

Primary cilia

Youn and Han (2018) noted that the primary cilia play key roles in signaling pathways and in the cell cycle. Within the brain, they were noted to be important in neurogenesis and in neuronal maturation.

They noted that the core of the primary cilium is composed of nine microtubule doublets that form a ring and that secondary cilia in addition to nine central doublets have an additional two doublets. The distal appendage that extends from the central structure referred to as an axoneme. The cilia attach to the cell membrane via a basal body that includes proteins centriolar proteins CEP164, CEP 170. Motor protein in cilia dyneins and kinesins hydrolyze ATP and allow movement along cilia. Cilia and microtubules extend from the basal body that contains the microtubule organizing center. Cilia extend outward from the basal boy in microtubules extend inward.

Wheway et al. (2018) reported that PTCH, the receptor for the sonic hedgehog gene (SHH), is located on ciliary membranes. Binding of SHH to PTCH releases PTCH from repressing SMO. SMO is the release, and this releases repression of GLI1 transcription factor that is then transported to the nucleus to activate gene expression. They described primary cilia as organelles that protrude from cellular surfaces particularly epithelial cells.

Ciliary gene defects leading to neurobehavioral or neurodevelopmental defects will be discussed further in a later chapter.

Microtubules

Microtubules composed of alpha and beta tubulin are responsible for transport within cells. The tau protein that forms microtubules is also designated as MAPT (microtubule-associated protein tau). Microtubules are also rich in transport proteins dyneins and kinesins (Hakanen et al., 2019; Wheway et al., 2018).

The centrosome located on the membrane forms the microtubule organizing center. Immediately prior to cell division, microtubules extend and attach to the centromeres of chromosomes as the cells prepare for separation of chromatids and cell division. Microtubules thus play important roles in cell division. Specific microtubular abnormalities that negatively impact cell division therefore represent causes of inadequate cell divisions that impact brain development. Microtubule defects leading to microcephaly will be discussed in a subsequent section on neurodevelopmental defects. Macrocephaly and related gene defects will be discussed further in the neurodevelopmental defects section.

Hakanen et al. (2019) reviewed cell polarity, cortical development, and malformations. They noted that asymmetric distribution occurs in cells that includes organelles, cytoskeletal elements, and signaling molecules. This asymmetric distribution impacts polarity. Polarity also influences cell shape. Apical basal polarity orientation is referred to as planar cell polarity.

They noted that neural progenitor cells and neurons were highly polarized particularly during development. This polarity impacted migration of neurons and directions of outgrowths from neurons including axons.

The planar cell polarity pathway (PCP) is critical for development and proteins particularly important in this pathway include FAT1 and FAT2 atypical cadherin proteins and DCHS (dachsous cadherin-related proteins). Other proteins involved in this pathway are CELSR1-3 (cadherin EGf receptors), VANGL1 (tetraspanin-related protein), PRICKLE (negative regulator of WNTcatenin signaling), and Frizzled FZD1 (form receptors for WNT). WNT pathway proteins, which are secreted signaling proteins, were reported to promote assembly of planar polarity proteins at particular positions.

The importance of the PCP pathway in neurogenesis has been illustrated by finding specific defects in proteins in this pathway in specific neural tube and brain defects. This will be described in the section on brain development.

Sonic hedgehog pathway and interactions with ciliary pathway genes and planar cell polarity genes

Evidence for optimal function of products of genes in these three pathways has been derived from demonstration of their importance in ensuring proper development of the brain and midline regions of the face. Deleterious mutations in genes in these pathways can lead to holoprosencephaly (Kim et al., 2019). This condition will be described in a subsequent section.

Nano and Basto (2017) emphasized the key roles of neuronal cell proliferation and cell polarity in brain development. They noted the importance of chromosome segregation spindle orientation. The considered the centrosome to be key in these processes. The centrosome constitutes the microtubule organizing center of cells. Centrosome dysfunction due to specific gene mutations has been found to be associated with a number of different disorders associated with abnormal brain size and structure.

Lasser et al. (2018) reviewed the functions of microtubules in the neuronal cytoskeleton and the role of microtubules in promoting neuronal migrations and in establishing connections. In the early formed neuronal cells, microtubules are essential for establishing polarity, and they impact migrations. They also noted growing evidence of impaired microtubule functions in specific neurodevelopmental disorders. These included disorders associated with impaired cortical formation including lissencephaly and polymicrogyria.

Lasser et al. described microtubule structures that include alpha and beta tubulin proteins that form heterodimers. These tubulins can undergo polymerization or depolymerization, and guanine nucleotides including GTP and GDP are essential to these processes.

Centrioles, centrosomes, and microtubules

Nigg and Raff (2009) noted that centrioles are structures essential for formation of centrosomes, cilia, and flagella. They defined the centriole as a complex microtubule-based structure and noted that centrosomes give rise to cilia. Centrioles also give rise to the spindle microtubules that are essential for cell division in many cell types. Throughout these processes, microtubules and microtubule motors kinesin and dynein act to transport essential products and cargo along microtubules. Microtubule-related structures and the centrosome played essential roles in neuronal cell division. Microtubule remodeling was noted to occur throughout neuronal morphogenesis. Lasser et al. noted that microtubule-associated products were essential for these processes. These products included MAP1A and MAP1B and tau. Tau protein was shown to increase polymerization and to regulate microtubule organization in later stages. Tau was reported to influence radial migration of neurons. Romaniello et al. (2018) noted the importance of tubulin gene expression in postmitotic neurons during migrations and differentiation.

3. Interneurons and their embryonic origins

Cortical interneurons are defined as neurons that contribute to neural networks. They are sometimes defined as relay neurons. Wamsley and Fishell reported that some cortical interneurons emerged from the pallium within the ventral telencephalon, and others emerged primarily in from sites that line the medial lateral ganglionic eminence and the caudal ganglionic eminence. Emergence of cortical interneurons was reported to be influenced by specific morphogens. They reported that when the interneurons became postmitotic, they undergo a long period of migrations when they invade the cortex. There they mature and establish synaptic contacts.

Wamsley and Fishell (2017) emphasized that electrical and chemical activity can be observed in progenitor cells and early neurons and that stimuli impact neuronal cell proliferation, their migration, and axon guidance. Spontaneous network activity was also recorded in early development and was thought to be necessary for development. When the migrating interneurons achieved their destination, they appeared to undergo changes in activity, and in mouse models, changes in electrical potential particularly in interneurons were particularly marked in the early postnatal stages and neurotransmitter release occurred. They also noted that there was evidence for cell death and death of cortical neurons and a decrease in GABAergic cells in postnatal life. They noted that stimulations of activity impact both excitatory neurons and interneurons and that stimuli triggered gene expression. Stimulated excitatory cells were reported to produce the neurotropic factor BDNF. BDNF also increased interaction of inhibitory synapses with excitatory synapses.

Meganathan et al. (2017) reported that the transcription factor NKX2-1 is important in differentiation of cortical interneurons and that this transcription factor interaction with chromodomain helicase 2 (CHD2) in modifying chromatin to ensure gene transcription activity necessary for cortical neuron development. It is interesting to note that deletion of CHD2 and mutations that impair CHD2 activity have been reported in specific neurodevelopmental disorders.

Lim et al. (2018) reported that the unfolding of specific transcriptional programs led to development and differentiation of different types of cortical interneurons, which were reported to differ not only in their morphologies but also in their physiology and synaptic connections.

4. Astrocytes

Gene expression in astrocytes

Vasile et al. (2017) reviewed gene expressed in astroglia and astrocytes; they included the following:

- GFAP glial fibrillary acidic protein, major intermediate filament proteins of mature astrocytes.
- ALDH1L1 aldehyde dehydrogenase 1 family member L1, important in NADH synthesis
- GLUL glutamate-ammonia ligase catalyzes the synthesis of glutamine
- AQP4 aquaporin 4, water-selective channels in the plasma membranes
- SLC1A2 solute carrier family 1 member 2, transporter protein
- SLC1A3 solute carrier family 1 member, transporter protein
- GJB6 gap junction protein beta 6

Other gene products subsequently found to be abundant in human astrocytes that have calcium signaling–related properties:

- RYR3 ryanodine receptor 3, functions to release calcium from intracellular storage
- MRY11 (IRAG) murine retrovirus integration site 1 homolog in endoplasmic reticulum
- RGN regucalcin, highly conserved, calcium-binding protein

Metabolism-related genes abundantly expressed in cortical astrocytes:

- APOC2 apolipoprotein C2 activates the enzyme lipoprotein lipase, which hydrolyzes triglycerides.
- AMY2B amylase alpha 2B hydrolyzes 1,4-alpha-glucoside bonds in oligosaccharides and polysaccharides.
- Astrocyte precursor cells have also been studied and were shown to express gene products that impacted cell proliferation and cell cycle function. In addition, they were found to express Gap junction proteins GJA1 and GJB1.

Astrocyte functions

Vasile et al. (2017) noted that data from several different studies provided evidence that astrocyte functions promote neuronal survival. Astrocyte secretion of the growth factor TGFB1 was reported to promote formation of synapses. Other proteins produced by astrocytes that can potentially promote neuron survival include secreted thrombospondins THBS2 and THBS4.

Astrocytes manifest gap junctions that facilitate exchange of ions, metabolites, and neuromodulators. Gap junction connections of astrocytes with processes from other cell types could potentially enhance cell survival.

Vasile et al. noted evidence that astrocytes produce specific transporter proteins that enable their uptake and recycling of neurotransmitter molecules. SLC1A3 (EAAT1) and SLC1A2 EAAT2 are members of a high-affinity glutamate transporter family. They are reported to function in the termination of excitatory neurotransmission in central nervous system (CNS).

Astrocytes were also shown to manifest calcium signaling.

Astrocytes in cognitive processes

Santello et al. (2019) reviewed possible role of astrocytes in cognitive processes. They emphasized the intricate branching of astrocytes and their interaction with neuronal synapses, blood vessels, and glial cells and the presence of gap junctions in facilitating these connections.

In recent decades, additional information has emerged regarding astrocyte function, including some evidence that astrocytes manifested calcium signaling and transmitted specific factors in response to signaling. Santello et al. discussed potential roles of astrocytes in synaptic plasticity.

Specific perisynaptic processes (PAPs) were found to express glutamate transporters that remove glutamate from the synaptic environment. PAPs were reported to be particularly active in cerebellar astrocytes and were shown to be active to a lesser degree in the CA1 region of the hippocampus.

Santello et al. reported that PAPs of astrocytes play active roles in synaptic plasticity. Serine is one specific substance released from astrocytes that impacts synaptic plasticity. Astrocytes were also reported to release L-lactate that is thought to be important in CA1 hippocampal synapses. L-lactate is derived from metabolism of glycogen in astrocytes.

Cannabinoid receptors are expressed on neurons and are postulated to play roles in long-term potentiation and long-term depression. Santello et al. noted that cannabinoid receptors are also expressed at lower levels on astrocytes.

5. Neurotrophic factors in development and survival of neurons

Skaper et al. (2018) reviewed neurotrophic factors and noted that they play roles in development and survival of neurons in the central and peripheral nervous systems. Four neurotrophins present in humans were reported to function through specific neurotrophic receptor kinases NTRK1, NTRK2, and NTRK3. Neurotrophins were also reported to activate a specific receptor, sometimes referred to as p75 and also designated NGFR nerve growth factor receptor.

The prototype neurotrophic factor is nerve growth factor (NGF). It is synthesized by neurons in the CNS. However, in the periphery, it is synthesized by other cell types. The second neurotrophic factor described (Skaper, 2018) was brain-derived neurotrophic factor (BDNF). Other important neurotrophic factors in humans include NTF3 and NTF4.

Skaper reported that specific polypeptide factors also have neurotrophic activity. These include the following:

- CNTF: ciliary neurotrophic factor, primarily active in brain
- GDNF: glial cell–derived neurotrophic factor
- IGF1, IGF2: active in many tissues, minimally expressed in brain
- FGF1, FGF2: fibroblast growth factor expressed in many tissues including brain
- TGFB1: transforming growth factor B1, expressed in many tissues including brain

The neurotrophins NGF, NT3, and NT4 and BDNF were reported to all interact with the neurotrophic receptor tyrosine kinases. Skaper noted that in addition to promoting neuronal cell growth and survival during development, neurotrophic factors were also reported to be important in promoting survival during postnatal life and were active in promoting neuronal survival following injury.

6. Microglia in brain health and disease

Salter and Stevens (2017) emphasized the diverse roles of microglia in brain health and also in brain disorders. Particularly important is the role that microglia play in sculpting neuronal circuits and facilitating neuroplasticity. Microglia were reported to constitute 10% of brain cells and to be highly branched cells with multiple mobile processes.

Salter and Stevens noted that microglia carry out surveillance of the brain parenchyma and that they actively participate in brain function. They also noted that advances in information on microglial function have come mainly from studies on animal models and particularly from studies on mice.

Microglia were found to originate from embryonic yolk sac progenitors, and key transcription factors involve in their generation included RUNX1 and CKIT (CD117). Microglia could be distinguished from macrophages on the basis of lineage-specific genes IRF8 (interferon regulatory factor 8) and PU.1 (transcription factor, also known as SPI1).

Salter and Stevens noted that several subpopulations of microglia developed in the CNS. Activation of a specific receptor CSF1R (colony-stimulating factor 1 receptor) was reported to be important in microglia development.

Microglia were reported to impact CNS development from midembryonic stage. Key functional roles of the microglia in the CNS included surveillance, promotion of neuronal plasticity synaptic pruning, programmed cell death, and phagocytosis. Activities of microglia were shown to be influenced by neuronal activity and neurotransmitters. Microglial function was dependent on ATP levels.

7. Brain growth

Pirozzi et al. (2018) reviewed processes involved in cortical expansion and brain growth. They noted that these include neural stem cell proliferation, neuronal migrations and organization, synaptogenesis, and apoptosis. They noted further that alterations in any of these steps could lead to changes in brain size, undergrowth leads to microcephaly, and overgrowth leads to macrocephaly. They noted that defects in genes involved in cell cycle, centrosome formation, spindle orientation, microtubule organization, and cytokinesis were associated with decreased brain volume and microcephaly.

Megalencephaly was associated with defects in the PI3K (phosphatidyl inositol kinase) and MTOR pathways involved in growth and proliferation and also with defects in the RAS MAP signaling pathway.

Postnatal brain development

Stiles and Jernigan (2010) noted that during the postnatal period, glial progenitor cells continue to be generated in the forebrain ventricular region and to migrate outward from there. They also reported that glial—neuronal interactions play roles in neural circuit organization. Neurogenesis continues to a limited degree postnatally.

Stiles and Jernigan noted that continued organization of the neocortex continues postnatally in response to input from experience, molecular signaling, and cross-regional activity.

Van Praag et al. (2000) emphasized the importance of enriched environment in promoted brain development and neuronal plasticity.

Studies by Greenough and colleagues (1987) revealed the important role of experience in brain development. They demonstrated that enriched environment promoted development and that this related in part to selection of some synapses for pruning and maintenance of synapse and connections that were activated.

Oligodendrocyte precursor cells

Birey et al. (2017) reported that oligodendrocyte precursor cells are a specific glial cell type and that they occur in gray and in white matter. Oligodendrocyte precursor cells can be distinguished by expression of a specific proteoglycan chondroitin sulfate proteoglycan 4 (CSPG4) (NG2). They noted that in recent decades new roles for oligodendrocyte precursor cells (OPCs) and for oligodendrocytes have been defined.

There is evidence that myelination processes induced by oligodendrocytes are activated by environmental stimulation learning and neuronal activity. Specifically, axonal myelination and myelin stability were reported to be responsive to neuron firing and electrical activity. There is evidence that electrical activity increases oligodendrocyte numbers and myelin thickness. White matter was documented to change in response to complex skill learning.

Key molecular factors involved in increasing myelination include myelin regulatory protein

MYRF and transcription factor SOX10. Another important myelination stimulating is the enzyme ectonucleotide pyrophosphatase phosphodiesterase (ENPP1), reported to play a role in lipid metabolism.

Oligodendrocyte plasticity leading to increased myelination was shown to occur in response to increased motor activity and increased sensory stimulation. Birey et al. noted that studies have also been carried out to determine if oligodendrocytes impact neuronal activity through direct interaction with neurons. There are reports that CSPG4 (NG2) glia secreted molecules into the extracellular matrix and that these molecules stabilize synapses. In addition, the CSPG4 glia were shown to produce neuromodulators including prostaglandin D2 synthase, Pentraxin 2, PTGD2, and prostaglandin D2 receptor 2 (PTGDR2).

PTGD2 (prostaglandin D2 synthase) functions as a neuromodulator and as a trophic factor in the CNS. Bargmann (2012) included in the neuromodulator category a range of molecules that act as signals to reconfigure neural circuits particularly in response to environmental and metabolic conditions.

Development of brain white matter

Dubois et al. (2014) reported results of studies in white matter development and maturation of connections through myelination. They defined elements of white matter as glial cell astrocytes oligodendrocytes, microglia, and fibers that connect different functional brain regions. These include commissural fibers that connect the two cerebral hemispheres and pass mostly through the corpus callosum and bidirectional projection fibers. Projection fibers were reported to pass through the thalamus, cortex, brain stem and spinal cord, and optic radiation. Fibers classified as association fibers also formed connections within hemispheres.

In considering patterns of fiber growth, Dubois et al. noted that following migration of a neuron to its ultimate position, it extends dendrites into gray matter and axons into white matter. Another important observation was that migration initially begins in the subplate beneath the cortex.

They also noted that the establishment of connections occurred primarily during the second trimester. In addition, there is evidence that overproduction is followed by pruning. Huttenlocher (1984) initially described the important synaptic pruning processes that occurred late in gestation and also in early postnatal life.

Myelination of axon fibers was reported to occur after axon pruning. Myelination is dependent on proliferation of oligodendrocytes and myelination of axons. The oligodendrocytes produce myelin that ensheaths the axons.

Dubois et al. noted that there are four stages of oligodendrocyte maturation. Immature oligodendrocytes are reported to be rich in galactocerebrosides; subsequently, levels of cholesterol and glycolipids increase in oligodendrocytes.

Myelination is reported to continue from the second half of pregnancy until late adolescence. Myelination occurs in different brain sites at different times and was reported to occur earlier in somatosensory pathways than in motor pathways. There is evidence that glial factors and electrical activity impact myelination. Myelination has been shown to dramatically increase nerve impulse conduction speed.

Myelin is reported to be composed of 40% water with lipids and proteins being the other main components. Specific proteins present in myelin include myelin basic protein (MBP), myelin oligodendrocyte myelin protein (OMG), and proteolipid protein 1 (PLP1).

Synaptic pruning

Synapses are key to neural processes (Fig. 1.2). However, there is evidence that synaptic pruning plays a key role in promotion of optimal brain function.

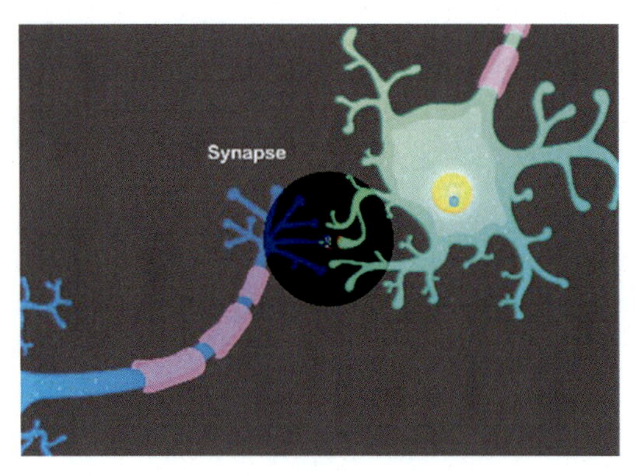

FIGURE 1.2 **Axon terminals, synapse, and release of neurotransmitters.** *From National Institute of Mental Health, n.d. National Institutes of Health, Department of Health and Human Services. https://images.nimh.nih.gov/LibraryImages/Medium/synapse.jpg.*

Schafer et al. (2012) reported clear evidence of participation of microglia in synaptic pruning. Synaptic pruning also involves participation of the complement cascade. Complement components were shown to be expressed in brain. Microglia were reported to be the only cells in the CNS that express the complement receptor C3R. Mice defective in complement components or complement receptor were shown to have defects in synaptic pruning. One role of microglia is to engulf and remove damaged and excess neurons. Schafer et al. noted that microglia may also act to promote neuronal cell death through the product of neurotoxin.

Studies on synaptic pruning

In a 2017 review, Neniskyte and Gross noted evidence for key role of synaptic pruning in nervous system development and increasing information on signaling pathways between glia and neurons. They noted that there is some evidence that synapses that are not activated are removed and the most active synapses are strengthened.

Many studies on synaptic pruning have been carried out in mice and in Drosophila. Specific proteins noted by Neniskyte and Gross to be implicated in synaptic pruning include complement C1q produced by neurons, and complement C3 and complement C3 receptors were shown to be produced by microglia. Astrocyte produced protein involved in synaptic pruning include receptor tyrosine kinase MERTK and MEGF10 (multiple EGF like domains 10) and POE that latter likely acts as a low-density lipoprotein receptor.

Neniskyte and Gross (2017) noted that human disorders postulated to be due to aberrant pruning including autism, schizophrenia, and epilepsy. They also noted that there is some evidence that epigenetic changes may constitute the bases for reported environmental factors that influence pruning.

8. Interactions of brain with peripheral immune system

Earlier dogma was that the brain was isolated from the peripheral immune system by the blood–brain barrier. Nutma et al. (2019) emphasized that there is now evidence for interaction between the immune systems of the CNS and the periphery. Within the nervous system, there is evidence of interaction between immune system cells, the microglia, and neurons.

The concept of separation of the CNS and peripheral immune systems was disproven through studies of the lymphatic system in meninges and vessel designated as glymphatics. Aspeland et al. (2015) and Louveau et al. (2015) described lymphatic vessels in the dura mater (the outermost brain covering), and they noted that immune cells from the CNS were trafficking into these vessels. In 2019, Ahn et al. reported that proteins and other large molecules from the CNS passed through the meningeal lymphatics into the peripheral nervous system.

Mäkinen et al. (2019) noted that there is evidence that the meningeal lymphatic vessels may decline in aging and that this decline may contribute to brain pathology. The growth factor

VEGFR3 was reported to stimulate the growth of meningeal lymphatic vessels.

There is also evidence that substances drained from the meningeal lymphatic system to peripheral lymph nodes can induce an immune response.

Mäkinen et al. (2019) emphasized that discovery of the mechanism by which material can be cleared from the nervous system to peripheral lymph nodes and can induce an immune response may lead to design of therapies.

Cortical layers in the developed brain

Layer I is also sometimes referred to as the molecular layer and has extensions of dendrites of pyramidal neurons and axons. In addition, glial cells are present. During development, Cajal—Retzius cells were present in this layer.

Layer II: In this layer, small pyramidal neurons occur; this layer is sometimes referred to as the external granular layer.

Layer III: This layer is referred to as the external pyramidal layer. It contains pyramidal neurons of different sizes and neurons of other types. In addition, it is rich in axons and connections between other cortical layers.

Layer IV: This layer is defined as the internal granular layer. It has different types of neurons and also contains cortical connecting processes including axons from the thalamus.

Layer V: Designated the internal pyramidal layer, it is the site of large pyramidal neurons including very large cells referred to as Betz cells and is rich in connecting axons. These include connections to the brain stem and spinal cords. Betz cells are primarily involved in motor control.

Layer VI: Referred to as the multiform layer, it is reported to have different types of neurons and processes that extend to subcortical structures. Afferents and efferents in the cortical layers are connected with other cortical regions or with subcortical sites. Connection pathways will be discussed later.

Different types of astrocytes in different cortical layers

Vasile et al. noted that in layer 1 of the cortex, intralaminar astrocytes occur that have processes that radiate tangentially and processes that radiate vertically. These astrocytes have round bodies and were found to be rich in mitochondria. Intralaminar astrocytes were reported to also occur in layers II and III.

Protoplasmic astrocytes were reported to be the most abundant astrocytes in layer II and were noted to have contacts with a large number of synapses.

Varicose projection astrocytes were reported to occur primarily in layers V and VI and were reported to have very long projections.

Fibrous astrocytes were reported to be particularly predominant in white matter in the cortex.

9. Glia—neuron interactions in neuroendocrine system

Clasadonte and Prevot (2018) reviewed glia—neuron interaction in the neuroendocrine hypothalamus. They emphasized the importance of hypothalamus neuroendocrine circuits in adaptive response and in maintenance of homeostasis. Clasadonte and Prevot also reviewed different types of glial cells and their functions. They reported that astrocytes, the most abundant glial cells, can be grouped into several distinct subgroups.

Other forms of glial cells that occur can have somewhat different function depending on their locations in the brain. The authors listed 7 distinct types of glial cells that were relevant to the neuroendocrine system and the functions of these cells. Astrocytes were reported to differ in morphology in different regions of the brain. They play important roles in energy, water, and ion homeostasis and can be activated by calcium. They can sense neurotransmitters and be activated by calcium and can release neurotrophic substances. Astrocytes exist in gray and white matter. Microglia represent components of the immunological system of the CNS. Clasadonte and Prevot reported that they differ in origin from other glial cell as they arise in the yolk sac and migrate to the brain. Microglia play important roles in synaptic pruning and modulation. NG2 glia were reported to be oligodendrocyte progenitor cells.

Clasadonte and Prevot noted that these cells are analogous to stem cells, and they are highly proliferative. Oligodendrocytes produce myelin, the insulation substance of axons, and play important roles in supplying metabolites to neurons. Radial glial cells are particularly important during brain development when their end feet are anchored in the ventricular surface and their elongated processes extend toward the out surface of the cortex and pia mater. Radial glial cell processes serve as guides for migrating neurons. Clasadonte and Prevot noted that after development, specialized cells related to radial glial cells persist in certain brain regions. Tanacytes are sometimes referred to as ependymoglial cells. They line the floor and walls of the third ventricle and extend processes to the hypothalamus and capillaries.

Clasadonte and Prevot emphasized that tanacytes participate in a number of different processes including metabolic neuroendocrine processes and neurogenesis. Pituicytes were reported to be abundant in the posterior pituitary zone and to play role in glia–neuron remodeling and to be involved in control of peptide secretion.

Hypothalamus and hypothalamus–neuro–hypophyseal system

Clasadonte and Prevot noted that the hypothalamus plays an important role in integrating information. The supraoptic nucleus and paraventricular nucleus contain neurons that secrete oxytocin and vasopressin. One population of magnocellular neurons secretes oxytocin, and a second magnocellular neuron population secretes vasopressin.

In a Zorn et al., 2017 review, Zorn et al. noted that stress is a risk factor for psychiatric disorders. They also noted that a number of studies on cortisol production in response to psychosocial stress have been carried out. However, they reported that sex and symptom level and other factors impact interpretation of the data. Another factor influences data related to the timing of cortisol measurements. Corticotropin-releasing hormone (CRH) is released from the hypothalamus paraventricular nucleus. It then promotes release of proopiomelanocortin (POMC) from the anterior pituitary, and CRH is processed to yield adrenocortical hormone and endorphin. Sandman (2018) reported that CRH is also produced by the placenta. The hippocampus is linked to a number of different brain structures.

Eichenbaum (2017) reviewed interactions between the hippocampus and the frontal cortex, and function of this linked system is important in memory especially episodic memory. He noted that one pathway involves direct projections from the hippocampus ventral region to the prefrontal cortex and that there are other indirect linking pathways. Surgical removal of the hippocampus for seizure therapy was reported by Scoville and Milner (1957) to lead to memory loss. However, indirect connections of the hippocampus to the prefrontal cortex could occur through numerous pathways so that some connection between the two cortices remained following surgery. Some of the hippocampus prefrontal cortex connections occur through the thalamus.

10. Specific brain regions and functions

The cerebral cortex comprises two hemispheres that include the following lobes, frontal, temporal parietal occipital, and limbic lobes. The hemispheres are separated from each other by a central sulcus. Specific gyri are identified in each lobe.

Cerebral hemispheres

The lobes of the left and right brain are characterized by gyri and sulci some of which have specific nomenclatures. The left and right frontal lobes extend from the central sulcus (sulcus of Rolando) to the front of the brain on the upper surface; on the lower surface, the frontal cortex ends laterally at lateral sulcus (sulcus of Sylvius) and the frontal cortex ends medially at the cingulate gyrus (Nolte, 1999).

The frontal lobe includes the primary motor cortex, the premotor cortex Broca's area, and prefrontal cortex. Broca's area was originally defined as an area important for speech. The prefrontal cortex was believed to be involved in analysis, insight, and planning. Currently, such functions are thought to be more widely distributed in the brain.

The parietal lobe is separated from the frontal cortex by the central sulcus (Rolando). Posteriorly, the parietal lobe is separated from the occipital lobe by a specific sulcus. Specific sulci separate the parietal lobe from the limbic lobe and the temporal cortex. In addition, specific sulci divide the parietal lobe into specific regions. Within the parietal lobe, the postcentral gyrus was determined to be important in processing tactile and proprioceptive input. The inferior region of the parietal lobe was designated as a region involved in language understanding. Other regions of the parietal lobe were defined and important for spatial orientation.

The temporal lobe has superior middle and inferior gyri. A specific area in the temporal lobe is designated as Wernicke region and was determined to be important in hearing.

Specific structures in the occipital lobe include the lateral gyri, lingual gyri, and parahippocampal gyri. The occipital lobe was defined as particularly important for vision.

The limbic lobe is a strip of cortex that lies between the corpus callosum and the frontal parietal and occipital lobes. The limbic lobe has cingulate and parahippocampal gyri. The limbic lobe is interconnected with the hippocampus to form the limbic systems designated as being important in emotional responses and memory.

The diencephalon

Structures included in this region include the thalamus, hypothalamus, epithalamus, and subthalamus. A region of the thalamus borders on the third ventricle. Information from the periphery travels through the thalamus to reach the cerebral cortices. Posteriorly the hypothalamus is above the optic chiasma. The hypothalamus is reported to process visual information and to be involved in limbic functions. The hypothalamus interacts with the pituitary gland. Structures in the epithalamus include the pineal gland.

The region designated as the brain stem includes the midbrain that is adjacent to the diencephalon the pons and the medulla. The pons is the brain region adjacent to the fourth ventricle. The medulla lies distal to the pons. The brain stem connects to the spinal cord. The cranial nerves articulate with the brain stem.

The cerebellum is adjacent to the fourth ventricle and comprises lateral hemispheres and a midline vermis. The cerebellum has multiple folds ridges and deep fissures that divide it into lobes and lobules. The cerebellum has more than 13 distinct lobules and distinct structures including the vermis, nodules, flocculus, and tonsil. Fibers pass into the cerebellum through the peduncle. The cerebellum is reported to have more neurons than the rest of the brain (Nolte, 1999). Specific layers in the cerebellar cortex include the outermost molecular layer that is rich in axons and dendrites. Under this is the molecular layer with large Purkinje cells. The next layer down is the thick granular layer that is dense with small granular cells.

Nuclei in the brain

The structures referred to as nuclei in the brain are gray matter structures. The basal ganglia are gray matter structures that if damaged lead to extrapyramidal symptoms including tremors, unusual movements, muscle spasm, and symptoms such as occurred in Parkinson's disease. The basal ganglia include the caudate nucleus, putamen, globus pallidus subthalamic nuclei, and substantia nigra with pars compacts and reticular regions.

The putamen and globus pallidus are sometimes referred to as the lentiform nucleus. The caudate and putamen together are sometimes referred to as the striatum. The basal ganglia have connections to other brain regions and are reported to be important in modulating cortical signaling.

Limbic system

Primary structures in the limbic systems include the hippocampus and the amygdala. The hippocampus is considered to be critical for memory. The amygdala lies adjacent to the thalamus and the putamen and receives and processes sensory input and had projections to both the putamen and the cerebral cortex. There is evidence that the amygdala is involved in emotional learning.

White matter tracts

White matter tracts form connections between different cerebral regions. These tracts include association fiber, commissural fibers, and projection fibers (Haines, 2002). Short association fibers form connections between gyri that are in close proximity to each other. Distant regions of the cortex are connected by long association fibers; these include the cingulum adjacent to the cingulate gyrus.

A tract that forms frontotemporal connections is referred to as the inferior longitudinal fasciculus. The superior longitudinal fasciculus connects frontal parietal and occipital cortices. Frontal and temporal cortices are connected by the arcuate fasciculus. The inferior frontooccipital fasciculus connects frontal and occipital regions.

Two small association bundles lie on either side of the neuron body structure known as the claustrum. A small bundle of association fibers occurs in proximity to the claustrum and is referred to as the capsule.

The corpus callosum is a large white matter structure that connects the left and right hemispheres of the brain. Specific regions within the corpus callosum are defined as the genu regions that connect the two frontal lobes. The fibers connecting the two occipital lobes pass through the splenium region of the corpus callosum.

Additional bundles of connecting fibers include the anterior commissure and the hippocampal commissure.

Projection fibers occur in the structure referred to as the internal capsule. The fibers of this structure form the corona radiata and include motor and sensory fibers. The internal capsule has five distinct regions. Fibers connect different regions of the cerebral cortex to each other and to the brain stem.

Characterizing functions of brain regions

Genon et al. (2018) noted that comprehensive conclusions regarding functions of specific brain regions in humans were not absolutely defined. They proposed that neuroimaging analyses under specific experimental conditions could provide greater insight.

They noted that Broca and Wernicke, working in the 19th century, analyzed function based on its deprivation when specific areas of the brain were injured. Wernicke also emphasized that the connections of a specific area with other areas impacted function.

Genon et al. emphasized that currently assessment of functional organization includes region assessment but also linkage of an area with other regions. Nevertheless, they noted that one area may be associated with a limited number of functions. Different methods used to assess function beyond lesion analyses include stimulation techniques, assessment of activation during specific tasks, and brain–behavior correlations.

Various stimulation approaches can include electrical current use and transcranial magnetic stimulation. Activation approaches with position emission tomography (PET) include monitoring proton activity primarily due to metabolic changes during performance of specific mental tasks.

The structure–behavior correlation approach involves analyzing effect of specific natural variants. For example, MRI is used to identify morphological differences, e.g., cortical thickness, gray matter volume, and correlating these parameters with behavior.

11. Genes with defects that lead to cortical malformations

- ARHGAP1B Rho GTPase activating protein 1B
- ASPM abnormal spindle microtubule assembly
- CDH1: 1
- SMO G protein–coupled receptor interacts with the patched protein and hedgehog receptor
- MEK/ERG mitogen-activated protein kinase
- PTEN: phosphatidylinositol-3,4,5-trisphosphate 3-phosphatase
- RASGAP family of GTPase-activating proteins
- TBC1D3 TBC1 domain family
- FGF: fibroblast growth factor
- GFR: fibroblast growth factor receptor
- GPSM1: G protein signaling modulator 1
- PDGF: platelet-derived growth factor
- PDGFR: platelet-derived growth factor receptor
- PAX6: paired box 6
- SMARCC1: member of the SWI/SNF family of proteins, impacts chromatin

12. Mitochondrial function, neurogenesis, and neurodevelopment

Khacho and Slack (2018) reviewed mitochondrial dynamics and roles in neurodevelopment. They noted long-standing evidence of mitochondrial function in contributing to the high energy required for neuronal activity. They noted new evidence regarding the important roles of mitochondrial function and dynamics in neural stem cells and in neurogenesis.

Mitochondrial dynamics include processes of fission and fusion that mitochondria undergo. These processes are dependent on GTP hydrolysis and GTPase activity. They noted that processes of fusion involve activity of mitofusin 1 (MFN1), mitofusin 2 (MFN2), and OPA1 mitochondrial dynamin-like GTPase (OPA1). The OPA1 gene product was also noted to be important in remodeling of the mitochondrial cristae.

Mitochondrial fission was shown to be dependent on the activity of DRP1 (DNM1L), a member of the dynamin superfamily of GTPases, and also on activity of additional gene products fission mitochondrial 1 (FIS1), MID49 (MIEF2) mitochondrial elongation factor 2, and MID51 (MIEF1) mitochondrial elongation factor 1.

13. Transcriptomics and neuroscience

Lein et al. (2017) noted that single cell transcriptome analyses were yielding important information on molecular cell types. They noted that additional efforts will be required to determine roles of different molecular cells types in determining physiological and behavioral functions. This would require information of the defined locations and positions of different cell types and their cellular connections.

Lein et al. noted that analyses of gene expression and analyses of protein by means of antibody staining provided insights into variations in products in gene expression in various brain regions but had not necessarily enabled analysis of gene expression in different cell types. They emphasized that great advances had been made in development of transcriptomic methods. Application of transcriptomic methods also facilitated analyses of developmental changes in cell type distribution and function.

They noted that our understanding of establishment and maintenance of cell type connectivity was also rudimentary. They noted further that transcriptomic analyses were important in determining brain architecture at different stages of development. Transcriptomic analyses could help illuminate molecular mechanisms involved in generation of specific brain diseases.

References

Ahn, J.H., Cho, H., Kim, J.H., Kim, S.H., Ham, J.S., Park, I., Suh, S.H., Hong, S.P., Song, J.H., Hong, Y.K., Jeong, Y., Park, S.H., Koh, G.Y., 2019. Meningeal lymphatic vessels at the skull base drain cerebrospinal fluid. Nature 572 (7767), 62–66. https://doi.org/10.1038/s41586-019-1419-5.

Aspelund, A., Antila, S., Proulx, S.T., Karlsen, T.V., Karaman, S., Detmar, M., Wiig, H., Alitalo, K., 2015. A dural lymphatic vascular system that drains brain interstitial fluid and macromolecules. J. Exp. Med. 212 (7), 991–999. https://doi.org/10.1084/jem.20142290.

Bargmann, C.I., 2012. Beyond the connectome: how neuromodulators shape neural circuits. Bioessays 34 (6). https://doi.org/10.1002/bies.201100185. PMID.

Birey, F., Kokkosis, A.G., Aguirre, A., 2017. Oligodendroglia-lineage cells in brain plasticity, homeostasis and psychiatric disorders. Curr. Opin. Neurobiol. 47, 93–103. https://doi.org/10.1016/j.conb.2017.09.016.

Clasadonte, J., Prevot, V., 2018. The special relationship: glia-neuron interactions in the neuroendocrine hypothalamus. Nat. Rev. Endocrinol. 14 (1), 25–44. https://doi.org/10.1038/nrendo.2017.124.

Dubois, J., Dehaene-Lambertz, G., Kulikova, S., Poupon, C., Hüppi, P., Hertz-Pannier, L., 2014. The early development of brain white matter: a review of imaging studies in fetuses, newborns and infants. Neuroscience 276. https://doi.org/10.1016/j.neuroscience.2013.12.044. Review (PMID).

Eichenbaum, H., 2017. Prefrontal-hippocampal interactions in episodic memory. Nat. Rev. Neurosci. 18 (9), 547–558. https://doi.org/10.1038/nrn.2017.74.

Genon, S., Reid, A., Langner, R., Amunts, K., Eickhoff, S.B., 2018. How to characterize the function of a brain region. Trends Cognit. Sci. 22 (4), 350–364. https://doi.org/10.1016/j.tics.2018.01.010.

Greenough, 1987. Experience and brain development. Child Dev. 58 (3), 539–559.

Guo, J., Higginbotham, H., Li, J., Nichols, J., Hirt, J., Ghukasyan, V., Anton, E.S., 2015. Developmental disruptions underlying brain abnormalities in ciliopathies. Nat. Commun. 6. https://doi.org/10.1038/ncomms8857.

Haines, D., 2002. Fundamental Neuroscience, second ed. Churchill Livingstone.

Hakanen, J., Ruiz-Reig, N., Tissir, F., 2019. Linking cell polarity to cortical development and malformations. Front. Cell. Neurosci. 13. https://doi.org/10.3389/fncel.2019.00244.

Huttenlocher, P.R., 1984. Synapse elimination and plasticity in developing human cerebral cortex. Am. J. Ment. Defic. 88 (5), 488–496.

Khacho, M., Slack, R.S., 2018. Mitochondrial dynamics in the regulation of neurogenesis: from development to the adult brain. Dev. Dynam. 247 (1), 47–53. https://doi.org/10.1002/dvdy.24538.

Kim, A., Savary, C., Dubourg, C., Carré, W., Mouden, C., et al., 2019. Integrated clinical and omics approach to rare diseases: Novel genes and oligogenic inheritance in holoprosencephaly. Brain 142 (1), 35–49. https://doi.org/10.1093/brain/awy290. PMID:30508070.

Kuijpers, M., Hoogenraad, C.C., 2011. Centrosomes, microtubules and neuronal development. Mol. Cell. Neurosci. 48 (4), 349–358. https://doi.org/10.1016/j.mcn.2011.05.004.

Lasser, M., Tiber, J., Lowery, L.A., 2018. The role of the microtubule cytoskeleton in neurodevelopmental disorders. Front. Cell. Neurosci. 12. https://doi.org/10.3389/fncel.2018.00165. eCollection 2018. (PMID).

Lein, E., Borm, L.E., Linnarsson, S., 2017. The promise of spatial transcriptomics for neuroscience in the era of molecular cell typing. Science 358 (6359), 64–69. https://doi.org/10.1126/science.aan6827.

Li, M., Santpere, G., Imamura, K.Y., Evgrafov, O.V., Gulden, F.O., et al., 2018. Integrative functional genomic analysis of human brain development and neuropsychiatric risks. Science 362 (6420). https://doi.org/10.1126/science.aat7615. PMID:30545854. pii: eaat7615.

Lim, L., Mi, D., Llorca, A., Marín, O., 2018. Development and functional diversification of cortical interneurons. Neuron 100 (2), 294–313. https://doi.org/10.1016/j.neuron.2018.10.009.

Louveau, A., Smirnov, I., Keyes, T.J., Eccles, J.D., Rouhani, S.J., Peske, J.D., Derecki, N.C., Castle, D., Mandell, J.W., Lee, K.S., Harris, T.H., Kipnis, J., 2015. Structural and functional features of central nervous system lymphatic vessels. Nature 523 (7560), 337–341. https://doi.org/10.1038/nature14432.

Mäkinen, T., 2019. Lymphatic vessels at the base of the mouse brain provide direct drainage to the periphery. Nature 572 (7767), 34–35. https://doi.org/10.1038/d41586-019-02166-7.

Meganathan, K., Lewis, E.M.A., Gontarz, P., Liu, S., Stanley, E.G., Elefanty, A.G., Huettner, J.E., Zhang, B., Kroll, K.L., 2017. Regulatory networks specifying cortical interneurons from human embryonic stem cells reveal roles for CHD2 in interneuron development. Proc. Natl. Acad. Sci. U S A 114 (52), E11180–E11189. https://doi.org/10.1073/pnas.1712365115.

Nano, M., Basto, R., 2017. Consequences of centrosome dysfunction during brain development. In: Advances in Experimental Medicine and Biology, vol. 1002. Springer New York LLC, pp. 19–45. https://doi.org/10.1007/978-3-319-57127-0_2.

Neniskyte, U., Gross, C.T., 2017. Errant gardeners: glial-cell-dependent synaptic pruning and neurodevelopmental disorders. Nat. Rev. Neurosci. 18 (11), 658–670. https://doi.org/10.1038/nrn.2017.110.

Nigg, E.A., Raff, J.W., 2009. Centrioles, centrosomes, and cilia in health and disease. Cell 139 (4), 663–678. https://doi.org/10.1016/j.cell.2009.10.036.

Nolte, J., 1999. The Human Brain: An Introduction to Functional Anatomy, fourth ed. Mosby.

Nutma, E., Willison, H., Martino, G., Amor, S., 2019. Neuroimmunology – the past, present and future. Clin. Exp. Immunol. 197 (3). https://doi.org/10.1111/cei.13279 (Review. PMID).

Penisson, M., Ladewig, J., Belvindrah, R., Francis, F., 2019. Genes and mechanisms involved in the generation and amplification of basal radial glial cells. Front. Cell. Neurosci. 13. https://doi.org/10.3389/fncel.2019.00381.

Pirozzi, F., Nelson, B., Mirzaa, G., 2018. From microcephaly to megalencephaly: determinants of brain size. Dialogues Clin. Neurosci. 20 (4), 267–282. https://doi.org/10.31887/DCNS.2018.20.4/gmirzaa. PMID.

Romaniello, R., Arrigoni, F., Fry, A.E., Bassi, M.T., Rees, M.I., Borgatti, R., Pilz, D.T., Cushion, T.D., 2018. Tubulin genes and malformations of cortical development. Eur. J. Med. Genet. 61 (12), 744–754. https://doi.org/10.1016/j.ejmg.2018.07.012.

Salter, M.W., Stevens, B., 2017. Microglia emerge as central players in brain disease. Nat. Med. 23 (9), 1018–1027. https://doi.org/10.1038/nm.4397.

Sandman, C.A., 2018. Prenatal CRH: an integrating signal of fetal distress. Dev. Psychopathol. 30 (3), 941–952. https://doi.org/10.1017/S0954579418000664.

Santello, M., Toni, N., Volterra, A., 2019. Astrocyte function from information processing to cognition and cognitive impairment. Nat. Neurosci. 22 (2), 154–166. https://doi.org/10.1038/s41593-018-0325-8.

Schafer, D.P., Lehrman, E.K., Kautzman, A.G., Koyama, R., Mardinly, A.R., Yamasaki, R., Ransohoff, R.M., Greenberg, M.E., Barres, B.A., Stevens, B., 2012. Microglia sculpt postnatal neural circuits in an activity and complement-dependent manner. Neuron 74 (4), 691–705. https://doi.org/10.1016/j.neuron.2012.03.026.

Scoville, W.B., Milner, B., 1957. Loss of recent memory after bilateral hippocampal lesions. J. Neurol. Neurosurg. Psychiatr. 20 (1), 11–21. https://doi.org/10.1136/jnnp.20.1.11.

Skaper, S.D., 2018. Neurotrophic factors: an overview. In: Methods in Molecular Biology, vol. 1727. Humana Press Inc, pp. 1–17. https://doi.org/10.1007/978-1-4939-7571-6_1.

Stiles, J., Jernigan, T.L., 2010. The basics of brain development. Neuropsychol Rev. 20 (4), 327–348. https://doi.org/10.1007/s11065-010-9148-4. PMID: 21042938.

Van Dam, T.J.P., Kennedy, J., van der Lee, R., de Vrieze, E., Wunderlich, K.A., Rix, S., Dougherty, G.W., Lambacher, N.J., Li, C., Jensen, V.L., Leroux, M.R., Hjeij, R., Horn, N., Texier, Y., Wissinger, Y., Van Reeuwijk, J., Wheway, G., Knapp, B., Scheel, J.F., Huynen, M.A., 2019. Ciliacarta: an integrated and validated compendium of ciliary genes. PloS One 14 (5). https://doi.org/10.1371/journal.pone.0216705.

van Praag, H., Kempermann, G., Gage, F.H., 2000. Neural consequences of environmental enrichment. Nat. Rev. Neurosci. 1 (3), 191–198. https://doi.org/10.1038/35044558.

Vasile, F., Dossi, E., Rouach, N., 2017. Human astrocytes: structure and functions in the healthy brain. Brain Struct. Funct. 222 (5), 2017–2029. https://doi.org/10.1007/s00429-017-1383-5.

Wamsley, B., Fishell, G., 2017. Genetic and activity-dependent mechanisms underlying interneuron diversity. Nat. Rev. Neurosci. 18 (5), 299–309. https://doi.org/10.1038/nrn.2017.30.

Wheway, G., Nazlamova, L., Hancock, J.T., 2018. Signaling through the primary cilium. Front. Cell Dev. Biol. 6. https://doi.org/10.3389/fcell.2018.00008.

Youn, Y.H., Han, Y.G., 2018. Primary cilia in brain development and diseases. Am. J. Pathol. 188 (1), 11–22. https://doi.org/10.1016/j.ajpath.2017.08.031.

Zeng, H., Sanes, J.R., 2017. Neuronal cell-type classification: challenges, opportunities and the path forward. Nat. Rev. Neurosci. 18 (9), 530–546. https://doi.org/10.1038/nrn.2017.85.

Zorn, J.V., Schür, R.R., Boks, M.P., Kahn, R.S., Joëls, M., Vinkers, C.H., 2017. Cortisol stress reactivity across psychiatric disorders: A systematic review and meta-analysis. Psychoneuroendocrinology 77, 25–36. https://doi.org/10.1016/j.psyneuen.2016.11.036.

CHAPTER

2

Neurotransmitters, neuromodulators, synapses

1. Neurons

Cell bodies contain nuclei; dendrites arise from cell bodies; the lengths and branching patterns of dendrites vary. Cell bodies also give rise to single projections, the axons, which can branch in their lower regions.

Synapses

At synapses information passes from one neuron to dendrites of another neuron. Information can also pass from the axon of one neuron to the cell body of another neuron. Synaptic vesicles are present in an axon region that is in contact with a dendrite.

Synaptic vesicles contain neurotransmitters (Fig. 2.1). Neurotransmitter receptors are involved in conversion of chemical signals to electrical signals. Pumps and ion channels also control electrical potential in neurons.

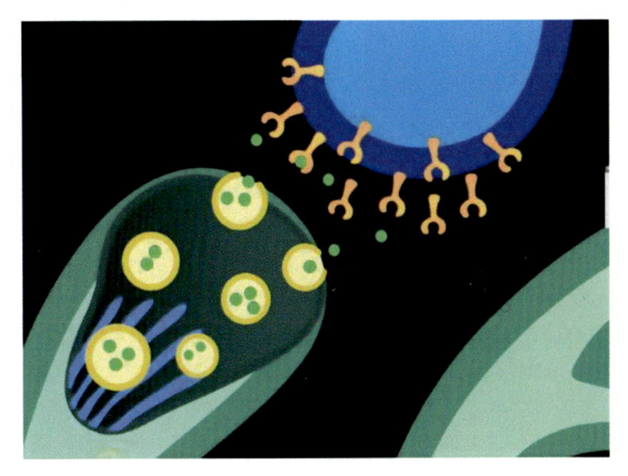

FIGURE 2.1 Neurotransmitter crossing neural synapse. NIMH image illustrating neurotransmitter release and neurotransmitter receptors. *From National Institute of Mental Health, National Institutes of Health, Department of Health and Human Services.*

Dendritic spines

Dendritic spines were first described by Ramon y Cajal in 1888. Studies by Gray in 1959 reported that dendritic spines contact presynaptic axons. In 2012, Rochefort and Konnerth (2012) reviewed information on dendritic spines, which arise as protrusions from dendrites of many types of neurons. Spines arise at many different degrees of density, and spines vary in size. Microscopy studies reveal four different morphological types of spines. There are thin spines including filopodia with minimal heads, thin spines with a long neck and small head, mushrooms spines with a long neck and large head, and stubby spines with no neck. Several different types of spines may occur on one neuron. Spines are associated with specific densities referred to as postsynaptic densities (PSDs). Actin fibers

https://doi.org/10.1016/B978-0-12-821913-3.00008-1

extend from the spines to the plasma membrane and also to the cytoskeleton. Studies by Yuste and Majewska (2001) revealed calcium within dendritic spines, and calcium dynamics were reported to be impacted by spine morphology. A surprising number of biochemical processes take place in spines. Rochefort and Konnerth emphasized the importance of calcium in spine function. They noted that influx of calcium into spines occurs via ionotropic receptors and through voltage-gated calcium channels. Both NMDA and AMPA receptors were found to be calcium permeable.

Dendritic spines and neuroplasticity

Nishiyama (2019) reviewed functions of dendritic spines and noted that structural changes in dendritic spines are critical for synaptic plasticity. He noted also that there is evidence that spine morphology and plasticity are altered in neurodevelopmental disorders.

Nishiyama emphasized the important role of the RAS GTPase signaling pathway in the regulations of dendritic spine actin cytoskeleton and protein synthesis. Dendritic spines were reported to play important roles in storing calcium ions and supporting the role of calcium in signaling. Dendritic spines contain PSDs, and they accommodate neurotransmitter receptors. Nishiyama noted that long-term potentiation and long-term depression of neural function were associated with alterations in dendritic spine size; potentiation was associated with increased spine size and decreased spine size occurred during long-term depression.

Nishiyama reported that advances in molecular and optical techniques facilitated analyses of spine functions and structural changes. Two-photon microscopy enables imaging of dendritic spine morphological changes in key components of spines and associated cytoskeleton components including microtubules, neurofilaments, and actin in response to neuronal stimulation. Calcium was also shown to accumulate in spines in response to activation. Fluorescence resonance energy transfer (FRET) is a technique that has been used to detect energy transfer related to signaling. Calcium can flow into spines through glutamate receptors; it then binds to calmodulin and activates calmodulin kinase complexes CAMK1A, CAMK2A, and CAMK2B that in turn activate downstream signaling mechanisms, e.g., RAS GTPases. Activated RAS GTPases regulate cofilin and actin in the cytoskeleton. CAMK2A and CAMK2B also have actin-binding activity.

Nishiyama emphasized the importance of dendritic spines in synaptic transmission and that there is evidence for dendritic spine impairment in certain neurodevelopmental conditions. In 1974, Huttenlocher (1974) reported defects in dendritic spines in children with cognitive impairment. Increased dendritic spine density in cases of autism was reported by Weir et al. (2018). Nishiyama noted that altered dendritic spine structure, including increased numbers of thin immature spines, was identified in Fragile X mental retardation.

Synapse formation

Südhof (2018) reviewed synapse formation and emphasized the importance of cell adhesion molecules and proposed that cell adhesion molecules are critical to the establishment and organization of synapses and for synaptic plasticity. Key elements in synaptic function include generation of neurotransmitter containing vesicles in the presynaptic region, exocytosis of these vesicles, and detection and uptake of the vesicle contents by receptors on the postsynaptic membranes. Südhof also drew attention to the fast information transfer system at synapses and noted that there is bidirectional transfer of information between synapses.

He noted that synapses development occurs during prenatal and postnatal periods and that synaptic pruning occurs and may extend into the third decade of life. It is important to note that developmental specific synapse formation is reported to be largely activity independent, but pruning may be related to activity levels. However, there is some evidence that neurotransmitter release does have some impact on synapse development.

Synapse formation and neural communication

In a 2018 review on synapse formation, Südhof emphasized two basic mechanisms of neuronal communication, fast point communication that is mediated by transmission at synapses and a slow signaling mechanism that depends on activities of a range of molecules including neuropeptides, monoamines, and endocannabinoids.

Synaptic transmission is dependent on formation and extrusion of synaptic vesicles from presynaptic structures, release of neurotransmitters from synaptic vesicles, and detection of neurotransmitters by postsynaptic receptors.

Südhof emphasized the abundance of cell adhesion molecules (CAMs) that play important roles at synapses. Cell adhesion molecules were reported to mediate precise apposition of pre- and postsynaptic sites. He noted that difficulties remained in distinguishing CAMs that play essential roles at synapses from those that had accessory functions.

Südhof noted that presynaptic terminals are neuronal. However, postsynaptic structures, though primarily neuronal, can also occur on oligodendrocytes. Axons form presynapses and were noted to extend over long distances. There is evidence that cellular organelles are sparse in axons and rough endoplasmic reticulum and ribosomes were reported to be sparse in axons.

There is evidence that release of neurotransmitters from the presynapse plays important roles in promoting postsynaptic development and organization. Südhof noted that synaptic pruning was important in prenatal and postnatal development, with evidence that synaptic pruning occurred into the third decade of life.

Südhof illustrated phases of synapse formation. Neurogenesis with development of axons and dendrites was reported to occur during embryonic and postnatal development. Presynaptic synapse for formation and postsynaptic specialization occurred during embryonic and postnatal development and occurred at lower rates throughout life and was activity dependent. Physiological synapse elimination was noted to occur throughout life but particularly during puberty. Synaptic dysfunction and pathological synapse elimination also occur. Südhof reviewed synaptic CAMs and their roles at synapses. He emphasized that specific CAMs played key roles at synapses and mediated connection between presynapses and postsynapses. Some presynaptic CAMs interacted with specific postsynaptic proteins, while other presynaptic CAMs interacted with a number of different postsynaptic CAMs. In addition, specific CAMs also interacted with proteins in the synaptic cleft. Importantly a number of presynaptic and postsynaptic CAMs have been reported to play roles in specific diseases.

Synapses and cellular adhesion molecules

Südhof focused on specific cellular adhesion molecules that play key roles in synapse formation. These include neurexin and specific ligands that bind to neurexins that include neuroligin, cerebellins, neuroexophilins, and five forms of leucine-rich repeat transmembrane neuronal proteins LRRTM1–LRRTM5; in addition, there are many other interconnections between pre- and postsynaptic membrane surfaces. Neurexins were also shown to bind to an interact with dystroglycans and latrophilins. Other important CAM interconnections between pre- and postsynaptic sites include netrins and neuronal pentraxins that interact with glutamate receptors, SynCAMs, cadherins, and teneurin. Südhof reported that teneurins play particularly important roles in embryogenesis. SynCAMs are thought to play roles in synaptic plasticity.

Cellular adhesion molecules implicated in developmental functions

- MDGA1 and MDGA2, MAM domain–containing glycosylphosphatidylinositol anchor 1
  - Dystroglycans
    - Latrophilins
    - Teneurins, transmembrane protein
    - PTPRD/F/S, protein tyrosine phosphatase receptor

- SynCAMs, synaptic cell adhesion molecules
- Cadherins, calcium-dependent cell–cell adhesion proteins
- Ephrin/Ephrin receptor, subfamily of receptor protein–tyrosine kinases

Cell adhesion molecules and interacting molecules with genetic association with neuropsychiatric diseases

- Neurexins, cell surface receptors that bind neuroligins to form Ca (2+)-dependent complexes
  - GRIK, kainate receptors; GRIK2, glutamate ionotropic receptor kainate type subunit 2
    - GLUD1 and GLUD2, glutamate dehydrogenases
    - MDGA1 and MDGA2, glycosylphosphatidylinositol (GPI)-anchored cell surface glycoprotein
    - Neuroligins, neuronal cell surface proteins
    - Dystroglycans, central component of dystrophin–glycoprotein complex that links the extracellular matrix and the cytoskeleton
    - LRRTMs, leucine-rich-repeat transmembrane neuronal proteins involved in synapse organization
    - GABAA receptors, gamma-aminobutyric acid type A receptor alpha subunits
    - Latrophilins; ADGRL1-4, adhesion G protein–coupled receptors
    - SLITRKs, integral membrane proteins with 2 N-terminal leucine-rich repeat (LRR) domains
    - ILRAPS, interleukin 1 receptor accessory protein

Presynaptic calcium channels and postsynaptic neurotransmitter receptors

Südhof noted that presynaptic calcium channels and postsynaptic neurotransmitter receptors form essential components in synaptic transmission. He reported that synapse formation occurs during embryogenesis and also for several years in postnatal life; synaptic formation declines slowly during the first three decades of human life. Significant synapse elimination occurs particularly prior to the third decade of life.

Studies of synapse elimination in the cerebellum revealed that secretion of semaphorins SEM3A, SEM7A, and progranulins may play roles in synapse elimination.

Synapses and astrocytes

Südhof noted growing evidence that astrocytes associate with synapses. The precise functional relevance of astrocyte–synaptic connections remains to be determined.

Postsynaptic regions of synapses: structures and functions

Postsynaptic densities

In 2011, Sheng and Kim (2011) reviewed the post-synaptic regions of synapses. They noted that excitatory synapses are present primarily on dendritic spines, while inhibitory synapses occur predominantly on dendritic shafts and on neuron cell bodies. Significant differences exist between excitatory and inhibitory synapses in the structure of their postsynaptic regions. The neurotransmitter receptors on excitatory synapses include primarily AMPA and NMDA glutamate receptors and also metabotropic receptors mGLUR, while inhibitory receptors include gamma-aminobutyric acid (GABA) receptors.

The postsynaptic regions of excitatory receptors were noted to be more elaborate than those of inhibitory receptors.

Postsynaptic regions of excitatory receptors

Sheng and Kim described a series of scaffold proteins that connect glutamate neurotransmitter on the postsynaptic membrane to cytoskeletal elements and to signaling molecules.

These scaffold proteins form a structure that is visible on electron microscopy as a membrane and that became known as the PSD.

Intense studies were carried out on the protein components of the PSD, on the scaffold proteins, and on the particular domains present in these proteins. The most abundant scaffold proteins were found to be PSD95 (also known as SAP90 and DLG4) and PDS93 (DLG2). These proteins have C terminal domains and guanylate kinase domain and are sometimes referred to as membrane-associated guanylate kinase proteins (MAGUKs).

Other proteins abundant in excitatory postsynaptic membranes are SYNGAP synaptic Ras GTPase activating protein and GKAP G kinase anchoring protein that has a guanylate kinase domain.

On the cytoplasmic side of the PSD, SHANK proteins occur; Sheng and Kim described Homer proteins that connect the mGLUR receptor with SHANK proteins. They also reported that the PSD is associated with cytoplasmic signaling molecules including CAM kinase, nonreceptor tyrosine kinases and phosphatases, and other small GTPases and GEFs guanylate exchange factors also occur in the PSD. The PSD is connected to F actin in the cytoplasm.

They noted that during different stages of development, the protein composition of the PSD changes. It is important to note that PSD function likely also plays roles in glutamate receptor incorporation that occurs in excitatory synapses during long-term potentiation and in removal of AMP receptors during long-term depression.

Components of the PSD also undergo post-translational modification including phosphorylation and palmitoylation. These modifications can lead to increased density of the PSD that occurs during long-term potentiation. Ubiquitination of PSD proteins can lead to loss of these proteins and decreases in density during long-term depression. Sheng and Kim emphasized increasing levels of evidence that mutations in PSD proteins played roles in the causation of neurodevelopmental disorders.

Postsynaptic regions of inhibitory synapses

Sheng and Kim (2011) emphasized that the postsynaptic specializations in inhibitory synapses were much less complicated than those of excitatory synapses. Inhibitory synapses include GABA-ergic receptors and glycine receptors. They reported that a major component of the postsynaptic regions of inhibitory receptors is gephyrin (GHPN) that was reported to be particularly important for glycine receptors. Gephyrin was reported to undergo phosphorylation induced by the kinase GSK3B. Other gephyrin-binding proteins have been identified. One gephyrin-binding protein neuroligin2 was reported to interact with neurogenin on GABAergic axons.Gephyrin was also reported to interact with proteins that link to factors in the cytoskeleton including proteins that bind to microtubules.

Sheng and Kim emphasized the delineation of structure and functions of the PSD components that are of critical important in understanding brain disorders.

Reviews in 2017 by Frank and Grant (2017) and in 2018 by Kaizuka and Takumi (2018) revealed that within PSDs that underlie excitatory synapses, there are molecular complexes that together are composed of more than 2000 proteins. Kaizuka and Takumi reported that classes of proteins that occur within the PSD include neurotransmitter receptors NMDARs, AMPARs, GRMs, membrane trafficking proteins, scaffold proteins, cell adhesion proteins, cytoskeletal proteins, and signaling proteins.

Specific families of proteins that occur in the PSD include MAGUK membrane-associated guanylate kinases and SH3 ankyrin repeat domain proteins SHANK 1, 2, and 3. Soler et al. (2018) reported additional protein families in the PSD and DLG discs-large proteins (DLG1-DLG5) that constitute members of the membrane-associated guanylate kinase (MAGUK) family, DLGAP discs-large associated proteins 1–4, and HOMER 1, 2, and 3 scaffold proteins.

Kaizuka and Takumi classified protein complexes by their positions. The most exterior PSD proteins included neurotransmitter receptor complexes and specific adhesion molecules, N-cadherin, neuroligins, and leucine-rich repeat transmembrane protein (LRRTP). The various proteins on the external membrane were linked to interior protein PSD95 and beta catenin. A number of proteins associate with PSD95 and PSD95 linked to cytoskeletal proteins. Key complexes of proteins associated with the PSD included neurotransmitter complex proteins 52 NMDAR protein, 34 AMPR complex proteins, and 83 ARC activity-regulated cytoskeleton-associated proteins.

Kaizuka and Takumi reported that in different brain regions, there were differences in the composition of PSD-related proteins. In addition, different degrees abundance of PSD proteins occurred, and there were differences in the degree of phosphorylation of PSD proteins.

Other important findings were that at least 200 of the PSD-associated proteins were reported to be targets of the FMRP-CYFIP1 complex. The fragile X mental retardation protein linked to the cytoplasmic FMR1 interacting protein 1 plays a role in translation of mRNA of specific gene products into proteins.

Kaizuka and Takumi also noted that 671 of the 1019 PSD-associated proteins had been reported to have variants implicated in autism causation. It is also important to note the PSD scaffolding proteins, e.g., discs-large (DLG), DLGAP, SHANK, and HOMER proteins, have also been reported to have variants implicated in the risk for schizophrenia (Soler et al., 2018).

Káradóttir and Kuo (2018) also reviewed aspects of neurotransmitter function that stimulate neuronal activity.

Neural and synaptic plasticity

Neural plasticity involves structural and functional changes that occur in the brain in response to simulation, neuronal activity, and development. Ismail et al. (2017) noted that neuronal plasticity involves modulation of gene expression, molecules, and cellular mechanisms that impact synaptic activity and neural circuitry.

Synaptic plasticity is defined as the capacity of synapses to strengthen or weaken and that these changes may be related to neurotransmitter receptor content. Südhof (2018) noted that there is some evidence that synaptic plasticity may be associated with structural changes in synapses. Synaptic activity may therefore be determined by synaptic utilization. Synaptic plasticity is known to play important roles in learning and memory.

SynGO: synapse knowledge base

Koopmans et al. (2019) noted that synaptic plasticity and synaptic use–dependent changes underlie cognitive processes including perception, sensory processing, memory formation and retrieval, and attention and learning. Furthermore, impaired synaptic function has been shown to contribute to brain disorders, referred to as synaptopathies. In this category, they included disorders such as autism spectrum disorder (ASD), attention deficit hyperactivity disorder (ADHD), schizophrenia, and specific late onset disorders including Parkinson's disease and Alzheimer's disease.

The SynGO knowledge base includes a comprehensive list of gene products that function in synapses, the specific molecular function, and location of each of these. Koopmans et al. documented 1112 genes that encode products involved in synaptic functions (Smith and Robinson, 2019).

Mitochondria also contribute to synaptic function; however, the mitochondrial genes were not included in SynGO since they are listed in a separate database and annotated by Smith and Robinson (2019).

In SynGO, the gene products are listed according to their locations in the synapse, e.g., presynapse interior and endosomes, presynaptic membranes. Synaptic cleft proteins include transsynaptic adhesion proteins and signaling molecules. Postsynaptic components listed include postsynaptic membranes, receptors associated with postsynaptic membranes, the postsynaptic specialized region that underlies the postsynaptic membrane, postsynaptic cytosol that includes organelles endosomes, endoplasmic reticulum, and cytoskeleton spine generating apparatus. In the SynGO database, the biological functions of the products of 1112 genes are also documented.

Structural features of genes that encode products expressed in synapses

Koopmans et al. (2019) documented that genes that encode products expressed in the synapse have average length that is twice as long as the average length of other genes. Gene transcripts expressed in the synapse were 1.7 times longer than gene transcripts expressed elsewhere. The isoform diversity and number of MRNA isoforms of synapse expressed genes was higher than the number of isoforms of genes expressed elsewhere. Proteins present in synapses were also found to undergo more extensive posttranslational modifications.

Through studies of nucleotide variants in healthy individuals in the population, Koopmans et al. discovered that fewer damaging mutations occurred in synaptic genes than in other genes, indicating that synaptic genes were intolerant of damaging mutations.

Koopmans et al. also reviewed previously reported data on genome-wide association studies (GWAS) for specific disorders including ADHD, ASD, schizophrenia, and bipolar disorder and GWAS data on three continuous human traits such as educational attainment, intelligent coefficient (IQ), and height. They analyzed the GWAS data for variants in genes that encoded products expressed in the synapse.

The most significant enrichment for variants in synapse-expressed genes was documented in ADHD in presynaptic membrane active genes and in genes that encoded products related to synaptic assembly. In autism cases, there was significant association of variants in genes that encode structural constituents of the presynapse and synaptic active zone genes. Medium levels of association of variants in gene expressed in processes in the postsynaptic region were found in bipolar disease. No highly significant association of synaptic genes occurred in schizophrenia.

With respect to the control data sets on continuous traits, medium levels of association were found between postsynaptic region expression genes and educational attainment. Medium levels of association were documented between IQ levels and variants that encode products involved in synapse organization and synaptic adhesion between pre- and postsynapse.

2. Brain membranes and neuronal functions

Kim et al. (2014) reported that phophatidyl serine is the major form of phospholipid in neural tissues. Specific signaling molecules were shown to interact with phosphatidyl serine in neuronal membranes. Lipids in neuronal membranes were also reported to interact with synaptic vesicles and to play roles in synaptic vesicle exocytosis

Joensuu (2020) that the majority of studies undertaken on neuronal mechanisms involved in cognition, memory, and learning, concentrated on proteins, genetic, and epigenetic mechanisms. They emphasized that phospholipids are the chief components of neuronal membranes and noted that specific phospholipases play roles in membrane remodeling and that there is evidence for protein phospholipid interactions in neural functions. Lipids are known to be particularly abundant in myelin. Concentrations of specific lipids vary between white and gray matter in the brain.

Ion channels

In a 2013 review, Imbrici et al. (2013) noted the importance of ion channels in generating neurotransmitter functions, neuronal action potential, and neuronal firing. They also noted that in several neurological diseases, ion channel mutations have been identified.

Imbrici et al. reviewed potential roles of calcium, sodium, and potassium channels in bipolar disease, schizophrenia, and autism. Rare mutations in the calcium channel CACNA1C lead to Timothy syndrome in which autistic features occur. Rare mutations in the gene that encodes the sodium ion channel SCNA1C may play roles in causation of epilepsy. However, it is more common for ion channel variants to act in concert with other gene variants in the cause of neuropsychiatric diseases.

Features of voltage-gated ion channels

Imbrici et al. noted that entry of calcium into electrically excitable cells is mediated in part by voltage-gated calcium channels. The calcium channels are composed of pore-forming subunits and regulatory subunits. Ten genes in the CACNA1 class encode pore subunits.

Voltage-gated sodium channels play important roles in neurons and glia. A cluster of four sodium channel genes map to human chromosome 2q34.3.

Potassium channels are important in repolarizing neurons and in maintaining resting potential in neurons.

Brini et al. (2017) reviewed neural calcium signaling and emphasized the importance of calcium in neural signaling and synaptic activity. They considered mechanisms of influx of calcium into neurons that occurs through plasma membrane receptors and through specific voltage dependent ion channels.

Calcium plays important roles in synaptic signaling and in neuronal signaling. Calcium enters into the cytoplasm and also enters into specific intracellular organelles including endoplasmic reticulum and mitochondria. Within mitochondria, calcium plays important roles in energy metabolism. Levels of calcium within cells are closely regulated, and this regulation is dependent on calcium sensors, calcium-binding proteins, proteins that function as calcium pumps, and proteins that function as calcium exchangers.

Transmission of calcium across the plasma membrane of cells is mediated in part through voltage-gated calcium channels composed of different subunits alpha, beta, gamma, and delta. Calcium can also enter neurons through receptors that are activated by neurotransmitters, e.g., glutamate. Activation of purinergic receptors P2XR can also activate calcium entry into neurons.

A mechanism that is involved in entry of calcium into intracellular organelles is referred to as store-operated calcium entry. This process involves specific calcium-binding proteins, including the stromal interaction molecules STIM1 and STIM2 that act as calcium sensors and interact with ORAI1 protein subunits; ORAI1 is calcium release-activated calcium modulator 1.

Other channels that mediate calcium entry include arachidonic acid channels TREK and TRAAK (Bauer et al., 2018) and TRP channels (transient receptor potential cation channel) subfamily A members and store-operated calcium channels (SOCs) that are reported to occur in the endoplasmic reticulum and Golgi (Lopez, et al., 2020).

Specific mechanisms exist to promote extrusion of calcium from organelles and extrusion of calcium out of cells into the extracellular environment.

Brini et al. (2017) noted importance of plasma membrane calcium ATPase (PMCA) and plasma membrane sodium calcium exchanger NCX.

A specific calcium ATPase exists in the endoplasmic reticulum and is referred to as the SERCA pump. In mitochondria, the uniporter MCU acts to remove calcium. In neuronal cells, calcium can also move through the ryanodine receptors and the inositol 1.4.5 triphosphate receptors. Mitochondria play important roles in buffering calcium levels in cells. In addition, calcium is required to facilitate activity of specific mitochondrial enzymes involved in energy generation, e.g., enzymes in the tricarboxylic acid cycle.

Brini et al. also noted the calcium levels impact activity of the mitochondrial F1F0 ATPase. Calcium levels in mitochondria are also impacted by connections between mitochondria and endoplasmic reticulum. The protein MCU promotes entry of calcium into mitochondria. Regulators of MCU functions include MICU1, MICU2, MICU3 (mitochondrial calcium uptake 1, 2, 3), and EMRE (solute carrier family 8 member B1). Efflux of calcium from mitochondria is mediated by NCLX (SCL8B1 solute carrier family 8 member B1).

Brini et al. noted that calcium-binding proteins that are important in nervous system function include proteins with EF hand–binding domains. The EF hand is a helix–loop–helix calcium-binding motif in which two helices pack together at an angle of approximately 90°. The two helices are separated by a loop region where calcium actually binds. Important proteins with EF hand motifs in the nervous system include calmodulin, calbindin, calretinin, and parvalbumin. These proteins undergo conformation changes on binding to calcium, and these conformation changes then facilitate their binding to other proteins.

Calcium binding and conformation change facilitate calmodulin binding to calmodulin kinases and phosphatase binding of calcineurin.

Neuronal calcium sensors are another class of EF hand calcium-binding proteins, and Brini et al. noted that these are encoded by 14 different genes in mammals and that the different gene products are expressed in different classes of neurons. Important neuronal calcium sensors include NCS1 neuronal calcium sensor 1 and calsenilin.

Brini et al. noted that calcium plays important roles in control of neuronal-specific processes including release of neurotransmitters. Calcium promotes the fusion of neurotransmitter bearing synaptic vesicles specifically with synaptotagmins that interact with synaptobrevin and SNAP25 (synaptosome-associated protein 25) that eventually plays roles in docking with the postsynaptic membrane and neurotransmitter release.

Brini et al. noted that calcium within cells is also transferred to the nuclear envelope membrane and ultimately to the nucleoplasm where it plays important roles in gene transcription. Calcium was reported to target CREB-binding protein and the CREB transcription factor.

Calcium and neuronal pathologies

Brini et al. noted that calcium signaling is impacted in a number of different neurodegenerative disorders. Impact on calcium is apparently secondary to the underlying primary pathologies in these disorders.

3. Neurotransmitters and neuromodulators

Individual neurons synthesize a variety of neuroactive substances including neurotransmitters and neuromodulators, reviewed by von Bohlen und Halbach and Dermietzel (2002) in 2002. They noted that astrocytes also contain neurotransmitter transporters and can facilitate terminations of action of neurotransmitters.

Neuroactive molecules were found to bind to specific target receptors, and frequently, more than one type of receptor were found to bind to a specific type of neurotransmitter. Although neurotransmitters were initially defined as molecules released by neurons, they are now known to also be released by other types of cells, e.g., cells in the immune system. In addition, cells other than neurons have receptors for neurotransmitters, e.g., cell in the immune system.

Categories of neurotransmitters

- Biogenic amines, including catecholamines
- Acetyl choline, serotonin
- Catecholamines: norepinephrine, epinephrine, dopamine

Amino acids

- Glutamate
- Glycine
- GABA

Neurotransmitter receptors

Much work has been done on families of receptors that bind neurotransmitters since Rall and Sutherland (1958) in 1958 described neurotransmitter receptor binding and its downstream effects on second messengers, including adenyl cyclase. In 1994, Rodbell and Gillman received the Nobel Prize for their work on neurotransmitter receptor activation and its downstream effects on guanosine phosphates (see Lefkowitz, 2012).

In summarizing neurotransmitter-related receptors, von Bohlen und Halbach and Dermietzel in 2002 distinguished six main types of receptors active in the central nervous system: ionotropic receptors, metabotropic receptors, G protein–coupled receptors, guanylate cyclase receptors, tyrosine kinase receptors, and cytokine receptors (chemokine receptors).

Ionotropic receptors

Arrival of electrical stimulus through the axon to the presynaptic terminal leads to several events. Triggering of calcium channels on the presynaptic membrane channel simulates synaptic vessel exocytosis and neurotransmitter release into the synaptic cleft (Mochida, 2019).

Binding of specific neurotransmitters to particular receptors on the postsynaptic membrane can facilitate passage of ions through the receptors. Such receptors are referred to as ligand-gated ion channels. This receptor activity was noted to have short-term electrical effects.

Metabotropic neurotransmitter receptors

In these receptors, coupling of the neurotransmitter ligand to the receptor leads to second messenger activation. Second messengers include cyclic adenosine monophosphate, cyclic guanosine monophosphate, and subsequently activation of inositol phosphates and diacylglycerol.

In addition, von Bohlen und Halbach and Dermietzel noted that activation of metabotropic receptors may lead to tyrosine kinase activity and phosphorylation. They also noted that metabotropic neurotransmitter receptors could act as G protein–coupled receptors.

Serotonin receptors

Two major serotonin receptors were reported to be expressed in brain. These include 5HT1AR and 5HT2AR. Carhart-Harris and Nutt (2017) noted that these two receptors are among at least 14 different 5HT serotonin subtypes that occur in humans. Stimulations of these receptors by ligand triggers G protein receptor activity.

The 5HT2AR receptor was reported to be expressed in the brain cortex and to be particularly highly expressed in the associative cortex that includes the default mode network. 5HT2AR expression was considered to be excitatory.

In reviewing expression of 5HT1AR, Carhart-Harris and Nutt noted that these were expressed in many brain regions and particularly in the hippocampus. They noted that the

5HT1AR receptors functioned as inhibitory receptors. There was then evidence that 5HT1AR and 5HT2AR had opposite effects on stimulation.

The 5HT1AR stimulation was reported to moderate anxiety and stress, while 5HT2AR activation was reported to increase anxiety and environmental vigilance. However, Carhart-Harris and Nutt proposes that plasticity in the 5HT2AR response and 5HT2AR response was reported to have positive effects on learning and cognition.

G protein–coupled receptors

In 2012, the Nobel Prize in Chemistry was awarded for work on elucidation of the structure and function of G protein–coupled receptors to Robert Lefkowitz and Brian Kobilka (see Clark, 2013. G protein receptors include a large family of receptors that can be activated by a number of different ligands. Huang and Thathiah (2015) reported that 90% of all G protein–coupled receptors are expressed in brain. They contribute to a range of different functions.

G protein–coupled receptors have a domain that extends into the extracellular space, a seven-loop transmembrane domain, and an intracellular domain that extends into the cytoplasm.

In 1998, Selbie and Hill reported evidence that stimulation of G protein–coupled receptors could activate signaling of adenyl cyclase and could also lead to activation of a number of other signaling pathways including the phosphatidyl inositol pathway and tyrosine kinase pathways including SRC and GRB2. Increased protein-coupled receptor ultimately led to increased activity in the MAP kinase signaling pathway.

Small G proteins coupled to G protein receptors

In a review in 2010, Ye and Carew (2010) noted the importance of RAS family of small G protein receptor–coupled proteins. They noted that when small G proteins were in the GTP-bound state, they were active and could bind to downstream effectors. They noted that the small G protein family includes RAS, RHO and RHEB, and ARF families of proteins. They noted that the RAS family of proteins were important in memory function through modification of synaptic strength.

In a 2013 review of disorders resulting from pathogenic mutations in member of the RAS family of small G proteins, referred to as RASopathies, Rauen (2013) noted the important connections of RAS protein function and the downstream MAP kinase (ERK) signaling pathways.

G proteins signaling and regulation

Syrovatkina et al. (2016) reviewed G proteins signaling and regulation. G protein are bound to G protein receptors. Many ligands including neurotransmitters and hormones bind to and activate G protein receptors, and G proteins bound to these receptors are involved in transmitting signals from these activated receptors.

A large number of genes are involved in encoding the G protein subunits, alpha, beta, and gamma. Each of these actually includes a family of subunits. G-alpha includes at least 15 different subunits. Some subunits are identified with subscripts indicating their functions. GNAS is a subunit with stimulatory properties; GNAI subunits have inhibitory properties, and GNAT subunits are involved in transduction. GNAO subunits function in the olfactory system; GNAQ subunits are particularly abundant in brain. There are also at least 5 different G protein beta subunits (GNB) and at least 12 different G protein gamma subunits (GNG).

Syrovatkina et al. reviewed the functions of G proteins. They noted that a trimeric form of G protein with alpha, beta, and gamma subunits, with guanosine diphosphate (GDP) bound to the alpha subunit, exists as an inactive protein complex. When the G protein receptor is activated by ligand binding, GDP is released, and the alpha subunit binds to GTP. This binding results in disassociation of the G protein trimer. The GTP-bound alpha subunit can initiate downstream signaling, and the beta and gamma G protein dimers can initiate signaling in a separate pathway.

Ultimately, GTP signaling is terminated through GTPase activity. The released alpha subunit subsequently binds to GDP and the trimeric G protein, with alpha beta and gamma subunits, forms. The binding of guanine nucleotide to alpha is reported to take place at a RAS-like domain within alpha.

Various studies continue to explore that nature of the interactions of the G protein trimer with the G protein receptor.

Specific proteins have been identified that regulate the binding of G protein alpha to GDP. These include RGS14 regulator of G protein signaling and GPSM1, G protein modulator that impacts the interactions between the G protein subunits.

In exploring the downstream signaling activities of G proteins, Syrovatkina et al. noted that the G protein alpha subunit (GNAS) was reported to stimulate adenyl cyclase activity and generation of cyclic adenosine monophosphate (AMP) that activates protein kinase A. Activated gamma G protein subunit was reported to stimulate phospholipase C activity and generation of phosphatidylinositol triphosphate (IP3) and diacylglycerol. IP3 was reported to open calcium channels, and diacylglycerol was reported to activate protein kinase C.

Mutations in specific G protein subunits have been reported to lead to specific diseases. Specific mutation in GNAS encoded on chromosome 20q13.32 was reported to lead to pseudohypoparathyroidism (Albright Hereditary Osteodystrophy). Malerba et al. (2019) and Poke et al. (2019) reported that specific pathogenic mutations in specific GNB subunits had been reported to lead to cardiac arrhythmias and also with epilepsy.

Signaling components downstream of receptors

Protein kinase A

Protein kinase A also known as adenosine 3′-5′ monophosphate (cyclic AMP)–dependent protein kinase has catalytic and regulatory subunits. Different studies carried out over a number of years have revealed that protein kinase A is the primary target of cyclic AMP. Specific anchoring proteins referred to as AKAPs were reported to bind the protein kinase A subunits to defined molecules (Logue and Scott, 2010).

Neuronal activity can lead to stimulation of cyclic AMP activity. Neurotransmitter activity and stimulation of G protein activity can trigger activity of adenyl cyclase. Passage of calcium through receptors and release of calcium from intracellular stores (e.g., in endoplasmic reticulum) can trigger adenyl cyclase activity. This can lead to generation of protein kinase A and protein kinase activity. Leslie and Nairn (2019) reviewed cyclic AMP effect in the brain and noted that many studies confirmed the importance of cyclic AMP activity in brain functions. However, they emphasized the dichotomy of cyclic AMP functions. In addition to promoting protein kinase production and kinase function, cyclic AMP also regulated the activity of phosphatases PP1 and PP2A.

Leslie and Nairn emphasized that a proper balance between kinase and phosphatase activity is essential to brain function. Cyclic AMP plays key roles in maintaining the balance through its influence not only on protein kinase A but also through its impact on activity of phosphatases including protein phosphatases PP1 and PP2A.

Phosphoinositide signaling in the nervous system

Dickson (2019) reviewed phosphoinositide signaling in the nervous system and noted that phosphoinositides occur on the cytoplasmic side of membranes. Components of phosphoinositide include fatty acid chains that project into the membranes, a glycerol group, and an inositol ring that occur in the cytoplasm.

Transformations occur in which phosphotidylinositol (PtdIns) undergoes sequential increases in phosphorylation to generate PtdIns4P, PtdIns(4,5)P2, and PtdIns (3,4,5)P3. The increasing levels of phosphorylation are catalyzed by specific lipid kinases. In humans, six different genes encode phosphatidyl inositol kinases. Phosphotidylinositol phosphates can also undergo dephosphorylation by various enzymes. The enzyme PTEN plays an important role in dephosphorylation of PtdIns (3,4,5)P3.

Dickson documented specific nervous system processes in which phosphotidylinositols are involved. These included functions in membranes, including ion channel functions and functions in the cytoplasm including endocytic trafficking, endocytic fusion, and autophagy.

In a 2017 review of critical signaling cascades in the nervous system, Borrie et al. (2017) presented information on contributions of the RAS, MAP kinase, and phosphotidylinositol. The PI3K pathway is also linked to another kinase AKT in downstream signaling in cells. The kinase AKT is recruited by binding to PtdIns (3,4,5)P3. AKT serves to block activity of TSC1, TSC2, and RHEB and thus enhance activity of MTOR. There is evidence that these pathways play important roles in function in the central nervous system, and factors that disrupt these pathways lead to intellectual impairment.

Neuronal receptors

Tyrosine kinase receptors

Tyrosine kinase receptors include an external N-terminal segment and an internal C terminal segment. Receptor activation stimulates tyrosine kinase activity of the internal domain and phosphorylation of specific molecules.

Dopamine receptors

Five different dopamine receptors that function in the brain have been studied; the five dopamine receptors are noted to be G protein–coupled receptors and to impact adenyl cyclase activity.

DRD1 5q35.2: This is reported to be the most abundant dopamine receptor in the CNS.

DRD1 receptors are reported to regulate neuronal growth and development, mediate some behavioral responses, and modulate dopamine receptor D2-mediated events.

DRD2 11q23.2: Alternative splicing of this gene results in two transcript variants encoding different isoforms.

DRD3 3q13.31 is expressed in the limbic system of the brain, which is associated with cognitive, emotional, and endocrine functions.

DRD4 11p15.5: The DRD4 gene contains a polymorphic number (2–10 copies) of tandem 48 nucleotide repeats.

DRD5. 4p16.1: This receptor is expressed in neurons in the limbic regions of the brain. It is reported to have 10-fold higher affinity for dopamine than the D1 subtype.

Beaulieu et al. (2015) reported that the dopamine receptors differ in their function; however, overall, they play roles in impacting attention, cognition, reward pathways, and sympathetic and hormonal systems.

DRD1 and DRD5 then form one type of dopamine receptors, while DRD2, DRD3, and DRD4 that constitute second-type agonists and antagonists for the two types of receptors may differ. However, there is evidence that dopamine receptors may use a variety of different downstream signaling mechanism. In addition, subunits of the different dopamine receptors form heterodimers. Medications used in treatment of psychiatric disorders often target the G protein receptors. Newer medications will hopefully target some but not all of the G protein receptor downstream signaling pathways (Komatsu et al. 2019).

Key enzymes in the synthesis of dopamine from tyrosine include tyrosine hydroxylase and DOPA decarboxylase (DDC). This protein catalyzes the decarboxylation of L-3,4-dihydroxyphenylalanine (DOPA) to dopamine. There is evidence that dopamine synthesized in the cytoplasm must be readily taken up into vesicles to prevent its decay by monoamine oxidase. Dopamine is released form vesicles through activation of presynaptic membranes by calcium signaling. Dopamine released into the synaptic cleft may trigger activity of the dopamine receptors. Excess released dopamine not taken up by receptors is removed by the dopamine reuptake system including dopamine transporters that restore excess dopamine to the presynapses (von Bohlen und Halbach and Dermietzel, 2002). There is evidence that differences in dopamine receptors types exist in different brain regions.

The enzyme catechol-O-methyltransferase (COMT) plays a major role in degradation of dopamine and also of other catechol neurotransmitters epinephrine and norepinephrine. COMT catalyzes the transfer of methyl groups to the neurotransmitters.

The five dopamine receptors are separated into two groups based on their downstream activators and on responses to different activators. There are also differences in the types of dopamine receptors present in different brain areas.

Striatal dopamine system

Collins and Saunders (2020) reviewed the heterogeneity of the dopamine system in terms of anatomical location and function. They noted information concerning the location of dopamine in the cortical striatal regions that are associated with reward perception and are disrupted in disorders associated with altered motivation and learning impairments.

They focused on three levels of heterogeneity in the dopamine system. Earlier immunochemical studies established that dopaminergic neurons occurred in the midbrain in the retrorubral region, in the substantia nigra compacta, and in the ventral tegmental area. Subsequent studies drew attention to the occurrence of dopaminergic neurons in other regions, including the hypothalamus, retina, and olfactory regions.

Collins and Saunders particularly studied the ventral tegmental and substantial nigra dopamine neurons that projected to the striatum and also to the forebrain. The ventral tegmental region is located on the floor of the midbrain close to the midline. This region has been particularly associated with the reward system and with learning.

Within the striatum, GABAergic neurons were reported to also be abundant and to be modulated both by interneuron circuits and by dopamine release.

Dopamine neurons were reported to be most dense in the ventral striatum that projects to the midline thalamus, amygdala, hippocampus, and prefrontal cortex. Both D1 and D2 dopamine neurons occurred in the ventral striatum.

Collins and Saunders reported evidence that deficiency in dopamine signaling from the ventral tegmental area and the striatal region impaired learning and reward-determined behavior. They noted that dopaminergic neurons also released neuropeptides and that functional effects of this corelease required additional studies. Aspects of the circumstances and events leading to burst release of dopamine also required additional study.

Dopamine receptor subtypes

Klein et al. (2019) reported that five different subtypes of dopamine receptors occur and that all act as metabotropic receptors, leading to formation of second messengers. Dopamine receptors were reported to occur in the central nervous system, blood vessels, heart, adrenal gland, kidney, and retina. They noted that dopamine receptor subtypes D1 and D2 were expressed in the brain. Subtypes D1 and D5 were reported to be coupled to cyclic AMP release.

The genes encoding these subtypes had no introns. Subtypes D2, D3, and D4 were coupled to inhibition of cyclic AMP. The genes encoding these subtypes each had two to three introns.

Dopamine receptors were noted to occur in the central nervous system and also in blood vessels, heart, adrenals, kidney, and retina. Klein et al. identified specific sites of distribution in the brains. D1 receptor–like subunits were distributed in the caudate nucleus, putamen, striatum, nucleus accumbens reticulate structure, amygdala, frontal cortex, and olfactory bulb. D2 receptor–like subunits were distributed in the striatum, globus pallidus ventral tegmental area, hypothalamus, hippocampus amygdala, cortex, and pituitary.

Functions of different areas in which dopamine receptors occur

Caudate region is important in learning, storage, and processing of memories. Putamen is involved in complex feedback with caudate nucleus, nucleus accumbens, and globus pallidus and impacts movement. Striatum is involved in planning and modulations of movements and in cognitive processes. Substantia nigra pars reticulata has high levels of neuromelanin that conveys signals from basal ganglia. Amygdala is part of the limbic system, involved in emotion processing. The frontal cortex participates in consciousness, memory, and attention. The nucleus accumbens facilitates interaction between limbic and motor systems. Hippocampus is involved in storage of recent information. Globus pallidus is involved in movement regulation. The hypothalamus impacts metabolism and endocrine system. The amygdala is involved in processing of emotions and memory.

Gamma-aminobutyric acid and its receptors

GABA is synthesized from glutamate through the activity of glutamate decarboxylases GAD65 and GAD67. GABA can be inactivated through activity of GABA transaminase. GABA is released from inhibitory neurons. Excess GABA in the synaptic cleft or cytoplasm can be removed through activity of GABA transporters (GATs) now identified as solute carriers, in the SLC6A family (von Bohlen und Halbach and Dermietzel, 2002).

GABA uptake is reported to be dependent on binding of sodium (Na^+) and chloride (Cl^-). The chloride binding to GABA was reported to block the interaction with GABA transaminase. Sodium alters the electrochemical gradient and facilitates GABA uptake through different receptors. Two main classes of GABA receptors have been defined that GABAA (GABRA) receptors are reported to be ligand-activated chloride channels and defined as fast response receptors. GABAC receptors have also been identified (GABRR1), a ligand-activated chloride channel. Metabotropic GABA receptors that utilize G protein–coupled receptors include GABBR1 and GABBR2. Other GABA receptors have also been identified such as GABRD encoding delta subunits, GABRE encoding epsilon subunits, GABRG encoding gamma subunits, GABRQ encoding theta subunits, and GABR encoding pi subunits.

Sixteen different subunits including alpha, beta, and rho subunits can be involved in the formation of GABA receptors, and a number of different chromosomes serve as the locations of GABA receptor encoding loci.

- GABRA subunits and the chromosome locations of their encoding genes
- GABRA1 5q34 alpha1 subunit
- GABRA2 4p12 alpha2 subunit
- GABRA3 Xq28 alpha3 subunit
- GABRA4 4p12 alpha4 subunit
- GABRA5 15q12 alpha5 subunit
- GABRA6 5q34 alpha6 subunit
- GABRB1 4p12 beta1 subunit
- GABRB2 5q34 beta2 subunit
- GABRB3 15q12 beta3 subunit

- GABRD 1p36.3 delta
- GABRE Xq28 epsilon
- GABRQ Xq28 theta
- GABRR1 6q15 rho1
- GABRR2 6q15 rho2
- GABRR3 3q11.2 rho3

Metabotropic gamma-aminobutyric acid receptors (G protein—coupled gamma-aminobutyric acid receptors)

- GABBR1 6p22.1 GABABR1, GABA receptor type B
- GABBR2 9q22.3 GABABR2, GABA receptor type B

Fritzius et al. (2017) reported that specific subtypes of GABA receptors were found to associate with potassium channel subunits KCTD8, KCTD12, and KCTD16. The KCTD subunits were found to interact with the receptor and the G protein.

There is evidence that the subunit composition of GABA receptors differs in different parts of the brain (Stefanits et al., 2018). Claxton and Gouaux (2018) analyzed the subunit composition of GABAA receptors. They noted that the receptors consisted of five subunits and that the channel propertied were influences by the subunit composition.

Chua and Chebib (2017) noted that a large number of therapeutic agents have been found to impact GABAA receptors.

Paine et al. (2020) reported that decreased GABAA signaling in studies on rats led to impaired sociability and increased anxiety.

Glutamate receptors

In 2012, Lüscher and Malenka (2012) reported that both AMPA and NMDA glutamate receptors are ionotropic receptors. Binding of neurotransmitter ligand to receptors leads to strong influx of sodium and only to minimal influx of potassium so that depolarization of the neuron results.

Ionotropic glutamate receptors of the *N*-methyl-D-aspartate (NMDA) type were noted to form multisubunit heteromeric protein complexes that form ligand-activated ion channels that play critical roles in synaptic plasticity and learning and memory.

- GRIN1 9q34.3 is a critical subunit of *N*-methyl-D-aspartate receptors.
- GRIN2A 16p13.2 is a glutamate ionotropic receptor NMDA-type subunit 2A.
- GRIN2B 12p13.1 is a glutamate ionotropic receptor NMDA-type subunit 2B.
- GRIN2C 17q25.1 is a glutamate ionotropic receptor NMDA-type subunit 2C.
- GRIN2D 19q13.3 is a glutamate ionotropic receptor NMDA-type subunit 2D.
- GRIN3A 9q31.1 is a glutamate ionotropic receptor NMDA-type subunit 3A.
- GRIN3B19p13.3 is a glutamate ionotropic receptor NMDA-type subunit 3B.

Gibb et al. (2018) reviewed NMDA glutamate receptors structures and functions, noting their critical role in excitatory synaptic transmission and their role in learning and memory. The NMDA receptors were noted to have a tetrameric structure and could be composed of different types of NMDA subunits. They reported that the subunit structure impacted the function of the receptors. There is evidence that the NMDA subunits bind glutamine but at specific sites could also bind glycine.

Skrenkova et al. (2019) reported the effects of mutations in the glycine-binding sites of NMDA receptors. They reported that multiple factors regulate both the number and type of NMDA receptors on neuronal surfaces. These factors include synthesis levels, endoplasmic reticulum processing, tracking of subunits to the cell membranes, and the rates of degradation. The activated NMDS receptor led to calcium infusion.

Rajani et al. (2020) reported that the NMDA receptors composed of four subunits could include different subunits and that the subunit composition altered the receptor ligand binding. Some receptor compositions led only to glutamine binding, and other compositions led to only glycine binding, while some receptor subunit compositions led to binding of both glutamine and glycine. They illustrated the ionotropic function of these receptors noted that ligand activation and receptor opening could lead to influx of calcium, influx of sodium, and efflux of potassium.

The influx of calcium was shown to lead to activation of calmodulin kinase II and activation of protein kinase A phosphatidyl inositol. Activation of calcineurin was associated with phosphatase activity.

AMPA receptor glutamate signaling

The AMPA receptor mediates fast glutamate signaling. The receptor is formed by tetrameric glutamate ionotropic receptor AMPA-type subunits (GRIA subunits GRIA1, 2, 3, and 4). In 2020, van Vugt et al. studied the contributions of NMDA and AMPA reactivity to working memory and determined that both receptor types contributed. Greger et al. (2017) emphasized that function heterogeneity of AMPARs is dependent to obtain a repertoire of subunits and auxillary subunits.

Kamalova and Nakagawa (2021) reported that the functions of the AMPARs are impacted by the receptor subunit composition. The pore-forming domain of each receptor has four domains. They noted evidence that the pore-forming domain could have homotetrameric structure or heterotetrameric structure.

Specific AMPAR regulatory proteins have been identified; these include TARPS transmembrane regulatory proteins and cornichons. TARP transmembrane regulatory proteins were reported to regulate AMPAR trafficking and also include calcium voltage-gated auxillary subunits CANG 1–8. In humans, two cornichons have been defined, CNIH1 and CNIH2, referred to as cornichon family AMPA auxillary proteins. Other AMPA auxillary regulatory proteins include GSGK and SHISA9 (CKAMP44).

Kamalova and Nakagawa noted that lipids may also play roles in determining AMPA complex structure.

AMPA receptor activity was reported to play a major role in synaptic plasticity.

Kainate receptors

It is important to note that kainic acid is neurotoxin that is produced by seaweed; it is, however, used in experimental systems to study neural activity and seizure induction. Kainate binds to ionotropic receptors that are also activated by glutamate.GRIK1, GRIK2, and GRIK3 are defined as subunits that form tetrameric ligand-activated ion channels that can be activated by kainic acid or by glutamate. They are expressed in the brain and in the adrenal gland.

Metabotropic glutamate receptors

These are G protein transmembrane receptors. Suh et al. (2018) reviewed trafficking of glutamate receptors, noting that these receptors undergo endocytosis. Receptor trafficking was influenced by posttranslation modification on the receptor protein subunits. They were reported to move in and out of the synaptic plasma membranes to modulate excitatory synaptic function. Three subclasses of metabotropic glutamate receptors have been defined that differ in their downstream effects. In animals, the receptors are named MGLU1–MGLU8.

In humans, the designation is GRM1–GRM8. Group 1 receptors include GRM1 and GRM5, and they are reported on activations to simulate G protein GNAQ that stimulate phospholipase c, leading to elevation of intracellular calcium and protein kinase C activation. Group 1 receptor activation increases intracellular calcium.

Group II metabotropic glutamate receptors include GRM2 and GRM3.

Group III receptors include GRM4, 6, 7, and 8, and they are coupled to GNAO and GNAI, leading to inhibition of adenyl cyclase and to protein kinase A inhibition. Intracellular calcium is reduced.

Activation of the GRM receptor could lead downstream to release of calcium from intracellular calcium stores.

4. Noradrenergic system

Norepinephrine is derived in the catecholamine synthesis pathway whereby phenyl alanine and tyrosine are converted to dopamine. Dopamine is then converted to noradrenaline (norepinephrine) through activity of the enzyme dopamine beta hydroxylase. Noradrenaline (norepinephrine) is then converted to adrenaline (epinephrine) in specific neurons and in the adrenal gland. Noradrenergic (A1) and adrenergic neurons A2 are mixed in the locus coeruleus.

von Bohlen und Halbach and Dermietzel in 2002 noted evidence that noradrenergic cells are present in the locus coeruleus in the pons of the brain stem, and it sends projections to the cerebral cortex, hippocampus, amygdala, thalamus, hypothalamus, and spinal cord. They emphasized that neuronal projections from the locus coeruleus have abundant collateral extensions.

The noradrenergic system is reported to impact alertness and attention. Norepinephrine and epinephrine interact with adrenoreceptors, which act as G protein–coupled receptors.

Three different types of adrenoreceptors have been described. The different receptor types differ in the intensity of their responses to agonists and in their responses to different antagonists.

Epigenetic interaction in neuronal activity

Belgrad and Fields (2018) reviewed epigenetic interaction in neuronal activity. They documented steps on neuronal function that began with patterned action potential followed by intracellular network activation and specific phosphorylation reactions of specific protein followed by dephosphorylation of other proteins. Action potential also impacts cytoplasmic calcium dynamics activation of transcription factors and regulation of epigenetic mechanism. The latter involves alteration in nucleosome caption and alterations in chromatin looping, histone, and DNA modifications.

Belgrad and Fields emphasized that action potential firing could lead to calcium-dependent and calcium-independent cytosolic signaling.

They emphasize that specialized gene expression in consequence of neuronal firing and signaling requires diverse epigenetic modifications. Importantly, unique neuronal activation patterns and unique epigenome interaction lead to transcription that is stimulus specific.

Belgrad and Fields noted that tests have been undertaken using specific forms of stimulation of neurons in model organism and in cultures neuronal cells. Optogenetic stimulation of channel rhodopsin represents one such study; channel rhodopsins function as light-gated ion channels. Another study utilized heat activation as a stimulus of TRP1 cation channels.

Calcium signaling functions in part through activation of protein kinase C and ERKMAP signaling that ultimately impact transcription factors. These processes include phosphorylation of CREB, a calcium responsive transcription factor. NMDA glutamate receptors are excitatory neurotransmitters reported to activate ERK signaling and calcium calmodulin kinase.

Specific environmental factors were reported to impact nuclear calcium signaling, and calcium was reported to bind nucleosomes and to impact the degree of nucleosome compaction. In addition, different forms of neuronal firing were reported to impact transcription factor binding to genomic sites.

The transcriptional effects of different histone and DNA modification were related to the specific position modified and the degree of modification.

Qureshi and Mehler (2018) reviewed epigenetic modification with particular reference to the nervous system. Key factors in epigenetic modifications. Key factors in epigenetic mechanisms include DNA methylation and hydroxymethylation, histone modifications, nucleosome repositioning and remodeling, and chromatin organization. Also important are noncoding RNAs including short and long nonprotein coding RNA and RNA editing.

Environmental stimuli, gene transcription, and neural activity

In 2018, Yap and Greenberg (2018) reviewed activity-dependent gene transcription in brain. They emphasized that neuronal function must adapt to changing environments. In addition, neurons mus3 encode short- and long-term memories. One adaptations method involved changes in synaptic properties; key aspects of this involve coupling of synaptic activity to nuclear function and gene expression.

Yap and Greenberg emphasized the importance of differentiating activity-regulated changes in gene expression from basal gene transcription. They also emphasized the importance of neuronal transcription and epigenome in bringing about behavioral adaptations to specific stimuli.

Early studies by Greenberg and Ziff (1984) documented that an early response to neuronal stimulation was expression of the FOS gene. These studies led to the conclusion that there was rapid communication of stimulated synapses with the nucleus, leading to expression of FOS transcripts. This transcription was then followed by transcription other genes, and this became known as the immediate early gene (IEG) program of response to stimulation.

The effect of the IEG program was to stimulate a subsequent pattern of gene expression referred to as the late response gene expression program.

Polyamines and ion channels

Polyamines are described as organic compounds with two or more amino groups. Moinard et al. (2005) defined ornithine decarboxylation and condensation processes that led to formation of polyamines putrescine, spermine, and spermine from ornithine and reported that polyamine homeostasis is tightly controlled.

Baroli et al. (2020) described polyamines as positively charged alkylamines essential for a number of different life processes in eukaryotes, including regulation of intracellular signaling. In the brain, polyamines were noted to be involved in modulation of ion channels and glutamate receptors. In addition, there are connections between polyamines, noradrenalin, and serotonin.

Bowie et al. (2018) reported that pores of ionotropic glutamate receptors are accessible to polyamines, spermine, and spermidine and that these large cations can block passage of smaller cation calcium and sodium. Baroli et al. reported that altered polyamine levels had roles in epilepsy.

Voltage-gated sodium and calcium channels

Catterall et al. (2020) reviewed sodium and calcium channels and their roles in transmembrane signaling, action potential initiation, and neurotransmission. In recent years, much information has been gathered on the consequences of defective functions of these channels.

Voltage-gated sodium channels (Nav) were reported to be rapidly activated and to initiate fast action potentials that were also rapidly terminated. The pore-forming structure of the sodium channel was reported to be formed by an alpha subunit and one or two beta subunits.

Voltage-gated sodium channel alpha subunits are encoded by at least 11 different genes. in humans. A cluster of alpha sodium channels occurs on chromosome 2 in the 2q24.2 region; this cluster includes SCN1A, SCN2A, SCN3A, SCN7A, and SCN9A. A smaller cluster of sodium alpha chain encoding genes occurs in the chromosome 3p22.2 region; this cluster includes SCN5A, SCN10A, and SCN11A. A cluster of SCNB subunits occurs on chromosome 11 in the 11q23.3—11q24.1 region; this cluster includes SCN2B SCN3B, and SCN4B.

In studies on sodium channels isolated from muscle and nerves, Catterall et al. reported that the alpha subunit sodium channels have 24 transmembrane segments. The sodium channel B subunits were reported to comprise single membrane spanning glycoprotein segments with a small intracellular domain.

Intense studies have been carried out on pharmacological substances that bind to calcium and sodium ion channels.

Calcium influx and postsynaptic signaling

Puri et al. (2020) reported that calcium influx on activation of NMDA receptors led to GTP-associated molecule exchanges that also involved Rac guanine nucleotide exchange factor and stimulated calmodulin kinases.

A different activation pathway was reported to follow stimulation of AMPA glutamate receptors where ligand activation was not associated with calcium influx. Activation of the AMPA receptor was followed by activation of RAS guanine nucleotide releasing factor and subsequent activation of the MAP kinase (ERK) signaling pathway.

Activation of both of these downstream signaling cascades ultimately activates gene expression.

Postsynaptic signaling

Sheng and Kim (2011) reviewed the postsynaptic organization of synapses. They noted that neurotransmitter receptor channels are embedded in a network composed of anchoring and scaffolding molecules, signaling molecules, and cytoskeletal component. Important postsynaptic components include PSD95 and calcium calmodulin kinase.

Important components of the PSD include cytoskeletal actin, kinases, phosphatases, and GTPases (RAS, RAP, RHO RAC, ARF) and regulators such as guanine nucleotide exchange factors, cell adhesion molecules, metabolism-related molecules, membrane trafficking promoters, and motor proteins chaperones.

Tomasetti et al. (2017) described the PSD as a specialized structure that included multiple different proteins that acted as scaffolds, adapters, regulators, and effectors of the signaling process. These proteins serve then to facilitate routing of signal from the postsynaptic surface region to the nucleus, and key effectors of routing including CAMK (calmodulin-dependent kinases) and MAPK (ERK) protein kinases.

Sudhof et al. (2018) emphasized the importance of signaling between pre- and postsynaptic regions of the synapse and signaling within these two regions of the synapse, and they noted the importance of CAMs.

Wild et al. (2019) described a synapse to nucleus signaling system that involved the activity-dependent transcription factor NFAT. Specifically, they analyzed L-type calcium channel signaling from the dendrite to the neuron soma. Wild et al. noted that many signaling pathways to the nucleus follow influx through L-type calcium channels. Neuron depolarization leads to activity of calcineurin. Calcineurin is defined as a calcium/calmodulin-dependent phosphatase. Calcineurin activity was reported to lead to activation of the transcription factor NFAT.

Perfitt et al. (2020) noted the importance of coupling of scaffolding complexes in the PSD and calcium conducting channels to downstream signaling output. They reported interaction of calcium/calmodulin-dependent protein kinase with SHANK3 a multidomain scaffold protein in the PSD. They noted that this interaction increased phosphorylation of the CREB transcription, and it subsequently increased expression of FOS a transcription modulator.

5. Neuromodulators and neuropeptides

The range of substances included in the neuromodulator category differs in different reviews. In some reviews, they are considered to include dopamine, serotonin, adrenaline, noradrenaline, histamine, and neuropeptides, and in other reviews, nitric oxide is also included as a neuromodulator.

Nitric oxide

In 1995, Kuriyama and Ohkuma reported that nitric oxide was synthesized through conversion of L-arginine to citrulline catalyzed by the enzyme nitric oxide synthase and the coenzyme NADH. Major sources of nitric oxide include endothelial cells of blood vessels, macrophages, and neurons. There is evidence that nitric oxide is synthesized in brain and that it plays major roles in the cerebral vasculature. The key biological effects in the brain include regulation of the cerebral blood flow. In 2010, Vincent reviewed nitric oxide and neurotransmission. Stimulation of NMDA glutamate receptor activity was noted to be associated with an increase in nitric oxide formation.

Small neuromodulators and neuropeptides may be synthesized in the brain, or they may reach the brain in the circulatory system or in the cerebrospinal fluid.

Neuropeptides

At least 55 genes in human are known to encode neuropeptides, and many of these neuropeptides are known to be derived from precursor molecules (ncbi.nlm.nih.gov/gene).

In 2012, van den Pol reviewed neuropeptide transmission in brain circuits. He noted that some neuropeptides are produced in specific brain regions, while others are more widely produced in brain. He also noted interactions between neuromodulator peptides and amino acid neurotransmitters and neuropeptide control of diverse processes. Some neuropeptides were noted to be named according to their functions. Examples included somatostatin that is released from the hypothalamic neurons. A number of neuropeptides are released from the arcuate nucleus of the hypothalamus.

Some neuropeptides that had brain effects were, however, noted to be released from organs and structures at some distance from the brain; van der Pol noted examples that included leptin, released from fatty tissue and ghrelin that is produced in the stomach.

An interesting fact noted by van den Pol was that although many peptides are too large to pass through the blood–brain barrier, this barrier is weak in some regions, and there peptides can pass through, and the medial eminence of the hypothalamus serves as an example of such a region.

Neuropeptides bind to specific receptors that may be close to the site of release, or they may be at some distance from the release site. Some hormones are released from a number of different sites in the nervous system, e.g., dynorphin is a neuropeptide that is released from sympathetic nerves, somatostatin is released in the brain and in digestive system (Kuriyama and Ohkuma, 1995).

Neuropeptides were reported to bind to G protein–coupled receptors Neuropeptide receptors were noted to be expressed at different positions on neurons, e.g., on neuronal cell bodies, dendrites, or axons. Also some neuropeptides were reported to interact with one major G protein–coupled receptor, e.g., GPR54, while others could interact with a number of different G protein–coupled receptors; some neuropeptides were reported to act on certain neurotransmitter receptors. Depolarization of calcium channels was shown to be associated with neuropeptide release.

Neuropeptides were noted to be particularly released from the hypothalamic paraventricular regions and nuclei and neurohypophysis of the pituitary. Neuropeptides in the latter region were often located in dense neurosecretory cores; van den Pol noted that the neuropeptides in other regions were often in smaller vesicles. Neuropeptide signaling was also noted to likely also impact glial cells.

Small molecule and neuropeptide transmission

Nusbaum et al. (2017) reviewed aspects of cotransmission of neural signals and the range of molecules involved. They noted that neurons release several different types of neurotransmitters, including acetylcholine, glutamate, glycine and gGABA, and biogenic amines including histamine, 5-hydroxytryptamine (serotonin), dopamine, noradrenaline, and related octopamine. Neuropeptides released by neurons include purine, adenosine, lipid-derived molecules, gas, and nitric acid.

They noted that signaling molecules released by neurons may bind to ionotropic receptors or to G protein–coupled receptors. Neuropeptides were reported to have primarily neuromodulatory effects. Nusbaum et al. emphasized that cotransmission of signals with different effects provided circuit flexibility.

Russo et al. (2017) reported that more than 100 neuropeptides were known and that families of neuropeptides have been described. Neuropeptides are defined as proteins that are produced and released by neurons and that act on neural substrates. Neuropeptides are released from vesicles and activate cell surface receptors.

Russo et al. also emphasized that neuropeptides are often released from vesicles as proproteins. Also, the proteins undergo post-translational modification and diverse products can be generated from the proprotein. One example is the proopiomelanocortin that can be processed to yield adrenocorticotropic hormone or beta endorphin.

Secretory vesicles that contain neuropeptides are often referred as dense core vesicles. Thus, in presynaptic neurons, these vesicles can be distinguished morphologically from clear vesicles that contain neurotransmitters. Neuropeptides are released by calcium-dependent exocytosis.

Russo et al. noted that within the central nervous system, neuropeptides act as neuromodulators, while in the periphery, they act as signaling molecules. Neuropeptides were reported to enhance or to dampen synaptic activity. Neuropeptides most clearly implicated in brain disorders include corticotrophin-releasing hormones and stress.

Specific classes of neuropeptides

- Opioid family: proenkephalin, proopiomelanocortin, prodynorphin, and orphanin
  - Vasopressin oxytocin family: vasopressin, arginovasopressin, and oxytocin
    - Gastrin cholecystokinin and gastrin families
    - Somatostatin and cortistatin families
    - Amide family
    - Neuropeptide Y and neuropeptide FF
    - Calcitonin family
    - Natriuretic family
    - Neuromedin
    - Glucagon secretin family
    - Corticotropin-releasing hormone and urocortin
    - Motilins and ghrelin
    - Galanin family
    - Relaxin

Neuromodulatory system

Avery and Krichmar (2017) reviewed neuromodulatory systems, noradrenergic, serotonergic, dopaminergic, and cholinergic and their roles in adaptation to environmental stimuli that reach the brain. They particularly emphasized neuromodulator interactions with the amygdala, frontal cortex, hippocampus, and sensory cortex and noted that these regions also interact with each other.

Dopaminergic system: considering dopamine as a neuromodulator

Dopamine was reported to be produced particularly in the ventral tegmental area and in the substantia nigra compacta. Release of dopamine was reported to be stimulated by input from the pedunculotegmental nucleus and lateral habenula.

Inhibition of dopamine release can occur as a result of signals from the striatum and from the cortical and limbic systems. The substantia nigra compacta gives rise to dopaminergic neurons that project to the dorsal striatum. Different dopaminergic receptors occur. Avery and Krichmar reported that D1R receptors are activated by phasic release of dopamine, while D2R receptors are activated by tonic dopamine release. However, exact effects of dopamine on receptors may differ in the striatum and cortex. Avery and Krichmar reported that studies have been carried out on roles of dopaminergic signaling on cognitive function and behavior. D2 dopamine receptors have been reported to play roles in reward processing, while D1 receptor stimulation was reported to play roles in working memory and attention.

Serotonergic system

Serotonin is produced in the raphe nuclei in the brain stem. The raphe nuclei in the brain stem are reported to give rise to projections that extend to the forebrain. There is evidence that serotonin activity in the forebrain influences decision-making and anxiety.

The amino acid tryptophan is essential for serotonin synthesis. Avery and Krichmar noted that acute depletion of tryptophan decreases levels of serotonin and impacts decision-making. Serotonin was reported to play a role in harm avoidance and in control of anxiety. Serotonin acts through binding to specific receptors, sometimes referred to as 5-hydroxytryptamine receptors.

Noradrenergic system

Noradrenergic neurons were reported to originate in the locus in the locus coeruleus in the brain stem and to project widely to cortical and subcortical brain regions. Avery and Krichmar noted that optimal sensory processing was reported to occur when noradrenaline levels were at optimal levels.

Cholinergic system

The cholinergic system and acetylcholine were reported to impact synaptic plasticity and synaptic transmission in many brain areas and to enhance adaptation to environmental stimuli. The central nervous system impact of acetylcholine was reported to differ in activity from that at neuromuscular junctions.

Avery and Krichmar concluded that it is difficult to define a specific function for each neurotransmitter type and that a specific neuromodulator may have different functions in different brain region depending partly on the receptors present in that region. They did note, however, that all neuromodulators played roles in attention and novelty detection.

Endogenous opioid systems

Benarroch in 2012 reviewed three families of endogenous opioid peptides such as beta endorphins, enkephalins, and dynorphins, and he described three families of receptors activated by opioids. The endogenous opioids were reported to be derived from large precursors, endorphin from proopiomelanacortin (POMC), enkephalins from preproenkephalins, and dynorphins from preprodynorphin.

Winters et al. (2017) reported that endogenous opioids regulate neuronal excitability. In 2020, Kissiwaa et al. reported that endogenous opioids inhibited glutamate release from certain synapses.

Endogenous cannabinoids

The endocannabinoid system involves two receptors CB1 (CNR1) and CB2 (CNR2). In 1992, Devane et al. first reported identification of a specific molecule synthesized in brains that acted as a ligand for cannabinoid receptors. Two specific endogenous ligands for cannabinoid receptors were subsequently identified: endocannabinoid substance and anandamide. Specific enzymes involved in the synthesis of endocannabinoids have been identified (Sugiura and Waku, 2002).

Cristino et al. (2020) reported that the CB1 receptor was located presynaptically in excitatory and in inhibitory neurons and that there was evidence that activation of the CB1 receptor impaired presynaptic calcium release and vesicular release of neurotransmitters GABA and glutamate.

They noted that there is also evidence that endocannabinoids may interact not only with CB1 and CB2 receptors but also with TRPV1 receptor transient receptor potential cation channel subfamily V member 1 that is involved in detection of painful stimuli. Postsynaptic CB1 receptors were reported to be present in external membranes of mitochondria and to potentially impact metabolism.

Cristino et al. reported that CB2 receptors were found to be located in microglia and to impact immune modulation. The endocannabinoid system was also reported to impact the sympathetic nervous system and gastrointestinal neuropeptide release. There is also evidence that cannabinoid receptors inhibit adenyl cyclase activity.

Neuropeptide Y

Reichmann and Holzer (2016) reported that neuropeptide Y has stress relieving and neuroprotective properties. NPY is highly expressed in the central nervous system in hypothalamus and limbic systems. The limbic system includes cortical and subcortical areas including the amygdala and nucleus accumbens, and the main feature of the limbic system is to serve as a connection between the cortex and subcortical regions. Five different G protein—coupled receptors were reported to bind neuropeptide Y.

Vasointestinal peptide

Vasointestinal peptide (VIP) was reported to function as a neuromodulator neuropeptide in a number of different tissues and organs. It is synthesized in the intestine. However, highest levels of VIP were reported to occur in the cerebral cortex hippocampus, amygdala, and hypothalamus (White et al., 2010).

Kisspeptin

Hellier et al. (2018) reported that the neuropeptide kisspeptin is produced by neurons in the hypothalamus in the rostral periventricular regions and in the arcuate nucleus. Kisspeptin-producing neurons were reported to project to the gonadotropin-releasing neurons in the arcuate neurons. Kisspeptin binds to a specific G protein—coupled receptor GPR54. Kisspeptin receptors occur in the gonadotrophin-releasing neurons. They were also reported to be present in fat tissues. The GPR54 receptor to which kisspeptin binds is encoded in human on chromosome 19p13.2 Trevisan et al. (2018) reported that specific mutations in GPR54 had been identified in individuals with hypotrophic hypogonadism. Transgenic mice with defects in the Kiss1 Gpr54 receptors were found to have hypotrophic gonads and impaired sexual maturation.

Trevissan et al. reported that kisspeptin is derived from preprokisspeptin. They noted that kisspeptin signaling through GPR54 increased calcium signaling and increased downstream MAPK signaling and promoted release of gonadotrophin-releasing hormone.

References

Avery, M.C., Krichmar, J.L., 2017. Neuromodulatory systems and their interactions: a review of models, theories, and experiments. Front. Neural Circ. 11 https://doi.org/10.3389/fncir.2017.00108.

Baroli, G., Sanchez, J.R., Agostinelli, E., Mariottini, P., Cervelli, M., 2020. Polyamines: the possible missing link between mental disorders and epilepsy (Review). Int. J. Mol. Med. 45 (1), 3–9. https://doi.org/10.3892/ijmm.2019.4401.

Bauer, C.K., Calligari, P., Radio, F.C., Caputo, V., Dentici, M.L., Falah, N., High, F., Pantaleoni, F., Barresi, S., Ciolfi, A., Pizzi, S., Bruselles, A., Person, R., Richards, S., Cho, M.T., Claps Sepulveda, D.J., Pro, S., Battini, R., Zampino, G., et al., 2018. Mutations in KCNK4 that affect gating cause a recognizable neurodevelopmental syndrome. Am. J. Hum. Genet. 103 (4), 621–630. https://doi.org/10.1016/j.ajhg.2018.09.001.

Beaulieu, J.M., Espinoza, S., Gainetdinov, R.R., 2015. Dopamine receptors – IUPHAR review 13. Br. J. Pharmacol. 172 (1), 1–23. https://doi.org/10.1111/bph.12906.

Belgrad, J., Fields, R.D., 2018. Epigenome interactions with patterned neuronal activity. Neuroscientist 24 (5), 471–485. https://doi.org/10.1177/1073858418760744.

Benarroch, E.E., 2012. Endogenous opioid systems: current concepts and clinical correlations. Neurology 79 (8), 807–814. https://doi.org/10.1212/WNL.0b013e3182662098.

Borrie, S.C., Brems, H., Legius, E., Bagni, C., et al., 2017. Cognitive Dysfunctions in Intellectual Disabilities: The Contributions of the Ras-MAPK and PI3K-AKT-mTOR Pathways. Annu. Rev. Genomics Hum. Genet. 18, 115–142. https://doi.org/10.1146/annurev-genom-091416-035332.

Bowie, D., 2018. Polyamine-mediated channel block of ionotropic glutamate receptors and its regulation by auxiliary proteins. J. Biol. Chem. 293 (48), 18789–18802. https://doi.org/10.1074/jbc.TM118.003794.

Brini, M., Carafoli, E., Calì, T., 2017. The plasma membrane calcium pumps: focus on the role in (neuro)pathology. Biochem. Biophys. Res. Commun. 483 (4) https://doi.org/10.1016/j.bbrc.2016.07.117 (Review.PMID).

Carhart-Harris, R., Nutt, D., 2017. Serotonin and brain function: a tale of two receptors. J. Psychopharmacol. 31 (9), 1091–1120. https://doi.org/10.1177/0269881117725915.

Catterall, W., Lenaeus, M., 2020. Gamal El-Din TM structure and pharmacology of voltage-gated sodium and calcium channels. Annu. Rev. Pharmacol. Toxicol. 60 https://doi.org/10.1146/annurev-pharmtox-010818-021757 (PMID).

Chua, H.C., Chebib, M., 2017. GABAA receptors and the diversity in their structure and pharmacology. In: Advances in Pharmacology, vol. 79. Academic Press Inc, pp. 1–34. https://doi.org/10.1016/bs.apha.2017.03.003.

Clark, R.B., 2013. 2012 Nobel Laureates in Chemistry. Proceedings of the National Academy of Sciences 110 (14), 5274–5275. https://doi.org/10.1073/pnas.1221820110.

Claxton, D.P., Gouaux, E., 2018. Expression and purification of a functional heteromeric GABAA receptor for structural studies. PLoS One 13 (7). https://doi.org/10.1371/journal.pone.0201210.

Collins, A.L., Saunders, B.T., 2020. Heterogeneity in striatal dopamine circuits: form and function in dynamic reward seeking. J. Neurosci. Res. 98 (6), 1046–1069. https://doi.org/10.1002/jnr.24587.

Cristino, L., Bisogno, T., Di Marzo, V., 2020. Cannabinoids and the expanded endocannabinoid system in neurological disorders. Nat. Rev. Neurol. 16 (1), 9–29. https://doi.org/10.1038/s41582-019-0284-z.

Dickson, E., 2019. Recent advances in understanding phosphoinositide signaling in the nervous system. F1000Res. 12 (8) https://doi.org/10.12688/f1000research.16679.1.

Frank, R.A., Grant, S.G., 2017. Supramolecular organization of NMDA receptors and the postsynaptic density. Curr. Opin. Neurobiol. 45, 139–147. https://doi.org/10.1016/j.conb.2017.05.019.

Fritzius, T., Turecek, R., Seddik, R., Kobayashi, H., Tiao, J., Rem, P.D., Metz, M., Kralikova, M., Bouvier, M., Gassmann, M., Bettler, B., 2017. KCTD hetero-oligomers confer unique kinetic properties on hippocampal GABAB receptor-induced K+ currents. J. Neurosci. 37 (5), 1162–1175. https://doi.org/10.1523/JNEUROSCI.2181-16.2016.

Gibb, A., Ogden, K., McDaniel, M., Vance, K., Kell, et al., 2018. A structurally derived model of subunit-dependent NMDA receptor function. J. Physiol. 596 (17) https://doi.org/10.1113/JP276093. PMID.

Gray, E.G., 1959. Electron microscopy of synaptic contacts on dendrite spines of the cerebral cortex. Nature 183 (4675), 1592–1593. https://doi.org/10.1038/1831592a0.

Greenberg, M.E., Ziff, E.B., 1984. Stimulation of 3T3 cells induces transcription of the c-fos proto-oncogene. Nature 311 (5985), 433–438. https://doi.org/10.1038/311433a0.

Greger, I.H., Watson, J.F., Cull-Candy, S.G., 2017. Structural and functional architecture of AMPA-type glutamate receptors and their auxiliary proteins. Neuron 94 (4), 713–730. https://doi.org/10.1016/j.neuron.2017.04.009.

Hellier, V., Brock, O., Candlish, M., Desroziers, E., Aoki, M., Mayer, C., Piet, R., Herbison, A., Colledge, W.H., Prévot, V., Boehm, U., Bakker, J., 2018. Female sexual behavior in mice is controlled by kisspeptin neurons. Nat. Commun. 9 (1) https://doi.org/10.1038/s41467-017-02797-2.

Huang, Y., Thathiah, A., 2015. Regulation of neuronal communication by G protein-coupled receptors. Fed. Eur. Biochem. Soc. Lett. 589 (14), 1607–1619. https://doi.org/10.1016/j.febslet.2015.05.007.

Huttenlocher, P.R., 1974. Dendritic development in neocortex of children with mental defect and infantile spasms. Neurology 24 (4), 203–210. https://doi.org/10.1212/wnl.24.3.203.

Imbrici, P., Camerino, D.C., Tricarico, D., May 2013. Major channels involved in neuropsychiatric disorders and therapeutic perspectives. Front. Genet. 4 https://doi.org/10.3389/fgene.2013.00076.

Ismail, F.Y., Fatemi, A., Johnston, M.V., 2017. Cerebral plasticity: windows of opportunity in the developing brain. Eur. J. Paediatr. Neurol. 21 (1), 23–48. https://doi.org/10.1016/j.ejpn.2016.07.007.

Joensuu, M, 2020. Phospholipases in neuronal function: A role in learning and memory? J. Neurochem. 153 (3), 300–333. https://doi.org/10.1111/jnc.14918.

Kaizuka, T., Takumi, T., 2018. Postsynaptic density proteins and their involvement in neurodevelopmental disorders. J. Biochem. 163 (6), 447–455. https://doi.org/10.1093/jb/mvy022.

Kamalova, A, Nakagawa, T, 2021. Kamalova A, Nakagawa T. AMPA receptor structure and auxiliary subunits. J. Physiol. 599 (2), 453–469. https://doi.org/10.1113/JP278701.

Káradóttir, R.T., Kuo, C.T., 2018. Neuronal activity-dependent control of postnatal neurogenesis and gliogenesis. Annu. Rev. Neurosci. 41, 139–161. https://doi.org/10.1146/annurev-neuro-072116-031054.

Kim, H.Y., Huang, B.X., Spector, A.A., 2014. Phosphatidylserine in the brain: metabolism and function. Prog. Lipid Res. 56 (1), 1–18. https://doi.org/10.1016/j.plipres.2014.06.002.

Klein, M.O., Battagello, D.S., Cardoso, A.R., Hauser, D.N., Bittencourt, J.C., Correa, R.G., 2019. Dopamine: functions, signaling, and association with neurological diseases. Cell. Mol. Neurobiol. 39 (1), 31–59. https://doi.org/10.1007/s10571-018-0632-3.

Komatsu, H., Fukuchi, M., Habata, Y., 2019. Potential utility of biased GPCR signaling for treatment of psychiatric disorders. Int. J. Mol. Sci. 20 (13) https://doi.org/10.3390/ijms20133207.

Koopmans, F., van Nierop, P., Andres-Alonso, M., Byrnes, A., Cijsouw, T., Coba, M.P., Cornelisse, L.N., Farrell, R.J., Goldschmidt, H.L., Howrigan, D.P., Hussain, N.K., Imig, C., de Jong, A.P.H., Jung, H., Kohansalnodehi, M., Kramarz, B., Lipstein, N., Lovering, R.C., MacGillavry, H., et al., 2019. SynGO: an evidence-based, expert-curated knowledge base for the synapse. Neuron 103 (2), 217–234.e4. https://doi.org/10.1016/j.neuron.2019.05.002.

Kuriyama, K., Ohkuma, S., 1995. Role of nitric oxide in central synaptic transmission: effects on neurotransmitter release. Jpn J. Pharmacol. 69 (1), 1–8. https://doi.org/10.1254/jjp.69.1.

Lefkowitz, R., 2012. Nobel Lecture. https://www.nobelprize.org/prizes/chemistry/2012/lefkowitz/lecture/.

Leslie, S.N., Nairn, A.C., 2019. cAMP regulation of protein phosphatases PP1 and PP2A in brain. Biochim. Biophys. Acta Mol. Cell Res. 1866 (1), 64–73. https://doi.org/10.1016/j.bbamcr.2018.09.006.

Logue, J.S., Scott, J.D., 2010. Organizing signal transduction through A-kinase anchoring proteins (AKAPs). FEBS J. 277 (21), 4370–4375. https://doi.org/10.1111/j.1742-4658.2010.07866.x.

Lopez, J., 2020. TRPC channels in the SOCE scenario. Cells 9 (1). https://doi.org/10.3390/cells9010126.

Lüscher, C., Malenka, R.C., 2012. NMDA receptor-dependent long-term potentiation and long-term depression (LTP/LTD). Cold Spring Harbor Perspectives in Biology 4 (6), 1–15. https://doi.org/10.1101/cshperspect.a005710.

Malerba, N., Nittis, D., Merla, G., 2019. The emerging role of Gβ subunits in human genetic diseases. Cells Vol. 8. https://doi.org/10.3390/cells8121567 (Review. PMID).

Mochida, S., 2019. Presynaptic calcium channels. Int. J. Mol. Sci. 20 (9) https://doi.org/10.3390/ijms20092217.

Moinard, C., Cynober, L., de Bandt, J.P., 2005. Polyamines: metabolism and implications in human diseases. Clin. Nutr. 24 (2), 184–197. https://doi.org/10.1016/j.clnu.2004.11.001.

National Institute of Mental Health, n.d. National Institutes of Health, Department of Health and Human Services.

Nishiyama, J., 2019. Plasticity of dendritic spines: molecular function and dysfunction in neurodevelopmental disorders. Psychiatr. Clin. Neurosci. 73 (9), 541–550. https://doi.org/10.1111/pcn.12899.

Nusbaum, M.P., Blitz, D.M., Marder, E., 2017. Functional consequences of neuropeptide and small-molecule co-transmission. Nat. Rev. Neurosci. 18 (7), 389–403. https://doi.org/10.1038/nrn.2017.56.

Paine, T.A., Chang, S., Poyle, R., 2020. Contribution of GABAA receptor subunits to attention and social behavior. Behav. Brain Res. 378 https://doi.org/10.1016/j.bbr.2019.112261.

Perfitt, T.L., Wang, X., Dickerson, M.T., Stephenson, J.R., Nakagawa, T., Jacobson, D.A., Colbran, R.J., 2020. Neuronal L-type calcium channel signaling to the nucleus requires a novel CaMKIIα-SHANK3 interaction. J. Neurosci. 40 (10), 2000–2014. https://doi.org/10.1523/JNEUROSCI.0893-19.2020.

Poke, G., King, C., Muir, A., de Valles-Ibáñez, G., Germano, M., Moura de Souza, C.F., Fung, J., Chung, B., Fung, C.W., Mignot, C., Ilea, A., Keren, B., Vermersch, A.I., Davis, S., Stanley, T., Moharir, M., Kannu, P., Shao, Z., Malerba, N., et al., 2019. The epileptology of GNB5 encephalopathy. Epilepsia 60 (11), e121–e127. https://doi.org/10.1111/epi.16372.

Puri, B.K., 2020. Calcium signaling and gene expression. In: Advances in Experimental Medicine and Biology, vol. 1131. Springer New York LLC, pp. 537–545. https://doi.org/10.1007/978-3-030-12457-1_22.

Qureshi, I.A., Mehler, M.F., 2018. Epigenetic mechanisms underlying nervous system diseases. In: Handbook of Clinical Neurology, vol. 147. Elsevier B.V, pp. 43–58. https://doi.org/10.1016/B978-0-444-63233-3.00005-1.

Rajani, V., Sengar, A.S., Salter, M.W., 2020. Tripartite signalling by NMDA receptors. Mol. Brain 13 (1). https://doi.org/10.1186/s13041-020-0563-z.

Rall, W., Sutherland, E., 1958. Formation of a cyclic nucleotide in tissue particles. J. Biol Chem 232, 1065–1076.

Ramon y Cajal, S., 1888. Estructura de los centros nerviosos de las aves. Rev. Trim. Histol. Norm. 1, 1–10.

Rauen, K.A., 2013. The RASopathies. Annu. Rev. Genom. Hum. Genet. 14, 355–369. https://doi.org/10.1146/annurev-genom-091212-153523.

Reichmann, F., Holzer, P., 2016. Neuropeptide Y: a stressful review. Neuropeptides 55, 99–109. https://doi.org/10.1016/j.npep.2015.09.008.

Rochefort, N., Konnerth, A., 2012. Dendritic spines: from structure to in vivo function. EMBO Rep. 13 (8), 699–708. https://doi.org/10.1038/embor.2012.102.

Russo, A.F., 2017. Overview of neuropeptides: awakening the senses? Headache 57, 37–46. https://doi.org/10.1111/head.13084.

Selbie, L.A., Hill, S.J., 1998. G protein-coupled-receptor cross-talk: the fine-tuning of multiple receptor-signalling pathways. Trends Pharmacol. Sci. 19 (3), 87–93. https://doi.org/10.1016/S0165-6147(97)01166-8.

Sheng, M., Kim, E., 2011. The postsynaptic organization of synapses. Cold Spring Harb. Perspect. Biol. 3 (12) https://doi.org/10.1101/cshperspect a005678. PMID: 22046028.

Skrenkova, K., Hemelikova, K., Kolcheva, M., Kortus, S., Kaniakova, M., Krausova, B., Horak, M., 2019. Structural features in the glycine-binding sites of the GluN1 and GluN3A subunits regulate the surface delivery of NMDA receptors. Sci. Rep. 9 (1) https://doi.org/10.1038/s41598-019-48845-3.

Smith, A.C., Robinson, A.J., 2019. MitoMiner v4.0: an updated database of mitochondrial localization evidence, phenotypes and diseases. Nucleic. Acids Res. 47 (D1), D1225–D1228. https://doi.org/10.1093/nar/gky1072.

Soler, J., Fañanás, L., Parellada, M., Krebs, M.O., Rouleau, G.A., Fatjó-Vilas, M., 2018. Genetic variability in scaffolding proteins and risk for schizophrenia and autism-spectrum disorders: a systematic review. J. Psychiatry Neurosci. 43 (4), 223–244.

Stefanits, H., Milenkovic, I., Mahr, N., Pataraia, E., Hainfellner, J.A., Kovacs, G.G., Sieghart, W., Yilmazer-Hanke, D., Czech, T., 2018. GABAA receptor subunits in the human amygdala and hippocampus: immunohistochemical distribution of 7 subunits. J. Comp. Neurol. 526 (2), 324–348. https://doi.org/10.1002/cne.24337.

Südhof, T.C., 2018. Towards an understanding of synapse formation. Neuron 100 (2), 276–293. https://doi.org/10.1016/j.neuron.2018.09.040.

Sugiura, T., Waku, K., 2002. Cannabinoid receptors and their endogenous ligands. J. Biochem. 132 (1), 7–12. https://doi.org/10.1093/oxfordjournals.jbchem.a003200.

Suh, Y.H., Chang, K., Roche, K.W., 2018. Metabotropic glutamate receptor trafficking. Mol. Cell. Neurosci. 91, 10–24. https://doi.org/10.1016/j.mcn.2018.03.014.

Syrovatkina, V., Alegre, K.O., Dey, R., Huang, X.Y., 2016. Regulation, signaling, and physiological functions of G-proteins. J. Mol. Biol. 428 (19), 3850–3868. https://doi.org/10.1016/j.jmb.2016.08.002.

Tomasetti, C., Iasevoli, F., Buonaguro, E.F., De Berardis, D., Fornaro, M., Fiengo, A.L.C., Martinotti, G., Orsolini, L., Valchera, A., Di Giannantonio, M., De Bartolomeis, A., 2017. Treating the synapse in major psychiatric disorders: the role of postsynaptic density network in dopamine-glutamate interplay and psychopharmacologic drugs molecular actions. Int. J. Mol. Sci. 18 (1) https://doi.org/10.3390/ijms18010135.

Trevisan, C., Montagna, e, Oliveira, Christofolini, D., Barbosa, C., et al., 2018. Kisspeptin/GPR54 system: what do we know about its role in human reproduction? Cell. Physiol. Biochem. 49 (4) https://doi.org/10.1159/000493406 (PMID).

van den Pol, A.N., 2012. Neuropeptide transmission in brain circuits. Neuron 76 (1), 98–115. https://doi.org/10.1016/j.neuron.2012.09.014.

von Bohlen und Halbach, O., Dermietzel, R., 2002. Neurotransmitters and Neuromodulators. Wiley-VCH Verlag GmbH.

Weir, R.K., Bauman, M.D., Jacobs, B., Schumann, C.M., 2018. Protracted dendritic growth in the typically developing human amygdala and increased spine density in young

ASD brains. J. Comp. Neurol. 526 (2), 262–274. https://doi.org/10.1002/cne.24332.

White, C.M., Ji, S., Cai, H., Maudsley, S., Martin, B., 2010. Therapeutic potential of vasoactive intestinal peptide and its receptors in neurological disorders. CNS Neurol. Disord. – Drug Targets 9 (5), 661–666. https://doi.org/10.2174/187152710793361595.

Wild, A.R., Sinnen, B.L., Dittmer, P.J., Kennedy, M.J., Sather, W.A., Dell'Acqua, M.L., 2019. Synapse-to-Nucleus communication through NFAT is mediated by L-type Ca^{2+} channel Ca^{2+} spike propagation to the soma. Cell Rep. 26 (13), 3537–3550.e4. https://doi.org/10.1016/j.celrep.2019.03.005.

Winters, B.L., Gregoriou, G.C., Kissiwaa, S.A., Wells, O.A., Medagoda, D.I., Hermes, S.M., Burford, N.T., Alt, A., Aicher, S.A., Bagley, E.E., 2017. Endogenous opioids regulate moment-to-moment neuronal communication and excitability. Nat. Commun. 8 https://doi.org/10.1038/ncomms14611.

Yap, E.L., Greenberg, M.E., 2018. Activity-regulated transcription: bridging the gap between neural activity and behavior. Neuron 100 (2), 330–348. https://doi.org/10.1016/j.neuron.2018.10.013.

Ye, X., Carew, T.J., 2010. Small G protein signaling in neuronal plasticity and memory formation: the specific role of Ras family proteins. Neuron 68 (3), 340–361. https://doi.org/10.1016/j.neuron.2010.09.013.

Yuste, R., Majewska, A., 2001. On the function of dendritic spines. Neuroscientist 7 (5). https://doi.org/10.1177/107385840100700508. PMID:11597098.

CHAPTER

3

Brain mapping

1. Introduction

Passingham et al. (2002) noted the limitations inherent in defining cortical regions functions on the basis of lesion studies. They stressed the advantages of application of brain mapping, particularly task-based brain mapping, to analyze functions of brain regions. On the basis of results of such studies, there is evidence that each brain region has a unique pattern of connections, and they proposed that each distinct area of the brain has unique extrinsic inputs and unique output connections, referred to as the connectional fingerprint. Passingham et al. noted that the concept of unique connectional fingerprints was initially supported by studies in primates and felines.

2. Normal brain and connectivity

There is a growing body of information on connections, networks, and functions in the brain. These have been investigated through a number of different techniques, primarily functional magnetic resonance imaging (fMRI), but also through electroencephalography and to a limited extent by magnetoencephalography (MEG). In addition to evidence of activity within and connections between brain regions during specific tasks, there is evidence of firing and connectivity in specific brain regions during rest. Core brain connectivity networks currently defined include the default-mode networks (DMNs), dorsal attention network, visual foveal network, and dorsal somatic motor network. The DMN and the frontoparietal network were reported to cooperate.

Connectivity and networks

In 2010, van den Heuvel and Pol reported that connectivity between specific distant brain regions can be assessed in resting-state functional magnetic resting-state imaging. These studies implied relationship in activation patterns of anatomically separated regions. Particular emphasis was initially placed on specific resting-state networks. These included a motor network, a visual network, and a network defined as the DMN described to include the medial frontal cortex, the inferior parietal cortex, and the temporal lobe, and a network that included bilateral temporal regions the insula and the anterior cingulate cortex; these regions overlapped with the attention network.

In considering possible functions of the DMNs, van den Heuvel and Pol noted that perhaps the activity served to keep functionality in the systems so that on activation the reaction times would be reduced. Petersen and Sporns (2015) referred to cognitive architectures as brain regions involved in performance of specific tasks or functions and noted that such regions are

Mechanisms and Genetics of Neurodevelopmental Cognitive Disorders
https://doi.org/10.1016/B978-0-12-821913-3.00004-4

sometimes referred to as networks. However, they stressed a more formal concept of networks, as regions having evidence of pairwise relationships between their elements. They addressed specific efforts undertaken to examine anatomically connected regions involved in a specific functional activity and noted that anatomical connections serve as the skeleton for passage of signals. They emphasized that the current notion was that cognitive functions depended on dynamic cortical and subcortical networks. Early studies analyzed the afferent and efferent connections between specific regions. Diffusion tensor imaging (DTI) and tractography focused on white matter connectivity and served to generate data on physical and anatomical connections. In addition to studies on long-range connections, more recent studies also provided information on short-range connections.

However, they noted that brain connectivity can be detected by fMRI (bold fMRI) in regions and that are anatomically connected and that functional connectivity could be detected during performance of a specific task. They noted a possibility that functional relationships between regions may reflect simultaneous changes in synaptic efficiencies. Specific studies have been carried out to examine coactivation of sites during a specific cognitive task.

Petersen and Sporns (2015) stressed that resting-state connectivity does not represent anatomical connectivity. It is important to note that a major DMN is now thought to be active not only during rest but also during quiet contemplation, during planning, and in contemplation of likely actions of others (theory of mind) and during meditation. The DMN is discussed further in the following in the context of its role in specific brain functions.

White matter connections

Over many years, anatomists and neurophysiologists have documented the physical pathways of major tracts and nerves in the nervous system and connections of the major tracts with specific brain regions. It is important to note that there are long-range connectivity tracts and short-range white matter connections.

Magnetic resonance imaging (MRI) of the brain was first used in a human in 1977. Becker (1993) noted that the technique was based on work of Isador Rabi who investigated properties of atoms exposed to strong magnetic fields. In the presence of a strong magnetic field emitted by a specific device, protons within the tissue align and then subsequently relax. These two processes enable the generation of images referred to as T1 and T2. The magnetic field causes protons to align in specific directions parallel or antiparallel, based on the properties of the proton. In the relaxation phase, the protons reassume their initial positions. In medical studies, parameters are geared to the properties of hydrogen that has a single proton. MRI has become the method of choice in medical imaging. fMRI is designed to detect the blood flow in a specific brain area during performances of a specific task. Detection is based on the contrast generated by blood oxygen levels as is referred to as BOLD fMRI.

Connectivity detected by diffusion weighted and diffusion tensor imaging

This is a modified form of MRI that is specifically designed to image passage of water molecules in a fibrous environment. This is carried out to image white matter that corresponds to connection fibers.

Specific white matter tracts

In addition to long white matter tracts, there are additional connections, e.g., commissural and associations connections and short axonal connections, Swanson et al. (2017) emphasized that cognition is supported by complex axonal connections that exist between regions. In addition, there are connections between hemispheres. These occur particularly in the corpus callosum.

Herbet and Duffau (2020) emphasized that new mapping techniques have led to new insights in anatomofunctional architecture and into brain functions (Fig. 3.1). They noted that previously brain functions were assigned to specific locations.

More recent studies emphasized spatiotemporal integration and greater connectivity through distributed networks. They also noted that interactions between regions can be reshaped over time, promoting neuroplasticity.

Herbet and Duffau initially reviewed the localization model in which a specific function was assigned to a specific brain area. They noted that the initial assignments of function to specific areas were based on lesion analysis. They suggested that these rigid interpretations led to simplistic models of cognition.

They noted that advances came particularly with noninvasive functional neuroimaging and information was also obtained during awake surgery through direct electrostimulation. Information on recovery of a specific function, despite specific brain lesion, led to concepts of neuroplasticity.

Herbet and Duffau noted that white matter tracts have received more attention during recent decades. There is new information on extensive connections between regions as DTI white matter analyses were applied.

Task-based fMRI studies capture metabolic activity in brain regions during performance of a specific task. A process referred to as effective connectivity analyzes how one region interacts with another during performance of a specific task.

3. Brain morphometry to gather information on structure

The purpose of brain morphometry is to utilize noninvasive neuroimaging techniques, including MRI, to gather information on overall brain structure and on features of specific brain structures within the brain (Good et al., 2001). Data can be gathered on size shape, volume and density of gray matter, patterns of gyri, white matter density, and connectivity. In addition, information can be obtained on ventricle size and cerebrospinal fluid volume, including axial and nonaxial cerebrospinal fluid volume. Data are then entered into specific data bases and comparisons drawn between similar or dissimilar cohorts for neuroinformatic analyses.

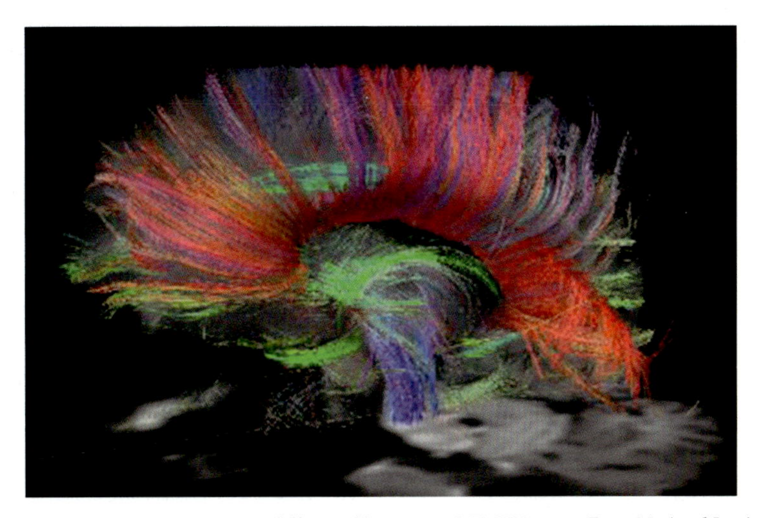

FIGURE 3.1 **Human brain connectome.** Note different fiber types. NIMH image. *From National Institutes of Health, Department of Health and Human Services. https://images.nimh.nih.gov/public_il/searchresults.cfm.*

Specific brain regions and connections

Brain nuclei

In the brain, structures referred to as nuclei are clusters of neurons connected to a specific neural tract. In the peripheral nervous system, a cluster of neurons connected to a specific tract is referred to as a ganglion.

Specific brain nuclei are in some cases named according to their position in the brain; in other cases, they are named according to the functions of the nerves that terminate in them. Other nuclei have the names of individuals who documented their positions or functions.

In 2014, Telford and Vattoth described use of MRI to study brain nuclei. Advances in MRI improved analysis of subcortical brain nuclei. Furthermore, recent studies had revealed the importance of subcortical brain nuclei in cognition and behavior.

Basal ganglia functional anatomy and connections

Bostan et al. (2018) reviewed the connections and function of the basal ganglia. They noted that the functions of the basal ganglia can be assessed through the connections.

The basal ganglia are subcortical nuclei positions at the base of the forebrain and above the midbrain. The basal ganglia region is sometimes referred to as the striatum. The dorsal striatum includes the caudate nucleus and the putamen. The ventral striatum includes the nucleus accumbens and olfactory nucleus. Other components of the basal ganglia include the globus pallidus, ventral pallidum, substantia nigra, and subthalamic nuclei. The substantia nigra has two components the pars compacta and the pars reticula. Bostan et al. documented the major connections of the basal ganglia.

Corticobasal ganglia connections and the thalamocortical circuit

The cerebral cortex sends afferent connections to the basal ganglia. These afferents pass to the striatum, including the putamen, caudate, ventral striatum, and subthalamic nuclei. Output signals of the basal ganglia to the cerebral cortex take place through the thalamus. These output projections are from the globus pallidus, substantia nigra, and the ventral pallidum. It is important to note that dopamine is synthesized in the substantia nigra.

Bostan et al. noted that the striatum has three territories, sensorimotor territory, an association-cognitive information processing territory, and a limbic territory involved in processing emotional and motivational information.

They also noted that efferents from the basal ganglia terminate in different regions of the thalamus.

Ultimately, efferents from the basal ganglia reach five different cortical areas. These connections then indicate that the basal ganglia influence not only the frontal cortex but also the parietal and temporal cortex. Importantly the basal ganglia influence not only movement but also cognitive and affective functions.

Bostan et al. note that this information has relevance for disorders in which the basal ganglia are affected. Defective basal ganglia function has been primarily associated with Parkinson's disease and dystonia. However, there is evidence that altered basal ganglia connections may be implicated in certain behavioral disorders. In Parkinson's disease and in Huntington's disease, initial dysfunction leads to motor changes; however, cognitive and affective changes subsequently arise. In disorders of the basal ganglia, signal from these structures may be excessive, leading to inhibition of thalamocortical circuit activity. In some disorders, there is decreased signal from the basal ganglia that lead to failure in suppression of unwanted movements.

In number of disorders that impact the basal ganglia, there is evidence of abnormal iron accumulation (Levi and Tiranti, 2019).

Basal ganglia defects have also been reported to lead to dyskinetic cerebral palsy (Aravamuthan and Waugh, 2016). Monbaliu et al. (2017) reported that dyskinetic cerebral palsy is due to nonprogressive lesions in the basal ganglia or thalamus. Motor system abnormalities have been attributed to dysfunction of the subcortical nuclei that include the thalamus and the basal ganglia.

Haber et al. (2016) reviewed corticostriatal connections. The striatum is included in the basal ganglia. Haber noted that concepts of basal ganglia functions have expanded in recent decades to include not only sensory motor functions but also goal-directed functions. They also defined connections between the cortex and dorsal striatum and the ventral striatum and noted anterograde connections from the frontal cortex and the thalamus to the thalamus to the dorsal and ventral striatum and also retrograde connections from the striatum.

Thalamus

This is midline gray matter structure with two halves located between the cerebral cortex and the midbrain. Rikhye et al. (2018) reviewed thalamic function. They noted that the thalamus has long been considered to play a role in cognition, but its precise impact was not known. They proposed that the thalamus is involved in cognitive computation.

Rikhye et al. defined the mammalian forebrain as composed of telencephalon that includes cortex and basal ganglia and diencephalon that includes thalamus and hypothalamus. In the course of evolution, the cortex, basal ganglia and thalamus were reported to undergo expansion and variation, while the hypothalamus revealed minimal changes.

They noted that differences exist between the cortex and the thalamus with the cortex noted to include both excitatory and inhibitory neurons while the thalamus gray matter structure contains primarily excitatory neurons.

Regions of the vertebrate thalamus include anterior, medial ventral, and posterior nuclei. Rikye et al. concluded that there was only a rudimentary understanding of the basis of the thalamus in cognitive function and that further analyses of thalamic circuits were necessary.

Hwang et al. (2017) undertook a study to define relationship of different thalamic nuclei to functional brain networks. Their study utilized resting-state fMRI data and graphic theoretic network analyses. They noted that the thalamus is widely connected to cortical regions. They noted further that while the thalamus was widely thought to primarily function as a relay station for sensory signaling, there is now evidence that it serves to integrate information through signaling and interactions with different cortical regions. They noted that thalamic nuclei have been classified in first-order and higher-order categories. First-order thalamic nuclei include the lateral geniculate nucleus and the ventral posterior nuclei to which ascending sensory pathways sent signal. The higher-order thalamic nuclei were reported to have connections to cortical areas. The higher-order thalamic nuclei included the pulvinar and medial dorsal nuclei.

Results of their study revealed that specific single thalamic subdivision had connections with multiple functional networks. Hwang et al. noted that the thalamus served to integrate information across networks.

Thalamic reticular region functions and dysfunctions

The thalamic reticular region, sometimes referred to as the thalamic nucleus, was noted to wrap round the thalamus, particularly on the ventral side, to form a capsule around the thalamus. This region was reported to primarily have contact with other thalamic nuclei. There are reports that this region may impact certain aspects of behavior.

In a 2018 review, Krol et al. emphasized the value of neural circuit information in relation to behavior and analysis of neural circuits in the context of genes that had been identified as having variants leading to or association with specific clinical endophenotypes. They focused their studies on the thalamic reticular nucleus (TRN) amd its role in sleep and attention.

Krol et al. noted that there is now clear evidence that both rare and common variations play roles in neurodevelopment and in neurodevelopmental and neurobehavioral disorders. Nevertheless, there is clear evidence for genetic heterogeneity in any one of these disorders. In addition, it seems that in the majority of cases, polygenic factors are causative. In addition, there is often overlap in symptoms in different disorders. In addition, there is often overlap across diseases. They also noted that not only were there overlaps in genetic factors related to different disease, but there were also overlaps in overlaps in certain functional studies such as EEG and overlaps in abnormalities detected in structural analyses including white matter structure or cortical thickness.

They emphasized importance of studies of biophysical mechanisms, including analyses of excitability and synaptic transmission and of correlation of findings of such studies with genetic studies. Krol et al. proposed therefore that correlative analyses be carried out of neuronal circuit endophenotypes and genetic expression studies. They noted that genes that had been to have deleterious alteration in certain neurodevelopmental disorders were noted to be normally expressed in the reticular nucleus of the thalamus. These included four genes, and the products and functions of these genes are listed in the following:

- PTCHD1, patched domain containing 1, acts as a receptor for signaling molecule.
  - SHH1 defects in this gene are associated with intellectual disability and autism.
    - CACNA1I calcium voltage-gated channel subunit alpha1 I, low voltage-activated, T-type, calcium channel, may be involved in calcium signaling in neurons.
- GM3 (GRM3 in humans), glutamate metabotropic receptor 3, linked to the inhibition of the cyclic AMP cascade.
- CHD2, chromodomain helicase DNA-binding protein 2, alters gene expression possibly by modification of chromatin structure, thus altering access of the transcriptional apparatus to its chromosomal DNA template.

Thalamic functions

Krol et al. (2018) reviewed specific thalamic functions. They noted that nuclei within the thalamus have excitatory neurons that transmit sensorimotor signals. In addition, nuclei in the thalamus play roles in integrating signals across different cortical networks. However, within the thalamus, inhibitory processes are also important. They note that the thalamic reticular nucleus constitutes the main source of inhibitory signals. Importantly the thalamic reticular nucleus (TRN) was shown to be rich in in GABAergic neurons. The TRN was also shown to have connections to a number of different targets within the thalamus. There is evidence that subnetworks from the TGN, under certain circumstances, suppress distracting sensory input. Such suppression is critically important for cognition and behavior.

Krol et al. proposed that elucidation of the function of the thalamic reticular nucleus had direct relevance for a number of neurodevelopmental disorders. They suggested that specific symptoms of such disorders may indicate TGN dysfunction. These include attention deficits, features of autism (ASD), schizophrenia (SCZ), and attention deficit hyperactivity disorder (ADHD).

Other symptoms that potentially relate to the thalamic reticular nucleus include sensory processing defects such as occur in ASD, ADHD, and SCZ. Sensory processing defects include hypersensitivity to auditory, visual, and tactile sensations.

Krol et al. noted that in 2018, direct testing of TRN function was not possible. However, they emphasized that it is possible to consider information on genetic expression in the TRN. In addition, animal models can be used to investigate how specific gene defects impair expression of genes in the TRN.

Through such studies, information was obtained on the importance of the gene products noted above PTCHD1, CACNA1I, and CHD2 on TGN function. They noted that another gene important in TGN function is ERBB4 protein, and defects in this protein have been reported in schizophrenia. Krol et al. emphasized that impairment in a specific circuit may give rise to a variety of clinical manifestations. They emphasized the importance of analysis of special circuits and documentation of endophenotypes that result from circuit disruption.

Cerebellothalamic tracts

Sakayori et al. (2019) defined projections from nuclei in the cerebellum to thalamic nuclei, particularly from the dentate nucleus in the cerebellum. Specific nuclei in the thalamus that receive cerebellar input include the ventroanterior and the ventrolateral nuclei.

Connections between basal ganglia, thalamus, and cerebellum

Hintzen et al. (2018) analyzed data from electrophysiological studies of interactions between thalamus, basal ganglia, and cerebellum and noted that connection pathways from these structures also converge on the cerebral cortex.

Specific structures they analyzed included basal ganglia components including the reticular part pars reticulata of the substantia nigra and the globus pallidus part of the basal ganglia. These structures were reported to interact with the thalamus and with deep cerebellar nuclei in animals and nonhuman primates.

The limbic system is also connected to the basal ganglia. This system is described in detail in the following. Bostan et al. also noted evidence for connections between the basal ganglia and the cerebellum. Projections to the cerebellum were noted to pass through the subthalamic nucleus of the basal ganglia. They reported that the basal ganglia and the cerebellum were interconnected at the subcortical level. The dentate nucleus of the cerebellum was reported to have projections to the striatum in the basal ganglia.

A new mapping technique involves analyses on the transneuronal transport of specific neurotropic viruses. These studies have been used to trace neuronal pathways in primates.

Functions of individual components of the limbic system

Components included in the limbic system differ to some degree in different texts. In terms of functions, this region is reported to be primarily involved in emotions and memory. Brodal (2004) noted the limbic system components to include the cingulate gyrus, hippocampal formation, septal nuclei, and amygdala. In some classifications, the mammillary bodies are also included in the limbic system.

Hippocampus

The hippocampus is a gray matter structure, and there are different determinations of the components of this structure. Classically key components were documented as CA1, CA2, CA3, and CA4 (CA cornu ammonis). Lee et al. (2017) reviewed hippocampus structure and functions and noted that hippocampus includes the dentate gyrus, subiculum, and CA regions. They emphasize that the hippocampus functions extended beyond memory and are related to cognitive skills, behaviors, and higher-level functions.

Lee et al. (2020) reported that substantial development of the hippocampus occurs in the prenatal period; however, further development occurs in the postnatal period. Later development likely includes substantial changes in circuit connections.

Rolls et al. (2019) proposed that the limbic system should no longer be considered as a single system. He made this proposal on the basis of findings of different connections and functions of the different components. Rolls noted that the cingulate cortex was so named since it hooks around the anterior end of the corpus callosum. Reynolds (2019) noted that the limbic system includes the cingulate gyrus, amygdala, hippocampus, thalamus, and hypothalamus. The cingulum tract conducts communication between different components of the limbic system.

The insula cortex

The insula cortex lies below the lateral sulcus, a groove that separates the temporal lobe from the frontal and parietal lobes. The insula cortex is also known as the island of Reil. This region is reported to be involved in sensory and emotional responses. Differences occur between connections of the anterior and posterior regions of the insula. Connection studies carried out with DTI revealed that the anterior region of the insula has connections with the anterior cingulate, frontal, frontoorbital, and anterior temporal region (Uddin et al., 2017). The posterior insula region was shown to be connected to the posterior temporal regions, parietal regions, and sensorimotor regions. A section in the middle of the insulas has connections that overlap with those of both the anterior and posterior insula regions. The insula function had not yet been definitively assessed.

Earlier reports indicated that the insula was involved in sensation of autonomic function, e.g., sensation from the cardiovascular, urinary, and gastrointestinal systems, and possibly involved in body temperature sensation. Uddin et al. (2017) noted that there is also evidence that the insula is involved in processing of sound. Isolated insula defects have been detected in patients with the central auditory defects. In fMRI studies, there is also evidence for involvement of the insula in recognition and interpretation of emotions of other individuals and also in empathy.

Uddin noted that the insula cortex together with the dorsal anterior cingulate and amygdala constitute part of the salience network. The salience network mediates switching between DMN and executive networks. There is also evidence that the insula may be involved in planning speech.

4. Developmental changes in connectivity

Brain imaging in early childhood

Gilmore et al. (2018) reviewed structural brain imaging and functional development in childhood. They noted that during the period from birth to 2 years, dynamic brain development occurs in humans. They noted further that after 2 years of age, developmental processes predominantly involve reorganization and remodeling of circuits and networks.

Resting-state functional networks were reported to develop rapidly after birth as rapid myelination of white matter connections occurred.

Gray matter in both cortical and subcortical regions was reported to expand during the first year of life. Brain volume was reported to be 35% of adult value at 2–3 months of postnatal life. Volume increased to be 80% of that of adult level by the second year of life. Cortical thickness was reported to peak by the second postnatal year; however, the surface area of the brain expanded in childhood and in early adolescence. In the newborn brain, the gyrification patterns were reported to be similar to those in adult brain.

DTI studies revealed that major tracts in the corpus callosum, longitudinal fasciculi, inferior and superior arcuate tracts, and cingulum were in place at birth.

Longitudinal fasciculi connect frontal occipital, parietal, and temporal. The arcuate tracts connect the inferior frontal lobe and caudal temporal cortex; the cingulum tract conducts communication between different components of the limbic system. Gilmore et al. (2018) noted, however, that both white matter maturation and microstructure changes occurred after birth.

Developmental changes in neural circuits

Turk et al. (2019) reviewed neurodevelopment process involved in the development of neural circuits. They noted that fetal MRI studies enabled investigation of brain connectivity during the prenatal period. Earlier studies indicated that connectome formation begins once neurons have completed migration. There is evidence that spontaneous firing and signaling processes support structural outgrowth and connectivity.

Short-range white matter connections, sometimes referred to as U-fibers, were noted to be important for connections between gyri. Oyefiade et al. (2018) reported that these U fibers undergo continuing development during childhood and adolescence.

Reynolds et al. (2019) used DTI in a study of 120 children aged between 2 and 8 years. They reported increasing functional anisotropy in white matter tracts over the 6-year period. They attributed this to increased myelination and axonal packing. Greatest changes were noted in the occipital and limbic connections. The limbic system includes the cingulate gyrus, amygdala, hippocampus, thalamus, and hypothalamus.

In 2016, Posner et al. described brain attention networks and their development in childhood. The attention network test was designed to examine different aspects of attention including alerting, orienting to visual stimuli and executive control.

Orienting, where the subject pays attention to a stimulus, was reported to involve frontal eye field, left and right frontal lobes, and areas in the superior and inferior parietal lobes. Attention was shown to involve the executive attention networks, sometimes defined as the cingular opercular network, and this includes the anterior cingulum sometime referred to as the operculum. The anterior cingulum surrounds the frontal portion of the corpus callosum. The reaction time in this test was shown to be related to efficiency of white matter connections. Importantly, Posner et al. demonstrated that training can increase the efficiency of the attention network.

Lee et al. (2020) reported that substantial development of the hippocampus occurs in the prenatal period; however, further development occurs in the postnatal period. Later development likely includes substantial changes in circuit connections.

Brain region—specific functions and connections

Movements and the motor cortex

Motor regions to consider include the primary motor cortex involved in execution of movements, the premotor cortex involved in spatial guidance, and the supplementary motor cortex involved in planning of movements and in coordinating movements of two sides of the body. Another related region is the somatosensory cortex that receives input from the cerebellum basal ganglia, pontine nuclei, and the red nucleus and exchanges information with the motor cortex.

The motor cortex in the precentral gyrus is connected with the pyramidal tracts (PTs) and sends signals ultimately to the spinal cord. The precentral gyrus lies in the posterior region of the frontal lobe and borders on the central gyrus sulcus that separates the frontal cortex from the parietal cortex. Laterally the precentral gyrus borders on the Sylvian sulcus that separates the frontal cortex from the temporal cortex.

Connections to the larynx and face are represented in the regions of the precentral gyrus that is closest to the Sylvian sulcus. Moving upward, this is followed by representation of the fingers, arm, trunk, and lower limb.

Fibers that arise in the pyramidal cells of the motor cortex make their way through the white matter and descend through the cerebral peduncle that attach the cerebrum to the brain stem and to the motor nuclei that are located in the brain stem. It is estimated that 80% of tracts from the pyramidal motor cortex cross over and became the crossed PT, while 20% of fibers b do not cross over and form the direct PT (Netter, 1972).

Economo et al. (2018) carried out studies on different types of PT neurons and their connections within the mouse brain. PT neurons were noted to be structurally heterogeneous and were shown to link the motor cortex with specific premotor cortex centers and with centers located in the brain stem and spinal cord. Projections to the thalamus from the PT neurons were also identified.

Papale and Hooks (2018) carried out studies of changes in the motor cortex that took place when specific new skills were learned. Their studies provided insights into plasticity in the motor cortex. They noted that major areas involved in skill learning had been obtained in part through imaging studies and physiological recordings. They reported that the precise factors and changes that occurred included changes in synaptic strength, changes in neuronal excitability, and changes in underlying circuitry.

They reported that studies in animals have demonstrated that representation within the M1 motor cortex associated with specific muscles increases as use of those muscles increases. In the case of certain skills, changes were also in the corticothalamo cortical pathways. In addition, in the motor cortex, local interneuron changes were documented.

5. Mapping sensory systems

Hearing

Appler and Goodrich (2011) reviewed details of the connections in the peripheral and central auditory circuits that have been enabled through the development of new techniques and access to the inner ear. In the inner ear, sound and fluid movement leads to hair cell movement in the organ of Corti and initiates hearing.

Further analyses have provided details on the arrangements of the inner and outer layers of hair cells. More details have been obtained in the neurons in the spiral ganglion that send axons to the hair cells and that also send connections to the eighth cranial nerve.

Type 1 and type II spinal ganglion neurons have been distinguished. These two different forms of neurons have been shown to differ with respect to the position of hair cells with which they connect, and in addition, the two different ganglion types have different firing rates.

Appler and Goodrich noted that understanding the roles of the spiral ganglion neurons has been an important goal of auditory neuroscience, given evidence that the different spiral ganglion neurons differ in the sounds to which they respond.

Details have emerged on steps in the differentiation of the components of the peripheral and central hearing systems. The peripheral hearing system emerges from the embryonic otic vesicle that responds to specific growth factors including fibroblast growth factor (FGF) and bone morphogenetic proteins (BMP) and WNT genes that encode signaling proteins. Sonic hedgehog signaling pathways are also implicated in differentiation of the otic vesicle. Both sensory and nonsensory structures develop from the otic vesicle. An important step in differentiation was segregation of the auditory and vestibular neurons and their connections.

Appler and Goodrich described details of differentiation of auditory neurons. These neurons become bipolar; they extend processes to the sensory epithelium. Processes from type I neurons extend process to cells in the inner layer of the sensory epithelium, while type II neurons extend processes further and contact the cells in the outer sensory epithelium regions. In a later stage, the sensory epithelium cells develop into hair cells. Hair cells in a specific position in the cochlea respond to sounds of a specific frequency, and sounds of different frequencies are conducted from hair cells to specific spiral ganglion neurons.

The eighth cranial nerve has two branches and is sometimes referred to as the vestibulocochlear nerve. The two branches are in one unit as the eighth nerve passes through the internal auditory meatus to enter the brain at the pontomedullary junction.

The spiral ganglion neurons are bipolar, and they send connections to the hair cells in the cochlea and also extend processes to the auditory component of the eight cranial nerve. In the brain, the auditory components of the eight cranial nerve that innervated the cochlea make complex with ventral and dorsal regions in the brain stem that then send projections to the cortex.

Appler and Goodrich described organization of tonotropic maps with components that separate high and low tone frequencies in the auditory pathway. They noted evidence that axons are topographically organized in the eighth cranial nerve. Auditory axons were reported to target tonotropically correct neurons. The spiral ganglion–derived axons were reported to end in large synaptic endings referred to as the end bulbs of Held.

Auditory cortex

In 2015, Hackett reviewed the auditory cortex and noted that every region of the brain has neurons that respond to sound. However, he emphasized that the auditory cortex in an anatomical sense included regions of the cortex that receive input from the medial geniculate body of the thalamus. This structure receives input from the auditory pathways that originate in the brain stem. The medial geniculate body (MGB) has input to other areas, particularly to the temporal lobe. Hewitt defined auditory-related brain regions as regions that received minor input from the MGB.

Five major auditory regions received information from the cochlea via the auditory nerves. These included complexes in the brain stem and in the thalamus. These regions processed information before transmitting it to the cortex.

Tonotropy was reported to be organized in the cochlear nucleus and in the auditory pathway, with high frequencies represented in one area and low frequencies represented in another area with a gradient of frequency representations between these two.

The medial geniculate body (MGB) was noted to have different regions, each with different inputs and outputs. Specific regions that constitute the auditory cortex include the superior temporal cortex, prefrontal and cingulate cortex, posterior parietal cortex, occipital cortex, amygdala, and striatum.

Yi et al. (2019) investigated the encoding of speech sounds in the superior temporal gyrus. They noted that there, neuronal populations distinguished acoustic phonetic features, including vowels and consonants and also pitch. They noted that responses in the superior temporal gyrus were modulated by learning.

Long et al. (2018) noted that myelination processes changed as the axons present in the eighth cranial nerves transition to the brains.

In reviewing steps in the central auditory conduction, they noted distribution of signal from the anteroventral cochlear nucleus bilaterally to the superior olivary complex of nuclei in the pons. From there, circuits were shown to propagate to the inferior colliculus in the midbrain and thence to the midgeniculate nucleus (MGN) of the thalamus. The MGN then sends signals to the auditory cortex and associated regions.

Importantly, Long et al. noted that myelination is a dynamic process and varies at different times of life. They reported that sensory deprivation alters myelination in segments of the central auditory cortex.

Auditory processing defects

Moore et al. (2010) carried out studies to gain insight into auditory processing disorder in children. They noted that at least 5% of children referred for audiology services did not have hearing loss; rather, their difficulties involved speech perception and they were given a diagnosis of auditory processing defects.

Moore et al. designed tests of sensory and nonsensory elements of auditory perception. Children were tested in quiet environments. Tests involved detection of sounds with different tones and speech testing with different vowels and consonant and also with non–language-related sounds. Phonologic memory tests were done using nonsense words. In addition, tests of cognition, including nonverbal IQ and Wechsler tests, were carried out.

Moore et al. reported that multivariate analyses of their study results revealed evidence of auditory processing defects that were correlated with auditory attention defects and with indications of reduced cognitive abilities.

6. Vestibular system

Cullen (2012) reviewed the vestibular system and noted that it included two separate components: the semicircular canals that were found to sense angular acceleration and the otolith systems that detected acceleration in all directions.

In 2020, Ramos de Miguel et al. reported that the function of the semicircular canals and the otolith function are complementary. The otolith structures include the sacculus and the utriculus, also sometimes known as the maculae. They noted that the otolith system is evolutionary older than the semicircular canal system.

They also reported that semicircular canals detected angular changes and accelerations, while the otolith system detected linear acceleration.

Information from the vestibular system was also noted to be integrated with visual and proprioception systems.

Key to the functions of the vestibular system are hair cells and endolymph. The hair cells, sometimes referred to as stereocilia, and sometimes referred to as kinocilia, are located int the sensory epithelium of vestibular system structures. The hairs that project from cell bodies are coated with a gelatinous substance. The hair cell cells bodies were reported by Rabbit (2019) to be anchored in the sensory epithelium by specific proteins tectorin and otogelin.

The endolymph fluid passing over the hair cells in response to specific movements causes deflection of the hair cells, and this deflection sends signals to nerve fibers of the vestibular nerve that have cell bodies in the vestibular ganglion, known as the ganglion of Scarpa, which is located in the internal auditory meatus.

Rabbit (2019) reported that the stereocilia have specific channels referred to as MET channels and mechanotransduction channels. These channels open in response to hair cell movement, and this leads to movement of potassium and calcium ions and to depolarization and excitation of the hair-associated nerve synapses.

The acoustic and vestibular branches of the eighth cranial nerve pass together through the internal auditory meatus and then diverge on reaching the brain.

Central vestibular system

Dieterich and Brandt (2015) reviewed the bilateral central vestibular system. They noted that from each vestibular nerve, both ipsilateral and contralateral pathways are derived. Nerve fibers then reach the brain vestibular nuclei at the level of the pons. Pathways then extend to the thalamus and the cortex.

Coordination of the vestibular system changes with motor activities is critical. Connections from the vestibular nucleus also pass to the hippocampus and parahippocampus where cognitive functions and spatial memory are processed.

Vision

Within the eye, rods and cones transmit signals to the dendrites of bipolar cells that transmit signals to ganglion cells that have axons that connect with the optic nerve. On reaching the position of the optic chiasma, cranial nerves each cross over to the opposite side and connect with the geniculate body of the thalamus. Within the geniculate body, there are different layers of cells. From the different layers, signal passes to different regions of the cortex.

Six different brain regions are included in the visual cortex. In the primary visual cortex, regions are V1 and V2. These are regions around the calcarine sulcus adjacent to the occipital lobe and include an area defined as the striatal cortex on the basis of its morphologic features. Some studies have revealed the signal passes from V1 to V2.

Other important regions of the visual cortex include regions in the temporal cortex, defined as middle and superior temporal regions. Youssofzadeh et al. (2015) studied signal propagation utilizing EEG and fMRI data. They identified signal passing within visual cortical regions in dorsal and ventral pathways. The V6 region was considered to be a key region in the dorsal pathway and to also have connections to the frontal lobe. The V3 region had connections to dorsal and ventral pathways. The V4 region was in the ventral pathway.

There is evidence that dorsal and medial pathways are each involved in specific aspects of the visual input. Yang et al. (2015) carried out studies on brain region activity in response to different visual tasks and confirmed that certain brain regions are preferentially activated by certain tasks. They reported that the V5 region responded to movement signals. Regions V3 and V4 were reported to be activated by shape differences. The region surrounding the right calcarine sulcus was shown to accept vision-related information that passed to other cortical regions. Brincat et al. (2018) reported that the V4 regions and the posterior infratemporal region were particularly involved in color perception.

Parietal lobe and primary somatosensory cortex

The parietal lobe is reported to integrate information on proprioception, touch, space navigation, and language processing.

The parietal lobe and postcentral gyrus form important parts of the somatosensory cortex. The inferior parietal lobe and portion of the temporal lobe are noted to be involved in language perception. Large regions of the parietal lobe play roles in perception of spatial orientation.

A segment of the parietal lobe forms the parietoinsular vestibular cortex PIVC, and an adjacent region is known as the parietal insular cortex (Frank and Greenlee, 2018). Thus, in humans two anatomically separate areas, the PIVC and the PIC, constitute the vestibular cortex.

Stimuli to the vestibular sensory cortex include active or passive head movement. This system may also be stimulated by electrical stimuli. Differences exist between the PIVC and PIC although both receive signals from the thalamus.

Frank and Greenlee noted that the PICV region is suppressed during visual processing, while the PIC region is activated during visual stimulation.

Touch and proprioception

Peripheral sensation is particularly dependent on mechanoreceptors in skin, muscle, and joints. Signals from peripheral receptors are conveyed by peripheral nerves to nuclei and the dorsal column of the spinal cord and then to the central nervous system.

Delhaye et al. (2018) reviewed proprioception and touch sensation in the cortex. The dorsal column tracts of the spinal cord convey signal to dorsal column nuclei in the brain stem. These include the cuneate nucleus, the gracile nucleus, and trigeminal nucleus. Signals from the upper body are conveyed to the cuneate nucleus; signals from the lower body are conveyed to the gracile nucleus. The trigeminal nucleus is reported to receive signal from the trigeminal nerve and the face and neck. Delhaye et al. noted that these nuclei also receive signals from the cortex that could potentially impact signals entering these nerves from the periphery.

Specific tracts conduct signals from the dorsal column nuclei in the brain stem to the thalamus. These include the medial lemniscus pathway and the trigeminal lemniscus. The ventroposterior regions of the thalamus primarily receive lemniscus pathway signals. Signals form the thalamus are then conveyed to different regions of the cortex.

The primary somatosensory cortex was reported to be located in the anterior body of the parietal lobe.

7. Speech

Hickok and Poeppel (2015) reviewed the neural bases for speech and its perception. They outlined tasks necessary in the learning of native language, the key purpose of which is to transform sounds to represent meaning. Key processes involved using the vocal tract (or hands and face in the case of sign language) to represent sounds or concepts. They noted that key processes involved auditory conceptual routes and auditory motor routes.

Hickok and Poeppel then presented details of the dual route model of speech processing. Key structures in these routes included the ventral stream that involves superior and middle portions of the temporal lobe that convert speech sounds to comprehension.

The dorsal route in speech processing was noted to involve the parietotemporal junction region and the posterior lobe of the frontal cortex. The dorsal route is involved in the production of representations of speech.

Hickok and Poeppel noted that neuroimaging studies have provided new insights into the mechanisms and central regions involved in speech perception.

Ventral stream

The ventral stream that is involved in deriving meaning from sounds is reported to be bilaterally organized in the brain. Processes that occur in the two hemispheres may not, however, be identical. Specific regions highlighted in phonologic processing include the region surrounding the superior temporal sulcus. Electrophysiological studies have also provided evidence that the left anterior temporal lobe is critical for sound intelligibility interpretation.

Dorsal stream

The dorsal stream is involved in the conversion of sound to action. Elements of this process were first described by Wernicke in 1874. Hickok and Poeppel noted that much progress has been made in identifying brain regions involved in sensorimotor integration. One domain identified on the basis of functional imaging is the left planum temporal region located deep to the Sylvian fissure at the parietotemporal boundary and designated as the SPT region. The SPT region was shown to correlate in activity with a region of the frontal cortex referred to as the posterior Broca region and known as the pars opercularis of the inferior frontal gyrus. Motor efferent activity was associated with a network that included the SPR, the pars opercularis and the premotor cortex. This network was anatomically and functionally involved in the sensorimotor integration required for speech-related vocal functions.

Hickok and Poeppel considered the dual stream model and its relevance to aphasia (speech disruption). They considered the condition known as Broca's aphasia and conduction aphasia to involve the dorsal stream. Wernicke's aphasia, word deafness, and transcortical sensory aphasia; they considered to be ventral stream defects. They noted that word deafness could also result from peripheral auditory defects.

Children and language skills

Lee et al. (2020) carried out detailed language analysis and neuroimaging, including T1-weighted MRI imaging, to study gray matter changes and also diffusion-weighted imaging (DTI) to examine whole brain networks and changes in children with language disorders.

Gray matter studies revealed specific changes in the right postcentral parietal gyrus, bilateral medial occipital gyrus changes, and changes in the superior occipital gyrus.

DTI studies revealed altered efficiency in a modular connectivity network that included the frontotemporal language areas, the hippocampus, and the cerebellum. Lee et al. noted altered efficiency in both long connections and short axonal connections.

8. Memory and cognition

Lee et al. (2017) emphasized that the hippocampus functions extend beyond memory and are related to cognitive skills, behaviors, and higher-level functions.

Memory and cognitive control networks

Gliebus et al. (2018) noted progress in understanding memory and emphasized that episodic memory depends on the medial temporal lobe and on interactions with large-scale brain networks including cognitive control networks.

They noted that semantic memory systems were less clearly defined but included lateral inferior temporal regions and that interconnected parietal regions may also be involved.

Vaz et al. (2020) reported that memory retrieval depended on the capacity to replay patterns of neural activity that occurred during the first generation of a specific memory. They noted that studies on space memory in rodents revealed firing of individual medial temporal neurons as they first navigated space. This neural firing pattern was proposed to be later replicated as the rats navigated space. They reported that replaying of memories was associated with a specific activity in the medial temporal lobe referred to as ripples.

Vaz et al. investigated memory-related ripple activity using subdurally implanted microelectrodes. In humans, these studies were done during the course of investigations on individuals with treatment-resistant epilepsy. The implanted electrodes were positioned adjacent to the middle temporal gyrus.

Their studies revealed that electrical spiking sequences generated during memory formation were retrieved when the memory was replayed. This electrical spiking activity was also associated with specific ripple oscillations.

In summary, their electrophysiological studies revealed that memory encoding involved cortical electrical spiking activity of a specific patterns and that this pattern was retrieved when the memory was recalled.

Temporal lobe, regions and functions that include memory and auditory cognition

The temporal lobe visualized from the lateral cerebral surface is separated from the frontal lobe by the fissure of Sylvius. The main gyri of the temporal lobe include the superior temporal gyrus, the middle temporal gyrus, and the inferior temporal gyrus that is separated from the occipital lobe posterior by a sulcus, and inferiorly, it is separated from the limbic lobe.

Sheldon and Levine (2018) reviewed functions and connectivity of the medial temporal lobe and the hippocampus and noted that these regions are involved in memory and other cognitive functions. Specific types of memory associated with the middle temporal lobe were reported to include autobiographical memories. Other functions include perception and imagination.

To gain information on the functions, Sheldon and Levine carried out studies and reported on connections of the middle temporal lobe and other brain regions.

Real-world episodic memory was reported to involve coupling of the middle temporal lobe, the medial prefrontal, and parietal regions. Significant connections between the temporal lobe, the hippocampus, and parahippocampal cortex were demonstrated to be involved in retrieval of autobiographical, spatial, and task-related memories. Connections with the parahippocampal cortex were particularly necessary for autobiographic tasks. Connections to the posterior parietal regions were necessary for spatial-related tasks. Conceptual tasks were reported to utilize connections to a number of different networks.

Memory ability and networks

Van Buuren et al. (2019) studied grades of memory performance and network activity across the whole brain. They determined that better memory performance occurred with stronger connections between the midtemporal lobe and the parietal cortex. Task-related memories were reported to involve activity between the DMN and the parietal network.

The DMN was initially described as the network active during rest and meditation. Regions included in the DMN are reported to include prefrontal ventral and dorsal regions, the posterior cingulate cortex, lateral temporal lobe, and parietal regions including posterior inferior regions and angular gyri.

The retrosplenial cortex is located behind the splenium of the corpus callosum and has been reported to be related to episodic memory, navigational, and spatial assessments. The angular gyrus within the parietal lobe is related to spatial cognition memory. Different investigators have defined subnetworks within the DMN.

Interactions are also reported to occur between the DMN and other networks; these include the frontoparietal also known as the execute control network. The frontoparietal network was shown to be involved in cognitive attention and memory and was reported to connect with the lateral prefrontal cortex, dorsal anterior cingulate cortex and precuneus in the occipital lobe, and the anterior inferior parietal lobe and inferior temporal lobe.

In summary, the temporal lobe can be defined as a region involved in processing sensory information and in generating memory.

9. Central autonomic nervous system

Cersosimo and Benarroch (2013) reviewed the central nervous system areas involved in body homeostasis and adaptation to changes. They listed the following forebrain regions as being involved in autonomic responses: temporal cortex, anterior cingulate cortex, amygdala, and specific regions of the hypothalamus.

The insula is located in proximity to the lateral sulcus, also known as the sulcus of Sylvius, which separates the frontal lobe from the temporal and parietal lobes. The insula is reported to play roles in autonomic functions and also in detection of specific sensation that arise in organs.

The dorsal region of the insula was reported to receive signals from the viscera including gastrointestinal tract and also signals from muscle and skin receptors. The insula was also noted to receive signals from the thalamus and signals from the cortex related to perception and emotional responses.

Cersosimo and Benarroch noted that fMRI studies indicated connections between the anterior insula and the anterior cingulate cortex. The anterior cingulate cortex was also noted to have connections to the amygdala, hypothalamus, brain stem, and frontal cortex. Through these extensive connections, the anterior cingulate cortex exerted control on autonomic functions. The amygdala was reported to receive sensory information.

Integration of autonomic and endocrine required for homeostasis and adaptation was noted to be carried in the hypothalamus and adjacent preoptic area, and the latter is reported to be particularly involved in thermal sensations.

The brain stem was reported to be involved in controlling parasympathetic and sympathetic response. Specific brain stem regions involved in responses were reported to include the preaqueductal gray matter (PAG) of the midbrain. Other brain stem regions involved in homeostasis included the nucleus of the solitary tract, the ventrolateral reticular formation, and the raphe of the medulla.

Intelligence and neuroimaging

Posner and Barbey (2020) traced the history of intelligence testing from the work of Binet (1917). They noted that tests were designed to assess general intelligence. Subsequently tests were designed to test specific domains of intelligence. Intelligence domains were defined by Gardner (1983). They included seven different domains: music–rhythmic, visual–spatial, verbal–linguistic, logical–mathematical, body–kinesthetic, interpersonal, and personal. Gardner also proposed that specific networks underlined the different domains. Cattell (1963) proposed existence of fluid and crystallized intelligence.

Posner and Barbey noted that neuroimaging has enabled analyses of specific networks associated with specific tasks and has assigned tasks to brain domains. They noted that general intelligence assessments initially concentrated particularly on the lateral prefrontal cortex. Network studies were used in intelligence assessment with particular focus on the frontoparietal network.

Later assessments focused on spatial overlap between networks. Newer networks were also identified and determined to play important roles, e.g., the cingular opercular network was associated with task performance.

Posner and Barbey noted that efficiency in a specific domain could be associated with efficiency of specific networks. Attention strength may operate across different domains of intelligence.

They noted further that during development, remote areas of the brain become connecting with long connections. Networks efficiency is dependent on the development of efficient connections and on myelination.

Roth et al. (2015) reported that the degree of methylation of brain-derived neurotrophic factor (BDNF) impacted learning. Furthermore, genetic variants of methylene tetrahydrofolate (MTHFR) reductase that increases methylation were reported by Voelker et al. (2017) to influence attention and learning skills in humans. There are reports that efficient methylation increases myelination, which may in turn increase connectivity. Moyon and Casaccia (2017) reported studies in DNA methylation and its importance in oligodendroglial cell activity and in myelination.

Posner and Barbey noted evidence that specific molecular factors impact learning. They emphasized that DNA methylation and activity of specific transcription factors are required to generate functioning oligodendrocytes to produce myelin to ensheath axons.

10. Brain mapping in psychiatric disorders

In 2017, Weinberger noted the relevance of transcriptional alterations in fetal life as playing roles in schizophrenic risk.

There is growing evidence that genetic loci with variants associated with psychiatric disorders encode gene products that play roles in the brain from prenatal the period on Cross-Disorder Group of the Psychiatric Genomics Consortium (2019). There is also growing evidence for polygenic risk factors in psychiatric disease and evidence that these risk factors can include pathogenic variants at the nucleotide and genomic copy number variants. The latter include particularly microdeletions at 22q11.2, 15q13.3, and 1q21.1.

Increasingly genetic studies in psychiatric disorders are being carried out in combination with brain mapping studies including neuroimaging. Studies are also being carried out to determine which cells in the brain normally express the products of genes that have been mapped to genomic loci associated with psychiatric disorders.

Liu et al. (2019) carried out genetic and genomic studies and neuroimaging studies on 502 Han Chinese patients with schizophrenia and 502 controls from the same population. Data from genome-wide association studies were used to derive a polygenic risk score for each individual. Structural and functional brain MRI was also carried out.

Liu et al. reported that individuals with a high polygenic risk score for schizophrenia had lower levels of hippocampal gray matter volume and lower levels of hippocampal medial prefrontal cortex connectivity.

They concluded that their findings may contribute to understanding of schizophrenia pathogenesis and may also contribute to schizophrenia.

Reinwald et al. (2020) developed rodent models of the genomic copy number variants that are reported to be associated with schizophrenia in humans, i.e., they deleted segments of the rodent genome that matched human 22q11.2, 15q13.3, or 1q21.1. On rodents with these deletions, they carried out functional and structural MRI diagnostic studies.

Rodents with deletions of the genomic segment that matched human 1q21.1 manifested microcephaly. In the rodents with deletions of the genomic segment that matched the human 15q13.3 genomic region, they identified reduced cerebellar volume; in addition, alterations in the striatal—auditory system were identified.

In rodents with deletions of the genomic segment that matched human 22q11.2, they reported cortical hypoconnectivity along with dopamine pathway hyperconnectivity; this rodent models manifested midbrain morphology abnormalities. They noted that the brain imaging in rodents with deletions that matched human 22q11.2 region simulated brain imaging abnormalities that had been reported in human individuals with schizophrenia.

Cell-type studies in psychiatric disorders

Skene et al. (2018) correlated information on the genes implicated on the basis of association studies in schizophrenia and information on gene expression patterns reported for specific brain cell types. Their efforts were therefore designed to determine if specific brain cell types were likely to be implicated in schizophrenia.

They also investigated which gene products were targets of specific antipsychotic medications and which brain cell expressed these products.

Specific cell types that particularly expressed gene products that map to genetic loci associated with schizophrenia included pyramidal cells in the hippocampus and medium spiny motor neuron cells. Furthermore, there was evidence that the molecular targets of useful antipsychotic medications also impacted products of those cells.

Medium spiny neuron cells are reported to be the major cell type in the striatum. They have dopamine receptors. One population has primarily D1 dopamine receptors, another population has predominantly D2 receptors, and a third population harbors both D1 and D2 dopamine receptors. The striatum has dorsal and ventral regions. The ventral striatum contains the nucleus accumbens and the olfactory tubercle. The dorsal striatum has the caudate nucleus and putamen, and these are separated by a white matter tract and the internal striatum. The medium spiny neurons in the dorsal striatum are reported to play roles in movements. The ventral region medium spiny neurons are reported to play roles in motivation, sense of reward, and sense of aversion.

Francis and Lobo (2017) noted that there is evidence that the ventral striatum that includes the nucleus accumbens is impacted in depression and stress. Studies in these disorders focused particularly on the medium spiny neurons and on the D1 and D2 dopamine receptors.

Torres (2016) reported structural brain abnormalities at different stages of schizophrenia. They analyzed morphometric brain MRI data accumulated on 101 individuals with schizophrenia, including 62 first-episode schizophrenia patients and 99 chronic cases of schizophrenia. They reported that the first-episode schizophrenia individuals had small volumetric defects in the insula, temporal lobe structures, and the striatum. In patients with chronic schizophrenia, extensive region gray matter losses were noted, and these losses were bilateral and included frontal cortices, anterior cingulate cortex, insula, and temporal cortex losses. They noted that they could not rule out that some of these changes were due to the cumulative effects of antipsychotic medications.

Striatum dopamine and dopamine receptors

In a 2015 review, Beaulieu et al. reported that in humans, five different genes encode dopamine receptors. Dopamine receptors were reported to signal through intracellular G proteins. However, they can also signal through other mechanisms. Products of the D1 and D2 dopamine receptors differ, and their mRNAs undergo different splicing processes, subsequently leading to the presence of an additional 29 amino acids in D2R. The D1R and D2R proteins are reported to have different signaling and different pharmacological properties.

Much work has been carried out on the striatum and dopamine and its receptors since medications that impact dopamine activity were first reported to be valuable in treatment of patients with psychosis; however, new relevant information continues to be reported.

Striatal dopamine receptors have received much attention in effort to unravel mechanisms, leading to psychiatric diseases. Olivetti et al. (2019) noted that negative manifestations of schizophrenia include impaired motivation. Positive manifestations include psychotic symptoms. Olivetti noted that increased dopamine release has been correlated with psychotic symptoms and that inhibitors of D2R dopamine receptors reduced these symptoms.

Studies of behavior and dopamine-related function have been intensively studied in rodents. However, in a 2018 review, Kesby et al. reported that important neuroanatomic differences exist between human and rodents with respect to the dopamine system. Also new information on networks associated with different regions of the striatum has been garnered. Kesby et al. emphasized that it is important to distinguish networks of the association striatum and networks of the sensorimotor striatum.

Kesby et al. reported that positron emission tomography (PET) imaging revealed that in patients with psychoses, there was increased dopamine synthesis in the association striatum. They noted that antipsychotics exerted remedial actions through reducing assess of dopamine to DR2 dopamine receptors.

They noted that specific networks also need to be considered. They drew attention to a network apparently involved in psychosis that leads to communication between the association striatum, the thalamus, and the prefrontal cortex. Other networks can feed into this network; they include a network that passes through the thalamus and originates in the hippocampus and amygdala.

Kesby et al. noted that antipsychotic medication does not improve cognitive functions. They also noted that different networks play roles in cognition and goal-directed action and behavioral flexibility. These can include the medial occipitofrontal cortex, prefrontal cortex, amygdala, and limbic structure.

Importantly the anatomic region of the association striatum differs from that of the other limbic striatum structures. The nucleus accumbens is an important component of the limbic striatum as is the dorsal limbic of the substantia nigra sometimes referred to as the ventral tegmentum.

Kesby et al. noted that it is possible that specific symptoms of schizophrenia may result from network defects. They also noted that brain imaging studies in schizophrenic patients revealed reduced volumes of the thalamic and caudate structures.

11. Neurodevelopmental delay, intellectual disability, brain imaging

Bélanger and Caron (2018a, b) reviewed evaluation of children with global developmental delay and children with intellectual disability (ID), in apposition statement of the Canadian Pediatric Society. They noted that the term global developmental delay (GDD) was applied to children below the age of 5 years and manifestations included in at least two of the following categories: speech and language, cognition, social, and/or personal skills activities of daily living.

The term intellectual disability was applied to children older than 5 years and included defects in three areas that include intellectual functioning, reasoning, and learning from experience; defects in communication and activities of daily living; and history of defects defined above during the developmental period before 5 years.

Central nervous system malformations were reported to occur in 28% of cases with developmental delay or intellectual disability. Belanger and Cohen noted that genetic factors contributing to causation were found to occur in 47% of cases with these disorders. They noted that important evaluations in these cases included clinical examination, metabolic, and genetic studies.

However, they noted that neuroimaging and MRI studies contributed to diagnosis only in a small percentage of children with GDD or ID. Neuroimaging studies were reported to have greater diagnostic utility in cases with microcephaly or macrocephaly, in cases with abnormal neurological features or epilepsy. In cases with epilepsy, electroencephalographic studies were noted to be important, and in some cases, specific EEG patterns were associated with specific disorders.

There are, however, specific neurodevelopmental disorders in which specific brain abnormalities occur that can be revealed on brain imaging studies, and these will be discussed in the following.

Joubert syndrome

This syndrome was reviewed by Valente et al., 2013. Specific brain abnormalities that occur in this syndrome include defects in genesis of the cerebellar vermis that lead to episodes of abnormal breathing and to movement abnormalities including ataxia. Intellectual disabilities may also occur. Valente et al. noted that hypoplasia of the cerebellar vermis was detectable on MRI studies, and in addition, the cerebellar peduncles were elongated and malrotated. Together, these structural abnormalities led to neuroradiological image referred to the molar tooth sign.

Joubert syndrome is in a class of disorders referred to as ciliopathies. Cilia function as cellular sensors and also play roles in intracellular migration of molecules and play roles in cell division. Cilia also play roles in cell migration including neuronal migration and in the projection of axons and dendrites from neurons. Cilia and functions are discussed in detail in Chapter 1 of this book. By 2013, 21 different genes were noted to have pathogenic mutations that led to Joubert syndrome.

Structural brain abnormalities associated with specific genomic copy number variants, microdeletion, or microduplications

Chromosome 22q11.2 deletion syndrome

Rogdaki et al. (2020) reviewed structural brain abnormalities identified in individuals with 22q11.2 deletion syndrome. This syndrome has heterogeneous clinical manifestations; it is, however, known to be associated with neurobehavioral problems.

Rogdaki et al. noted that 90% of cases were reported to result from de novo events. The deletion region is flanked by low copy repeat sequences, and deletions of three megabases of DNA between the largest repeat elements occur in most cases. Physical defects that occur in this syndrome can impact the heart, endocrine, and immune systems. Mild facial dysmorphism and cleft palate occur in some cases. Neurobehavioral manifestations can include autism-like behaviors and psychiatric manifestations including psychoses with variable degrees of severity.

Rogdaki et al. noted that there were earlier reports indicating reduced volumes of gray and white matter in individuals with 22q11.2 deletions. Reduced volumes of the frontal cortex and hippocampus were also reported.

Rogdaki et al. analyzed results of studies of 22q11.2 deletions on 988 individuals and 873 controls. Mean deviation in age of patients was 14.4 years, and mean age deviation in controls was 14.5 years. Analyses revealed that in the 22q11.2 deletions cases, there was widespread volumetric reduction in whole brain and volumetric reductions occurred in frontal and temporal lobes and in the hippocampus.

Two genes impacted by the 3 Mb deletion in 22q11.2 were reported to play roles in axonal differentiation regeneration and proliferation, and in neuronal migration, these are described in the following.

PI4KA (PI4CA, phosphatidylinositol 4-kinase alpha) catalyzes the first committed step in the biosynthesis of phosphatidylinositol 4,5-bisphosphate.

RTN4R (reticulon 4 receptor) is reported to mediate axonal growth inhibition and may play a role in regulating axonal regeneration and plasticity.

Analyses were carried out to determine the extent of altered imaging findings in individuals with the 3 MB 22q11.2 deletion. Frontal and temporal lobe gray matter thinning was identified, and extent of this was shown to have some degree of correlation with predisposition to psychosis.

Rogdaki et al. emphasized that heterogeneity existed in these individuals, both with respect to the extent of volumetric change and the degree of psychiatric manifestations.

15q11.2 BP1-BP2 region copy number variants and brain morphology

The Enigma working group committee (Van der Meer et al., 2019) reported that the 15q11.2

copy number variants constituted the most common pathogenic copy number variants. They analyzed data generated at different sites, from 45,756 individuals. The data were gathered from 204 15q11.2 deletion carriers, 366 individuals with duplications of this region, and 45,247 controls. Data analyses revealed that individuals with deletions of this region had lower brain surface area, and the lower surface area particularly impacted the frontal lobe the anterior cingulate, the precentral, and postcentral gyri. In addition, lower volumes of the nucleus accumbens occurred in the deletion carriers than in controls, and the differences were statistically significant. The normal nucleus accumbens is reported to have functions related to motivation, perceptions of reward, and pleasure. The frontal cortex, pre- and postcentral gyri, and cingulate region are known to play roles in cognition.

Duplication carriers showed larger surface areas and lower cortical thickness than controls.

Deletion carriers manifested lower performance on cognition tasks. Importantly the authors noted that cognitive task performance scores in duplication carriers were similar to those in the control group.

15q11.2 copy number variants impact 500 kb of DNA, and key genes encoded in these regions encode NIPA1, NIPA2, CYFIP1, and TUBGCP5. Studies on the functions of the products of these genes revealed that the NIPA1 and NIPA2 proteins function as magnesium transporters. Studies on the Cyfip protein in mice revealed that it play roles in size of presynaptic terminals and on presynaptic function Hsiao et al. (2016) reported that normal Cyfip1 function is essential for normal synaptic function.

Chromosome 15q11.2 duplication has been reported to be associated with autism. However, there is clear evidence of variable penetrance of effects of this copy number variant, and its significance is not always clear (Benitez-Burraco et al., 2017).

Chromosome 16p11.2 copy number variants

Hippolyte et al. (2016) carried out studies on 62 carriers of 16p11.2 deletions, 44 carriers of duplications of 16p11.2, and 71 intrafamilial controls. Cognitive assessments carried out included overall cognitive functioning, assessment of fine motor skills, language, memory, and executive function.

Deletion carriers and duplication carriers manifested cognitive impairment. The average full-scale IQ values were 72 for deletion carriers, 25 for duplication carriers, and 98 for intrafamilial controls.

Lower scores on tests of phonology, written language, and vocabulary were documented in 16p11.2 deletion carriers than in duplication carriers. Duplication carriers had scores similar to controls on these tests. Duplication carriers had higher scores than deletion carriers on tests of verbal memory.

Structural brain studies included whole brain morphometry studies. Gray matter volume differences occurred in the different groups. Overall, gray matter volume was increased in deletion carriers relative to controls, and in duplication carriers, gray matter volume was reduced relative to that in controls.

Copy number variants in chromosome 1q21.1

Bernier et al. (2016) carried out studies on 19 individuals with 1q21.1 deletions, 19 individuals with 1q21.1 duplications, and 23 familial controls. Clinical neurological assessment revealed borderline cognitive functions in both groups, and fine and gross motor impairments were also reported in both groups. Autism features were observed in the duplication carriers.

Brain studies revealed microcephaly in deletion carriers and macrocephaly in duplication carriers. Imaging studies in three of the duplication carriers revealed reduced corpus callosum volume, and later, ventricle volume was found to be increased. Deletion carriers did not differ from controls with respect to structural brain parameters.

Neuroimaging and brain phenotypes in autism spectrum disorder

Girault and Piven (2020) reviewed neurodevelopment in autism through studies in infants and young children. They reported that neuroimaging studies revealed five key alterations in autism. These included brain overgrowth, increased extraaxial cerebrospinal fluid, altered white matter densities, and aberrant structural and functional connectivity.

They noted that following evidence of heritability of autism, studies were designed to include younger children who were at increased risk for autism because of having older siblings with autism. Prospective studies on at-risk children revealed that clinical features of autism emerged during the latter part of the first year of life when clinical manifestations included lack of attention to faces and poor response to name. Clinical manifestations that emerged during the second year of life included disengagement with visual stimuli and minimal evidence of language development.

MRI studies on infants at risk for autism revealed that brain overgrowth was not present at birth but emerged during the latter part of the first year or during the second year of life. There was then evidence of increased brain volume in the frontal, temporal, and parietal lobes. Increased thickness in some cortical areas was noted by 3–5 years of life. Increased volume of specific subcortical structures emerged in 20% of cases, and these cases were particularly severely affected. Girault and Piven noted that these structures included the amygdala, caudate nucleus, globus pallidus, and putamen.

Diffusion imaging studies of white matter were carried out on a limited number of children with manifestations between 1 and 3 years of age and revealed increased maturity of myelination relative to controls and increased axonal density.

fMRI in toddlers with autism manifestations revealed weaker interhemispheric connectivity. An important neuroimaging finding was the increased volume of extraaxial fluid.

Girault and Piven considered possible neurobiological mechanisms implicated in autism causation. They noted that it was important to consider possible mechanism involved in brain expansion. It is important to note that evidence of increased brain expansion was not documented in birth histories of children who later developed autism. Possible factors in brain expansion include increased neuronal cell population expansion. More complex branching of neuronal processes could be involved. They particularly drew attention to the increased volumes of extraaxial fluid and noted that there are reports that this fluid contains growth factors. Another possible mechanism responsible for brain enlargement could be increased degrees of myelination, increased axon caliber, increased axon connectivity, and decreased synaptic pruning.

Girault and Piven noted that collectively genes implanted in autism encode products involved in a number of different cellular and molecular pathways. Courchesne et al. (2019) emphasized role of variants in genes that encode products that function in the phosphoinositol kinase/serine threonine kinase (PI3K/AKT) pathway and in the RAS/ERK (RAS/MAPK) pathway.

A number of investigators have noted the importance of the MTOR signaling pathway in autism. Deleterious variants in the TSC1 and TSC2 genes lead to the disorder tuberous sclerosis, and autism is a common manifestation of this disorder (Dickinson et al., 2019). The TSC1 and TSC2 genes encode products that indirectly reduce MTOR activity. Ganesan et al. (2019) also drew attention to downstream regulators in the mTOR pathway that impact cell proliferation, e.g., P70S76 (ribosomal protein S6 kinase), and eukaryotic translation initiation factors EIF4B and EIF4E.

Attention deficit hyperactivity disorder

Studies have been carried out to define genetic risk in ADHD and also to correlate genetic risk with specific behavioral manifestations of ADHD and with brain imaging findings.

Demontis et al. (2019) reported results of a genetic study that included genome-wide association studies. Studies included data from 29,183 individuals with ADHD and 35,191 controls. This study provided information on common genetic variants associated the disorder and also specific manifestations of the disorder.

Demontis et al. reported that 304 genetic variants were associated with disorders, and these variants were associated with 12 different regions in the genome. In addition, polygenic risk scores were determined in target samples. Demontis et al. also established correlations of ADHD risk loci with other genetically influenced conditions. Positive correlations were established between ADHD, neuroticism, depressive symptoms, and obesity.

Negative correlations were observed between ADHD and years of schooling and between ADHD and college completion.

The locus with the highest level of significance occurred on chromosome 1p, and 12 different genes occur in this region.

In all 12 different genome-wide ADHD-associated loci, specific loci were identified. Demontis noted that genes on chromosome 2, 7, and 10 were of particular interest. The FOXP2 encoding gene occurs in the associated region on chromosome 7. Variants in FOXP2 have previously been associated with ADHD. On chromosome 10, the ADHD-associated locus was in proximity to the gene that encodes SORC3, a transmembrane receptor reported as being important in neuronal development and in neuronal plasticity.

Gene loci with genome significance scores with ADHD also occurred on chromosomes 12 and 15. A gene that localizes in the significant region on chromosome 12 encodes DUSP6, a phosphatase that has been reported to influence dopamine neurotransmitter levels at the synapse. Demontis et al. noted that the chromosome 15 locus that segregates with ADHD encompasses the gene that encodes SEMA6D is located, SEMA6D is reported impact axon pathfinding, fasciculation and branching, and target selection.

Sudre et al. (2019) carried out analyses to correlate polygenic risk scores with ADHD clinical manifestations and with diffusion tensor brain imaging. Significant polygenic risk scores were associated with hyperactivity and impulsivity. DTI revealed that in individuals with significant ADHD polygenic risk scores, there was a specific white matter feature that involved altered axial diffusivity in the right anterior and right superior corona radiata and white matter correction fibers that radiate out from the cortex.

Sudre et al. also reported that the dorsal medial prefrontal cortex showed altered thickness that correlated with significant polygenic risk in ADHD.

Analyses of cognitive measures revealed evidence that significant polygenic risk scores were associated with reduced working memory.

References

Appler, J.M., Goodrich, L.V., 2011. Connecting the ear to the brain: molecular mechanisms of auditory circuit assembly. Prog. Neurobiol. 93 (4), 488–508. https://doi.org/10.1016/j.pneurobio.2011.01.004.

Aravamuthan, B.R., Waugh, J.L., 2016. Localization of basal ganglia and thalamic damage in dyskinetic cerebral palsy. Pediatr. Neurol. 54, 11–21. https://doi.org/10.1016/j.pediatrneurol.2015.10.005.

Beaulieu, J.M., Espinoza, S., Gainetdinov, R.R., 2015. Dopamine receptors - IUPHAR review 13. Br. J. Pharmacol. 172 (1), 1–23. https://doi.org/10.1111/bph.12906.

Becker, E.D., 1993. A brief history of nuclear magnetic resonance. Anal. Chem. 65 (6).

Bélanger, S., Caron, J., 2018a. Evaluation of the child with global developmental delay and intellectual disability. Paediatr. Child Health 23 (6), 403–419. https://doi.org/10.1093/pch/pxy093.

Bélanger, S.A., Caron, J., 2018b. L'évaluation de l'enfant ayant un retard global du développement ou un handicap intellectuel. Paediatr. Child Health 23 (6), 411–419. https://doi.org/10.1093/pch/pxy099.

Benítez-Burraco, A., Barcos-Martínez, M., Espejo-Portero, I., Jiménez-Romero, S., 2017. Variable penetrance of the 15q11.2 BP1-BP2 microduplication in a family with cognitive and language impairment. Mol. Syndromol. 8 (3), 139–147. https://doi.org/10.1159/000468192.

Bernier, R., Steinman, K.J., Reilly, B., Wallace, A.S., Sherr, E.H., Pojman, N., Mefford, H.C., Gerdts, J., Earl, R., Hanson, E., Goin-Kochel, R.P., Berry, L., Kanne, S., Snyder, L.G., Spence, S., Ramocki, M.B., Evans, D.W., Spiro, J.E., et al., 2016. Clinical phenotype of the recurrent 1q21.1 copy-number variant. Genet. Med. 18 (4), 341–349. https://doi.org/10.1038/gim.2015.78.

Binet, A., 1917. The Intelligence of the Feeble- Minded. William and Wilkins.

Bostan, A.C., Dum, R.P., Strick, P.L., 2018. Functional anatomy of basal ganglia circuits with the cerebral cortex and the cerebellum. Prog. Neurol. Surg. 33, 50–61. https://doi.org/10.1159/000480748.

Brincat, S.L., Siegel, M., von Nicolai, C., Miller, E.K., 2018. Gradual progression from sensory to task-related processing in cerebral cortex. Proc. Natl. Acad. Sci. U.S.A. 115 (30), E7202–E7211. https://doi.org/10.1073/pnas.1717075115.

Brodal, P., 2004. The Central Nervous System: Structure and Function, third ed. Oxford University Press.

Cattell, R.B., 1963. Theory of fluid and crystallized intelligence: a critical experiment. J. Educ. Psychol. 54 (1), 1–22. https://doi.org/10.1037/h0046743.

Cersosimo, M.G., Benarroch, E.E., 2013. Central control of autonomic function and involvement in neurodegenerative disorders. In: Handbook of Clinical Neurology, vol. 117. Elsevier B.V, pp. 45–57. https://doi.org/10.1016/B978-0-444-53491-0.00005-5.

Courchesne, E., Pramparo, T., Gazestani, V.H., Lombardo, M.V., Pierce, K., Lewis, N.E., 2019. The ASD Living Biology: from cell proliferation to clinical phenotype. Mol. Psychiatry 24 (1), 88–107. https://doi.org/10.1038/s41380-018-0056-y.

Cross Disorder Group, & Psychiatric Genetics Consortium, 2019. Genomic relationships, novel loci, and pleiotropic mechanisms across eight psychiatric disorders. Cell 179 (7), 1469–1482.e11. https://doi.org/10.1016/j.cell.2019.11.020.

Cullen, K.E., 2012. The vestibular system: multimodal integration and encoding of self-motion for motor control. Trends Neurosci. 35 (3), 185–196. https://doi.org/10.1016/j.tins.2011.12.001.

Delhaye, B.P., Long, K.H., Bensmaia, S.J., 2018. Neural basis of touch and proprioception in primate cortex. Compr. Physiol. 8 (4), 1575–1602. https://doi.org/10.1002/cphy.c170033.

Demontis, D., Walters, R.K., Martin, J., Mattheisen, M., Als, T.D., Agerbo, E., Baldursson, G., Belliveau, R., Bybjerg-Grauholm, J., Bækvad-Hansen, M., Cerrato, F., Chambert, K., Churchhouse, C., Dumont, A., Eriksson, N., Gandal, M., Goldstein, J.I., Grasby, K.L., et al., 2019. Discovery of the first genome-wide significant risk loci for attention deficit/hyperactivity disorder. Nat. Genet. 51 (1), 63–75. https://doi.org/10.1038/s41588-018-0269-7.

Dickinson, A., Varcin, K.J., Sahin, M., Nelson, C.A., Jeste, S.S., 2019. Early patterns of functional brain development associated with autism spectrum disorder in tuberous sclerosis complex. Autism Res. 12 (12), 1758–1773. https://doi.org/10.1002/aur.2193.

Dieterich, M., Brandt, T., 2015. The bilateral central vestibular system: its pathways, functions, and disorders. Ann. N. Y. Acad. Sci. 1343 https://doi.org/10.1111/nyas.12585 (PMID).

Economo, M.N., Viswanathan, S., Tasic, B., Bas, E., Winnubst, J., Menon, V., Graybuck, L.T., Nguyen, T.N., Smith, K.A., Yao, Z., Wang, L., Gerfen, C.R., Chandrashekar, J., Zeng, H., Looger, L.L., Svoboda, K., 2018. Distinct descending motor cortex pathways and their roles in movement. Nature 563 (7729), 79–84. https://doi.org/10.1038/s41586-018-0642-9.

Francis, T.C., Lobo, M.K., 2017. Emerging role for nucleus accumbens medium spiny neuron subtypes in depression. Biol. Psychiatry 81 (8), 645–653. https://doi.org/10.1016/j.biopsych.2016.09.007.

Frank, S., Greenlee, M.W., 2018. The parieto-insular vestibular cortex in humans: more than a single area? J. Neurophysiol. 120 (3) https://doi.org/10.1152/jn.00907.2017. Review.PMID:29995604.

Ganesan, H., Balasubramanian, V., Iyer, M., Venugopal, A., Subramaniam, M.D., Cho, S.G., Vellingiri, B., 2019. mTOR signalling pathway - a root cause for idiopathic autism? BMB Rep. 52 (7), 424–433. https://doi.org/10.5483/BMBRep.2019.52.7.137.

Gardner, H., 1983. Frames of Mind. The theory of Multiple Intelligences Basic Books.

Gilmore, J.H., Knickmeyer, R.C., Gao, W., 2018. Imaging structural and functional brain development in early childhood. Nat. Rev. Neurosci. 19 (3), 123–137. https://doi.org/10.1038/nrn.2018.1.

Girault, J., Piven, J., 2020. The neurodevelopment of autism from infancy through toddlerhood. Neuroimaging Clin. 30 (1) https://doi.org/10.1016/j.nic.2019.09.009. Review (PMID).

Gliebus, G.P., 2018. Memory dysfunction. Continuum (Minneap minn). Behav. Neurol. Psychiatry 24. https://doi.org/10.1212/CON.0000000000000619 (PMID).

Good, C., Ashburner, J., Frackowiak, R.S., 2001. Computational neuroanatomy: new perspectives for neuroradiology. Rev. Neurol. 157.

Haber, S.N., 2016. Corticostriatal circuitry. Dialogues Clin. Neurosci. 18 (1), 7–21. https://doi.org/10.1007/978-1-4614-6434-1_135-1.

Hackett, T.A., 2015. Anatomic organization of the auditory cortex. Handb. Clin. Neurol. 129, 27–53. https://doi.org/10.1016/B978-0-444-62630-1.00002-0. Elsevier B.V.

Herbet, G., Duffau, H., 2020. Revisiting the functional anatomy of the human brain: toward a meta-networking theory of cerebral functions. Physiol. Rev. 100 (3), 1181–1228. https://doi.org/10.1152/physrev.00033.2019.

Hickok, G., Poeppel, D., 2015. Neural basis of speech perception. Handb. Clin. Neurol. 129, 149–160. https://doi.org/10.1016/B978-0-444-62630-1.00008-1. Elsevier B.V.

Hintzen, A., Pelzer, E.A., Tittgemeyer, M., 2018. Thalamic interactions of cerebellum and basal ganglia. Brain Struct. Funct. 223 (2), 569–587. https://doi.org/10.1007/s00429-017-1584-y.

Hippolyte, L., Maillard, A.M., Rodriguez-Herreros, B., Pain, A., Martin-Brevet, S., Ferrari, C., Conus, P., Macé, A., Hadjikhani, N., Metspalu, A., Reigo, A., Kolk, A., Männik, K., Barker, M., Isidor, B., Le Caignec, C., Mignot, C., Schneider, L., Mottron, L., et al., 2016. The number of genomic copies at the 16p11.2 locus modulates language, verbal memory, and inhibition. Biol. Psychiatry 80 (2), 129–139. https://doi.org/10.1016/j.biopsych.2015.10.021.

Hsiao, K., Harony-Nicolas, H., Buxbaum, J.D., Bozdagi-Gunal, O., Benson, D.L., 2016. Cyfip1 regulates presynaptic activity during development. J. Neurosci. 36 (5), 1564–1576. https://doi.org/10.1523/JNEUROSCI.0511-15.2016.

Hwang, K., Bertolero, M.A., Liu, W.B., D'Esposito, M., 2017. The human thalamus is an integrative hub for functional brain networks. J. Neurosci. 37 (23), 5594–5607. https://doi.org/10.1523/JNEUROSCI.0067-17.2017.

Kesby, J.P., Eyles, D.W., McGrath, J.J., Scott, J.G., 2018. Dopamine, psychosis and schizophrenia: the widening gap between basic and clinical neuroscience. Transl. Psychiatry 8 (1). https://doi.org/10.1038/s41398-017-0071-9.

Krol, A., Wimmer, R.D., Halassa, M.M., Feng, G., 2018. Thalamic reticular dysfunction as a circuit endophenotype in neurodevelopmental disorders. Neuron 98 (2), 282–295. https://doi.org/10.1016/j.neuron.2018.03.021.

Lee, J.K., Johnson, E.G., Ghetti, S., 2017. Hippocampal development: structure, function and implications. In: Hannula D., Duff M. (eds). The Hippocampus from Cells to Systems. Springer, Cham. Chapter 3. https://doi.org/10.1007/978-3-319-50406-3_6.

Lee, M.H., O'Hara, N.B., Behen, M.E., Jeong, J.W., 2020. Altered efficiency of white matter connections for language function in children with language disorder. Brain Lang. 203 https://doi.org/10.1016/j.bandl.2020.104743.

Levi, S., Tiranti, V., 2019. Neurodegeneration with brain iron accumulation disorders: valuable models aimed at understanding the pathogenesis of iron deposition. Pharmaceuticals (Basel) 12 (1), 27. https://doi.org/10.3390/ph12010027. PMID: 30744104.

Liu, S., Li, A., Liu, Y., Yan, H., Wang, M., et al., 2019. Polygenic effects of schizophrenia on hippocampal grey matter volume and hippocampus-medial prefrontal cortex functional connectivity. Br. J. Psychiatry 1 (8). https://doi.org/10.1192/bjp.2019.127. PMID:31169117.

Long, P., Wan, G., Roberts, M.T., Corfas, G., 2018. Myelin development, plasticity, and pathology in the auditory system. Dev. Neurobiol. 78 (2), 80–92. https://doi.org/10.1002/dneu.22538.

The Group, van der Meer, D., Sønderby, I., Kaufmann, T., Walters, G., et al., 2019. Association of copy number variation of the 15q11.2 BP1-BP2 region with cortical and subcortical morphology and cognition. JAMA Psychiatry 30. https://doi.org/10.1001/jamapsychiatry.2019.3779 PMID.

Monbaliu, E., Himmelmann, K., Lin, J.P., Ortibus, E., Bonouvrié, L., Feys, H., Vermeulen, R.J., Dan, B., 2017. Clinical presentation and management of dyskinetic cerebral palsy. Lancet Neurol. 16 (9), 741–749. https://doi.org/10.1016/S1474-4422(17)30252-1.

Moore, D.R., Ferguson, M.A., Edmondson-Jones, A.M., Ratib, S., Riley, A., 2010. Nature of auditory processing disorder in children. Pediatrics 126 (2), e382–e390. https://doi.org/10.1542/peds.2009-2826.

Moyon, S., Casaccia, P., 2017. DNA methylation in oligodendroglial cells during developmental myelination and in disease. Neurogenesis (Austin) 4 (1), e1270381. https://doi.org/10.1080/23262133.2016.1270381. eCollection 2017. PMID: 28203606.

Netter, F., 1972. The ciba collection of medical illustrations: nervous system. In: Ciba Collections of Medical Illustrations, vol. 1. Ciba.

Olivetti, P.R., Balsam, P.D., Simpson, E.H., Kellendonk, C., 2019. Emerging roles of striatal dopamine D2 receptors in motivated behaviour: implications for psychiatric disorders. Basic Clin. Pharmacol. Toxicol. https://doi.org/10.1111/bcpt.13271.

Oyefiade, A.A., Ameis, S., Lerch, J.P., Rockel, C., Szulc, K.U., Scantlebury, N., Decker, A., Jefferson, J., Spichak, S., Mabbott, D.J., 2018. Development of short-range white matter in healthy children and adolescents. Hum. Brain Mapp. 39 (1), 204–217. https://doi.org/10.1002/hbm.23836.

Papale, A.E., Hooks, B.M., 2018. Circuit changes in motor cortex during motor skill learning. Neuroscience 368, 283–297. https://doi.org/10.1016/j.neuroscience.2017.09.010.

Passingham, R.E., Stephan, K.E., Kötter, R., 2002. The anatomical basis of functional localization in the cortex. Nat. Rev. Neurosci. 3 (8), 606–616. https://doi.org/10.1038/nrn893.

Petersen, S.E., Sporns, O., 2015. Brain networks and cognitive architectures. Neuron 88 (1), 207–219. https://doi.org/10.1016/j.neuron.2015.09.027.

Posner, M.I., Barbey, A.K., 2020. General intelligence in the age of neuroimaging. Trends Neurosci. Educ. 18 https://doi.org/10.1016/j.tine.2020.100126.

Posner, M.I., Rothbart, M.K., Voelker, P., 2016. Developing brain networks of attention. Curr. Opin. Pediatr. 28 (6), 720–724. https://doi.org/10.1097/MOP.0000000000000413. PMID: 27552068.

Rabbitt, R.D., 2019. Semicircular canal biomechanics in health and disease. J. Neurophysiol. 121 (3), 732–755. https://doi.org/10.1152/jn.00708.2018.

Ramos de Miguel, A., 2020. The Superiority of the Otolith System. https://doi.org/10.1159/000504595.

Reinwald, J.R., Sartorius, A., Weber-Fahr, W., Sack, M., Becker, R., Didriksen, M., Stensbøl, T.B., Schwarz, A.J., Meyer-Lindenberg, A., Gass, N., 2020. Separable neural mechanisms for the pleiotropic association of copy number variants with neuropsychiatric traits. Transl. Psychiatry 10 (1). https://doi.org/10.1038/s41398-020-0771-4.

Reynolds, J., Grohs, M., 2019. Global and regional white matter development in early childhood. Neuroimage 196, 49–58. https://doi.org/10.1016/j.neuroimage.2019.04.004.

Rikhye, R.V., Wimmer, R.D., Halassa, M.M., 2018. Toward an integrative theory of thalamic function. Annu. Rev. Neurosci. 41, 163–183. https://doi.org/10.1146/annurev-neuro-080317-062144.

Rogdaki, M., Gudbrandsen, M., McCutcheon, R.A., Blackmore, C.E., Brugger, S., Ecker, C., Craig, M.C., Daly, E., Murphy, D.G.M., Howes, O., 2020. Magnitude and heterogeneity of brain structural abnormalities in 22q11.2 deletion syndrome: a meta-analysis. Mol. Psychiatr. https://doi.org/10.1038/s41380-019-0638-3.

Rolls, E.T., 2019. The cingulate cortex and limbic systems for action, emotion, and memory. Handb. Clin. Neurol. vol. 166, 23–37. https://doi.org/10.1016/B978-0-444-64196-0.00002-9. Elsevier B.V.

Roth, E.D., Roth, T.L., Money, K.M., SenGupta, S., Eason, D.E., Sweatt, J.D., 2015. DNA methylation regulates neurophysiological spatial representation in memory formation. Neuroepigenetics 2, 1–8. https://doi.org/10.1016/j.nepig.2015.03.001.

Sakayori, N., Kato, S., Sugawara, M., Setogawa, S., Fukushima, H., Ishikawa, R., Kida, S., Kobayashi, K., 2019. Motor skills mediated through cerebellothalamic tracts projecting to the central lateral nucleus. Mol. Brain 12 (1). https://doi.org/10.1186/s13041-019-0431-x.

Sheldon, S., Levine, B., 2018. The medial temporal lobe functional connectivity patterns associated with forming different mental representations. Hippocampus 28 (4), 269–280. https://doi.org/10.1002/hipo.22829.

Skene, N., 2018. Genetic identification of brain cell types underlying schizophrenia. Nat. Genet. 50 (6), 825–833. https://doi.org/10.1038/s41588-018-0129-5.

Sudre, G., Frederick, J., Sharp, W., Ishii-Takahashi, A., Mangalmurti, A., Choudhury, S., Shaw, P., 2019. Mapping associations between polygenic risks for childhood neuropsychiatric disorders, symptoms of attention deficit hyperactivity disorder, cognition, and the brain. Mol. Psychiatry. https://doi.org/10.1038/s41380-019-0350-3.

Swanson, L.W., Hahn, J.D., Sporns, O., 2017. Organizing principles for the cerebral cortex network of commissural and association connections. Proc. Natl. Acad. Sci. U.S.A. 114 (45), E9692–E9701. https://doi.org/10.1073/pnas.1712928114.

Telford, R., Vattoth, S., 2014. MR anatomy of deep brain nuclei with special reference to specific diseases and deep brain stimulation localization. Neuroradiol. J. 27 (1), 29–43. https://doi.org/10.15274/NRJ-2014-10004.

Torres, U., 2016. Patterns of regional gray matter loss at different stages of schizophrenia: a multisite, cross-sectional VBM study in first-episode and chronic illness. Neuroimage Clin. 12, 1–15. https://doi.org/10.1016/j.nicl.2016.06.002.

Turk, E., van den Heuvel, M.I., Benders, M.J., de Heus, R., Franx, A., Manning, J.H., Hect, J.L., Hernandez-Andrade, E., Hassan, S.S., Romero, R., Kahn, R.S., Thomason, M.E., van den Heuvel, M.P., 2019. Functional connectome of the fetal brain. J. Neurosci. 39 (49), 9716–9724. https://doi.org/10.1523/JNEUROSCI.2891-18.2019.

Uddin, L.Q., Nomi, J.S., Hébert-Seropian, B., Ghaziri, J., Boucher, O., 2017. Structure and function of the human insula. J. Clin. Neurophysiol. 34 (4), 300–306. https://doi.org/10.1097/WNP.0000000000000377.

Valente, E.M., Dallapiccola, B., Bertini, E., 2013. Joubert syndrome and related disorders. Handb. Clin. Neurol. 113,

1879–1888. https://doi.org/10.1016/B978-0-444-59565-2.00058-7. PMID: 23622411.
Van Buuren, M., Wagner, I.C., Fernández, G., 2019. Functional network interactions at rest underlie individual differences in memory ability. Learn. Mem. 26 (1), 9–19. https://doi.org/10.1101/lm.048199.118.
van den Heuvel, M.P., Pol, H.E.H., 2010. Exploring the brain network: a review on resting-state fMRI functional connectivity. Eur. Neuropsychopharmacol 20 (8), 519–534. https://doi.org/10.1016/j.euroneuro.2010.03.008. Epub 2010 May 14.
Vaz, A.P., Wittig, J.H., Inati, S.K., Zaghloul, K.A., 2020. Replay of cortical spiking sequences during human memory retrieval. Science 367 (6482), 1131–1134. https://doi.org/10.1126/science.aaz3691.
Voelker, P., Sheese, B.E., Rothbart, M.K., Posner, M.I., 2017. Methylation polymorphism influences practice effects in children during attention tasks. Cogn. Neurosci. 8 (2), 72–84. https://doi.org/10.1080/17588928.2016.1170006.
Weinberger, 2017. Neurodevelopment and schizophrenia. Schizophr. Bull. https://doi.org/10.1093/schbul/sbx118.
Yang, Y.L., Deng, H.X., Xing, G.Y., Xia, X.L., Li, H.F., 2015. Brain functional network connectivity based on a visual task: visual information processing-related brain regions are significantly activated in the task state. Neural Regen. Res. 10 (2), 298–307. https://doi.org/10.4103/1673-5374.152386.
Yi, H.G., Leonard, M.K., Chang, E.F., 2019. The encoding of speech sounds in the superior temporal gyrus. Neuron 102 (6), 1096–1110. https://doi.org/10.1016/j.neuron.2019.04.023.
Youssofzadeh, V., Prasad, G., Fagan, A.J., Reilly, R.B., Martens, S., Meaney, J.F., Wong-Lin, K.F., 2015. Signal propagation in the human visual pathways: an effective connectivity analysis. J. Neurosci. 35 (39), 13501–13510. https://doi.org/10.1523/JNEUROSCI.2269-15.2015.

CHAPTER

4

Brain plasticity

1. Evolution, synaptic activity, plasticity, and cognition

Hardingham et al. (2018) reviewed the evolutionary advances in transcriptional responses to synaptic activity. They attributed the advanced cognition that developed in humans to expression of activity-related genes and to regulatory regions in DNA that respond to signals. Hardingham et al. emphasized that changes in gene expression primarily drove evolutionary cognitive advances and that gene expression changes were more important than protein coding differences in driving evolution. Specific regulatory regions important in cognitive advances included promoters and enhancers.

Studies revealed that signal-related electrical changes regulated transcription of genes and led to functional and structural changes in neurons. In addition, gene expression changes that followed synaptic stimulation led to altered metabolism, neuronal survival, and synapse stabilization.

Detailed transcriptome studies carried out on rodents revealed that several hundred genes could be defined as activity-related genes. In some cases, genes identified as activity-related genes in mouse studies were found to have homologous forms that acted as activity-responsive genes in humans. In this context, evolutionarily conserved activity responsive genes that encode produce involved in calcium signaling, e.g., voltage-gated calcium signaling genes, were shown to be particularly important. Hardingham et al. noted that the impact of calcium on gene expression depended on presence of calcium-binding sites in promoters or in enhancers of specific transcription factors.

Other important elements that were shown to respond to synaptic stimulation include the cyclic AMP-responsive element protein CREB1 and MEF2 (myocyte enhancer factor 2). They emphasized that many of the immediate early response genes encoded transcription factors including FOS, EGR1, and NPAS4 transcription regulator.

Hardingham et al. emphasized that excitation transcription coupling is highly conserved in evolution. This excitation transcription activity plays roles in synaptic plasticity, dendritic outgrowth, synaptic connections, and connection maturation.

They noted that studies of comparisons of neuronal stimulation effects in mouse and humans revealed quantitative differences in responses. In addition, in humans as compared with mice, a number of additional genes showed evidence of response to synaptic stimulation. Divergence was also documented in DNA sequence in promoters and sequences in enhancer elements.

Hardingham et al. concluded that in the course of evolution changes emerged in the signal-regulated elements that are located within

https://doi.org/10.1016/B978-0-12-821913-3.00011-1

gene regulatory regions, changes were particularly documented in promoter architecture.

2. Dendritic spines and neuroplasticity

Nishiyama (2019) reviewed functions of dendritic spines and noted that structural changes in dendritic spines are critical for synaptic plasticity (Fig. 4.1). He noted also that there is evidence that spine morphology and plasticity are altered in neurodevelopmental disorders. Nishiyama emphasized the important role of the RAS GTPase signaling pathway in the regulations of dendritic spine actin cytoskeleton and protein synthesis.

Dendritic spines were reported to play important roles in storing calcium ions and supporting the role of calcium in signaling. Dendritic spines contain postsynaptic densities, and they accommodate neurotransmitter receptors.

Nishiyama noted that long-term potentiation (LTP) and long-term depression (LTD) of neural function were associated with alterations in dendritic spine size, and potentiation was associated with increased spine size and decreased spine size occurred during LTD.

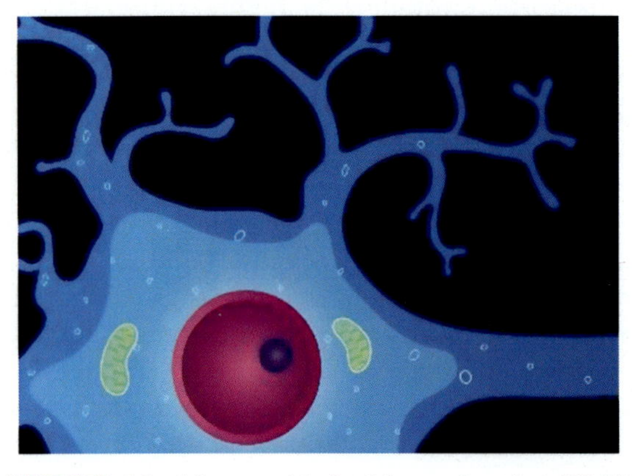

FIGURE 4.1 Neuron with dendrites and nucleus NIMH image. Note also presence of mitochondria essential for energy provision. *From http://National Institute of Mental Health, National Institutes of Health, Department of Health and Human Services.*

Nishiyama reported that advances in molecular and optical techniques facilitated analyses of spine functions and structural changes. Two-photon microscopy enables imaging of dendritic spine morphological changes in key components of spine and associated cytoskeleton components including microtubules, neurofilaments, and actins, in response to neuronal stimulation. Calcium was also shown to accumulate in spines in response to activation.

Fluorescence resonance energy transfer (FRET) is a technique that has been used to detect energy transfer related to signaling.

Calcium can flow into spines through glutamate receptors; it then binds to calmodulin and activates calmodulin kinase complexes CAMK1A, CAMK2A, and CAMK2B that in turn activate downstream signaling mechanisms, e.g., RAS GTPases. Activated RAS GTPases regulate cofilin and actin in the cytoskeleton. CAMK2A and CAMK2B also have actin binding activity.

In a 2017 review, Salter and Stevens emphasized the diverse roles of microglia in brain health and also in brain disorders. Particularly important is the role that microglia play in sculpting neuronal circuits and facilitating neuroplasticity. Microglia were reported to constitute 10% of brain cells and to be highly branched cells with multiple mobile processes.

In 2018, Bennet et al. noted that brain connectome studies had largely assumed stability of established connections. However, studies on neuroplasticity indicated plasticity of connections and indicated experience-based plasticity and rewiring. They noted that different approaches to understanding plasticity in the context of learning and behavior have been undertaken. Cellular changes and modifications in synaptic strength were the focus of some studies, while other studies focused on alterations in wiring.

Other aspects of rewiring and plasticity that have been established include changes in dendritic spine density, increased axon sprouting, formation of new connections, and decomposition or elimination of functional connectivity.

Plasticity mechanisms were thought to come into play when activity is altered over longer periods. Plasticity mechanisms are reported to include changes in neuronal plasticity, synaptic plasticity, or rewiring. Bennet et al. noted that rewiring modifies neural circuit architecture.

Synaptic plasticity

Earlier studies led to distinction of different forms of synaptic plasticity; Hebbian plasticity referred to changes in synaptic strength and could include increased synaptic strength based on neuronal activity. Homeostatic plasticity was the process that counterbalanced Hebbian plasticity.

Hebbian plasticity was thought to include changes in neurotransmitter release and changes in neurotransmitter receptor sensitivity or alterations in ion channel activity. Homeostatic plasticity was considered to be triggered by certain sensors that detected levels of synaptic stimulation. Homeostatic plasticity measures were proposed to impact synaptic efficiency and could include a reduction in dendritic arborization.

Neural plasticity and synaptic plasticity

Von Bernhardi et al. (2017) defined neural plasticity as "the capacity of the nervous system to modify itself functionally and structurally in response to experience and injury." They noted that in addition to adaptability, stability of function is also important in the nervous system.

Von Bernhardi et al. noted that different processes involved in plasticity include neurogenesis, cell migrations, excitability, neurotransmission, generations of new connections, and modification of existing connections. Remodeling can include synaptic formation of synapses, elimination or expansion of dendritic arborization, axonal sprouting, axonal retraction, or pruning of axons. In addition to structural changes, functional changes contribute to plasticity and can include changes in strength of electric signals generated and their transmission or decreases in signal generation or transmission.

Hebbian plasticity was defined as a change in neuronal activity following stimulation. LTP was noted to represent an example of Hebbian plasticity. Another form of neuronal plasticity is homeostatic plasticity that acts to counteract Hebbian plasticity.

Homeostatic plasticity involved changes in ion channel activity, alterations in neurotransmitter release, and alterations of sensitivity of neurotransmitter receptor responses.

Von Bernhardi et al. noted that neuronal intracellular elements could impact plasticity, particularly homeoplastic plasticity.

3. Memory, engram cells, and circuits

Tonegawa et al., 2015 reviewed evidence that memory is stored in specific cells designated as engram cells and that these cells are associated with circuits. They noted that Semon in 1904 and 1909 proposed that memory was a physical entity. Semon's theories were reviewed by Schacter et al. (1978). Tonegawa et al. noted that Semon also referred to engraphy, a modification produced by experience and stimulus, and ecphory that referred to storage and possible processes of memory retrieval.

Tonegawa et al. proposed reuse of Semon's terms taking into account more recent scientific developments. They proposed that the term engram be used to describe the physical and chemical changes that result from learning and that Engram cells include neurons activated by learning and that could be reactivated by stimuli related to the original stimuli. Engram pathways would then include neuronal cells that had been stimulated and their connections. Engram components included contents of an engram cell, and engram complexes referred to engram cells and their connections.

Tonegawa et al. noted that it was important to include concepts of an engram pathway since the engram does not necessary lie in a specific location.

Penfield and Rasmussen (1950), through studies on specific patients, obtained evidence for location of sites of a specific memory within the brain. Electrical stimulation of a specific area led a patient to recall a specific memory.

Tonegawa et al. noted that memory has been classified into different types, e.g., declarative, nondeclarative, explicit, or implicit memory.

Molecular methods applied to memory analysis include analysis of transcription of immediate-early genes. Other approaches include optogenetics and pharmacogenetics.

Flavell and Greenberg (2008) examined expression of specific immediate-early genes CFOS and Zif268 (EGR1) to identify cell activated during memory. Specific alterations in gene expression through stimulation signaling and CREB expression were shown to make cells more likely to be recruited as engram cells.

Studies revealed that synaptic activity and LTP were essential in establishing memory. Blocking of NMDA glutamate neurotransmitter receptor activity reduced long-term synaptic potentiation and reduced memory.

In a 2018 review article, Josselyn and Franklind noted that establishment of an engram likely also involves enhancement of connections between neurons that were activated in the process of establishing memory. At a subsequent time, the same excitation could activate memory. They noted further that specific regions are involved in forming an engram in response to a specific type of stimulus and furthermore that only a subset of neurons in a specific region are involved in establishing an engram. A subset of neurons was identified as important in responding with fear in the auditory tone foot shock experiments.

Josselyn and Franklind explored possible mechanisms that led to selection of a defined number of neurons in a specific area for memory storage. Early insights into this process include discovery of the important role of calcium signaling and overexpression of transcription factor CREB cyclic AMP response element. Neurons that overexpress CREB in response to signaling were shown to be recruited to establish the engram and to determine its size.

The ARC (activity-regulated cytoskeleton-associated protein) gene promoter was shown to have CREB-binding sites and MEF transcription factor–binding sites. Overexpression of CREB was proposed to stimulate LTP in a subset of neurons and that this led to enhanced potential for memory in these cells.

Memory consolidation

Tonegawa et al. (2018) reviewed brain networks involved in memory consolidation. They noted that recent studies had provided information on brain networks involved in system consolidation of memory and the roles of engram cells. They defined memory consolidation as the transformation of labile memory to a more stable long duration state.

They noted that processing of experience is accompanied by alterations in synaptic plasticity and changes in connections or possibly formation of new connections and the generation of episodic memory. Tonegawa et al. noted that the hippocampus is particularly important in episodic memory.

Optogenetic techniques have been utilized to trace neuronal stimulation to specific cells that are activated and processes that led to the identification of engram cells. Furthermore, there is evidence that these cells are subsequently reactivated, leading to memory retrieval.

Tonegawa et al. noted that additional processes take place resulting in more stable and long-term changes, and they referred to the processes as consolidation. Different theories concerning processes involved in system consolidation have been advanced.

Insights into brain regions involved in memory were initially obtained through studies of patients with specific memory deficits that followed damage to specific brain regions. Damage to the temporal lobe regions that includes the hippocampus was found to lead to retrograde amnesia (inability to recall past events). In addition, in part through animal studies, information was gathered on whether specific forms of memory tended to be stored in specific brain regions. Regions involved in contextual fear memory in humans included the cingulate cortex, the hippocampus, amygdala, and the entorhinal cortex; there was evidence that the medial prefrontal cortex was involved in the retrieval of older memories.

Tonegawa et al. noted that new techniques revealed that FOS-expressing neurons were particularly important in memory generation. In studies on mice, active engram cells were shown to have dense dendrites, while silent engram cells had fewer dendrites. Engram cells followed over a period of time and were shown to transition from active cells to silent cells.

Ghandour et al. (2019) used engram imaging and calcium imaging to study learning impact in the hippocampus. They identified subassemblies of engram cells. They proposed that subassembles of engram cell each represented different pieces of information.

4. Protein synthesis relevance to neuronal activity

Sonenberg and Hinnebusch (2009) reviewed translation of MRNA to protein. They noted the importance of the AUG translation start codon by methionyl tRNA and binding of the preinitiation complex and 40S ribosomal subunit. The preinitiation complex includes eukaryotic protein synthesis initiation factors that recognize the M7-guanine cap at the 5′ end of mRNA.

For mRNA translation to proceed, unphosphorylated EIF2 and RNA are transferred to the ribosome, and this transfer is dependent on guanosine triphosphate.

Hinnebusch et al. (2016) reported that the 5'untranslated region of mRNA and the EIF2 translation initiation factors exerted the control of protein synthesis that was implicated in learning and memory.

Hearn et al. (2016) reported that there are a number of different steps that lead to repression of mRNA translation and repression of protein synthesis. A key step in this repression is phosphorylation of serine 51 in the EIF2a subunit of the translation initiation complex. They noted that this phosphorylation can be catalyzed by different kinases including PKR kinase also known as EIF2AK2 and two other kinases that are apparently active under specific conditions including amino acid deficiency and oxidative stress.

Kashiwagi et al. (2019) reviewed the binding of translation initiation factor subunits EIF2A and EIF2B, and this binding was negatively impacted by phosphorylation of EIF2A. The stress-induced kinase EIF2A phosphorylation pathway was designated as the integrated stress response. Following phosphorylation of serine 21, the phosphorylated EIF2A was shown to inhibit a specific function of the EIF2B subunit, namely a guanine nucleotide exchange reaction that is essential for translation to proceed.

Hearn et al. reported that a small molecule screen led to identification of a class of small molecules that bind to EIF2a and result in impairment of phosphorylation. This impairment of phosphorylation meant that protein synthesis was no longer inhibited. These small molecules were designated as ISRIBs (integrated stress response inhibitors).

Importantly in rodent studies, ISRIB administration was shown to improve memory consolidation. The target of the small molecule was the interaction between EIF2a and the regulatory subunit encoded by EIB2B3.

Zyryanova et al., (2018) reported that the ISRIB molecule bound at the interface of EIF2B and the EIF2B3 encoded regulatory subunit and that this attenuated the nucleotide exchange activity of EIF2B.

In a 2019 review, Hegde and Smith noted that substantial progress has been made during the past half century in defining the underpinnings of memory formation and consolidation. With respect to transcription, key developments included elucidation of the roles of transcription factors including CREB and other factors encoded by immediate-early response genes particularly in the nuclear regulator 4 category, e.g., NR4A2 and NR4A3. Regulation of transcription was also found to be linked to CRTC1 a CREB coactivator. LTP of synaptic activity was shown to be linked to NR4A activity. Other important factors in expression of the immediate-early response genes include SRF serum response factor that was essential for expression of CFOS and EGR1.

Appropriate translation of mRNA through activity of EIF2A and related factors at the 5′ end of mRNA has also been shown to be important in regulating synaptic plasticity and memory.

Epigenetic factors have also been linked to potentiation and memory formation. Histone acetylation was also shown to play an important role in memory formation and consolidation.

Benito and Barco (2010) reported that ion channel activity and CREB regulated neuronal excitability. Inhibition of CREB was shown to reduce neuron excitability.

Other investigators identified memory-activated cells in specific brain areas. These studies were conducted in animals and insects. If pain was induced by foot shock leading to foot retraction and the animal was exposed to a specific odor at that time and then later the animal was exposed to the same odor, this elicited the foot retraction response.

In analysis of memory, much attention has been paid to the lateral nucleus of the amygdala to the CA1 region of the hippocampus also to the dentate gyrus, the piriform cortex, and the insular cortex. The amygdala has six regions, and the lateral nucleus region was specifically associated with fear. The dentate gyrus is a region within the hippocampus. The piriform cortex is located in proximity to the insular cortex and in the temporal lobe, and it includes the amygdala, is thought to be related to smell, and is sometimes referred to as the olfactory cortex.

The insular cortex is located deep in the lateral sulcus. The lateral sulcus separates the temporal lobe from the frontal and the parietal lobe. The insular cortex was reported to be important in taste.

The retrosplenial cortex is defined as being involved in spatial memory; it is located behind the splenium of the corpus callosum.

The pathway from early neuronal stimulation to early gene response and gene expression

Yap and Greenberg noted that neuronal stimulation is accompanied by calcium influx into the neuron. Calcium influx can occur through voltage-gated ion channels or through specific neurotransmitter receptors AMPA or NMDA glutamate receptors. The calcium influx then stimulates activity of RAS mitogen protein kinases and calmodulin kinase (CAMK) that stimulate preexisting transcription factors that then apparently stimulate CFOS and immediate-early response gene transcription in the nucleus.

FOS expressed in the nucleus then combines with JUN to form a protein complex known as activation protein complex 1 (AP1). The AP1 then triggers expression of the late response genes LRG.

Late response genes were reported to encode proteins involved in dendritic growth and spine maturation. Yap and Greenberg (2018) noted that complete identification of the late response genes had not yet been achieved.

Activation complexes in combination with pioneer transcription factors alter chromatin formation of late response genes. The AP1 complex (sometimes referred to as the FOS complex) was shown to interact with SWI/SNF chromatin remodeling complex.

Yap and Greenberg considered additional aspects of neuronal activity and the epigenome. Neuronal activity was shown to impact not only chromatin remodeling but also nuclear histones through activation of histone deacetylase and the NURD complex. The NURD complex is composed of proteins with both ATP-dependent chromatin remodeling and histone deacetylase activities.

Yap and Greenberg noted that questions still arise as to how behavioral information is stored. Possible mechanisms include epigenetic changes and modification induced by neural activity. Indirect evidence has been obtained in that inhibition of DNMT methyltransferase and impaired methylation of DNA may impact memory consolidation.

5. Alterations in synapses and postsynaptic regions in learning and memory

Synaptic plasticity involves LTP and LTD that impact synaptic transmission and are thought to be key processes in learning and memory Lüscher and Malenka (2012). There is, however, evidence that many changes in postsynaptic regions are required for learning and memory. Li et al. (2016) reviewed these processes. They include alterations in voltage-dependent ion channels, dendritic spine changes, and protein phosphorylation reactions, and they include changes in gene expression and protein synthesis.

Li et al. (2016) particularly concentrated on analyses of phosphorylation changes in the postsynaptic density proteins in excitatory synapses. Phosphorylation changes are primarily due to activities of protein kinases and protein phosphatases. Phosphorylation processes were postulated to play key roles at synapses and to be involved in processing, integrations, and storage of information.

Li et al. undertook studies to define kinases in the postsynaptic density (PSD). They determined that the PSD was particularly rich in four families of protein kinases, listed in the following:

- AGC family of protein kinases PKA, PKC, and PKG
  - STE family MAP3K, MEK
    - CMGC family includes CDK, GSK3 and CLK
    - CAMK calcium calmodulin kinases

Li et al. noted that the protein kinases also undergo phosphorylation within their proteolytic domains.

Phosphorylations occur particularly in the core scaffold proteins of the postsynaptic density; these include DLG (discs-large MAGUK proteins), DLGAPs (DLG-associated proteins), and SHANK protein. Phosphorylations can also occur in receptor proteins and in channel proteins and in SYNGAP1 (synaptic Ras GTPase activating protein 1).

A key finding relevant to potentiation was that phosphorylation led to generation of more connections between the components of the postsynaptic density. Li et al. documented categories of genes that were particularly shown to undergo phosphorylation during LTP.

- CAMK2A, calcium/calmodulin-dependent protein kinase II alpha
  - CAMK2B, calcium/calmodulin-dependent protein kinase II beta
    - CNKSR2, connector enhancer of kinase suppressor of Ras 2
    - DLG2/DLG3/DLG4, discs-large MAGUK scaffold protein 2/3/4
    - DLGAP1/2/3/4, DLG-associated protein 1/2/3/4

- GRIN2B, glutamate ionotropic receptor NMDA-type subunit 2B
- GRIN2C, glutamate ionotropic receptor NMDA-type subunit 2C
- MAPK1, mitogen-activated protein kinase 1
- PRKCG, protein kinase C gamma
- PRKCB, protein kinase C beta
- SYNGAP1, synaptic Ras GTPase activating protein 1

Dendritic structural plasticity

Forrest et al. (2018) reviewed neuronal circuits and noted that these are refined throughout development and into adult life. Following early development, neuronal circuits are modified by activity-dependent mechanisms and in response to stimuli, including those from the environment. Dendritic arbors and dendritic spines were noted to particularly change in response to activity stimulation.

Forrest et al. noted that neuropsychiatric disorders were thought to arise in part from disruption of synaptic circuits. They also noted that genetic risk factors in neuropsychiatric diseases have been shown to interact in part with factors that control structural plasticity of synapses, excitatory and inhibitory.

Disruption leading to neuropsychiatric disorders could occur in presynaptic regions, but it appeared to occur particularly in postsynaptic regions. There was evidence that neuropsychiatric disorders particularly impact postsynaptic component of excitatory neurons.

Forest et al. focused on dendrites and spines of pyramidal neurons and their structural plasticity and activity-dependent remodeling. They noted that pyramidal neurons, the main excitatory neurons of the neocortex, developed complex dendritic outgrowths during development. As the dendrites matured, they developed protrusions, the dendritic spines. The complexity of the dendritic outgrowths and spines determined the excitatory output of synapses.

Forest et al. noted that later in development, the numbers of dendrites stabilized. Dendritic spines varied in their structures and included filopodia and mushroom-like structures. Mature spines were reported to house neurotransmitter receptors and postsynaptic densities.

Developmental progression of dendrites

Forrest et al. noted that dendrites developed in the human neocortex between 17 and 25 weeks of gestation. Dendritic spine development coincided with the period during which cortical thickening occurred. In the perinatal and postnatal period, the numbers of spines increased. Pruning of spines occurred between 1 and 2 years. They noted that childhood was the period during which refinement of dendritic arbors and spines and adaptations of synaptic circuits occurred. In mature neurons, dendritic changes were noted to be much less dynamic.

Plasticity, associated with changes in spine size, impacted the signaling components. Plasticity and dendritic modification were noted to be related to experience. Morphologic modification was therefore considered to constitute a structural basis for learning and memory.

Forrest et al. considered synaptic changes in the light of Hebbian plasticity and homoplastic plasticity. Hebbian plasticity was correlated with presynaptic and postsynaptic activity. Hebbian LTP was reported to be associated with increases in stability and numbers of dendritic spines. Details of changes during LTP include actin polymerization, increased AMPA neurotransmitter receptor trafficking, and expansion of the postsynaptic density.

During Ltd., dendritic spine shrinkage occurred, associated with decreased actin polymerization and reduced AMPA receptor trafficking.

Forrest et al. noted that homeostatic synaptic plasticity is considered to be the process that limits excitability. This is accomplished by altering AMPA receptor expression.

Dendritic spines in neuropsychiatric disorders

Changes in the morphology of spines and size of dendritic spines have been reported to occur in histological studies of cases of neuropsychiatric disease.

6. Chromatin regulation and neuronal plasticity

In a 2018 review, Gallegos et al. emphasized that neuronal maturations and structural and functional plasticity are dependent upon regulation of gene transcription. Studies in recent decades have shed light on epigenetic processes active in control of functions in postmitotic neurons and indicate that dynamic changes in gene expression take place during information processing.

Gallegos et al. drew attention to neuroepigenetics and roles of related processes in determining neuronal plasticity. These epigenetic processes include DNA modification, histone changes, changes in chromatin structure, and activity-related changes in positioning of transcription enhancers relative to promoter regions of specific genes.

Gallegos et al. noted that early epigenetic changes occur in the processes of generation of postmitotic neurons and their migration and positioning in the cortex and during generations of their synaptic connections. Subsequently, both intrinsic and extrinsic factors influence the gene regulation required for neuronal function.

DNA methylation

Gallegos et al. noted that three DNA methyl transferases DNMT1A, DNMT3A, and DNMT3B are expressed in the developing brain. Specific patterns of DNA methylation lead to repression of expression of specific genes as cells differentiate to defined types. This methylation was shown to occur primarily at gene promoter regions and to occur as cytosine–guanine (CpG) methylation. They noted that promoters to be repressed were also targeted by polycomb repressive complex PRC2 that led to trimethylation of lysine in histone H3 (H3K27me3). PRC2 repression can be reversed under certain circumstances.

There is evidence that postnatal neurons accumulate methylation at nucleotides other than CG; these nucleotides are referred to as CH and include cytosine linked to adenine, thymine or cytosine, most of cytosine linked to adenine, and the methylation of CA nucleotide mCpA occurs most often in gene bodies (Lister et al., 2013).

Gallegos noted the possibility that experience-related neuronal activity impacts the degree of methylation of CpA nucleotides. There is evidence that mCpA modification occurs particularly in the bodies of long genes reported to play roles in neuronal functions (Zylka et al., 2015).

Studies on the neurological disorder Rett syndrome that results from mutations that impair the function of MECP2 have revealed that the normal MECP2 impacts methylation at CpA sites in long genes Leonard et al. 2017.

Gallegos et al. emphasized the important roles of enhancer elements in gene transcription and the interactions of enhancers, transcription factors, and chromatin rearrangements.

Romanoski et al. (2015) noted the important role of enhancers in gene expression. Enhancers may be located at some distance from genes that they regulate and that they recruit transcription factors and also coregulators that modify chromatin.

Heinz et al. (2015) reviewed enhancer sequences, structures, and functions and noted the occurrence of three states of enhancers: poised, primed, and active. Active enhancers were noted to have specific modifications.

Gallegos et al. referred to evidence that active enhancers were enriched for histone H3 lysine 27 acetylation (H3K27ac) and for specific monomethylation H3K4me 1 and that they bound p300

CREB-binding protein and p300 (EP300 transcriptional coactivator). Primed enhancers lacked H3K27ac. There is evidence that different enhancers are active in different cell types in the developing brain.

Chromatin organization also plays important roles in regulating gene expression; chromatin folding, and chromatin looping are particularly important in determining contact between enhancers and promoters of specific genes. The CCCTC-binding factor CTCF and cohesin play important roles in chromatin looping. In addition, connections of chromatin to nuclear lamina impact gene expression.

Gallegos et al. emphasized the important roles of epigenetic processes and chromatin regulation in dynamic regulation of the gene expression program in postmitotic neurons in response to environmental stimuli. Environmental stimuli lead to neuronal activity–regulated gene transcription and generation of new gene products. Neuronal activity leads to activation of CREB target genes and stimulates the transcription of FOS and ARC genes.

Gallegos et al. noted that differentiation and gene expression are also impacted by the incorporation of certain variant forms of histone into chromatin, e.g., H2AZ and H3.3. Deposition of these variant histones is regulated by neuronal activity.

The stimulus-regulated transcription of genes FOS and ARC has been shown to be due to looping of chromatin between specific upstream enhancers and the promoter regions of those genes.

Gallegos et al. concluded that there is clear evidence that in the postnatal and adult brain, chromatin regulatory processes play important roles in gene expression in postmitotic neurons.

7. Glia and synaptic pruning

In a 2017 review, Neniskyte and Gross reported that glial cells have been shown to play key roles in synaptic pruning. Synaptic pruning was noted to play roles in maturation of synaptic function. They noted that there is also evidence that insufficient pruning or excess pruning can lead to neurodevelopmental disorders. There is growing evidence for roles of neuron–glial signaling pathways in bringing about synaptic maturation.

Neniskyte and Gross noted evidence of two life stages when synaptic pruning particularly occurred. These stages were first demonstrated in mice. The first stage takes place in early postnatal life, and a second stage occurs during adolescence. The early stage of pruning is thought to play important roles in refining sensory circuits, while the second stage is thought to be particularly important to guide behavior and impulse control.

Synaptic pruning was also reported to occur at neuromuscular junctions.

Studies on the rodent cerebellum revealed that each Purkinje cell initially has contact with, and synapses with, four or more climbing fibers. These were subsequently pruned to leave contact with only one fiber. During the period of pruning, the Purkinje cells were shown to be surrounded by Bergmann glia. In addition, during the period of synaptic pruning, there was increased lysosomal activity. Neniskyte and Gross also noted evidence that Purkinje cell synaptic pruning required the complement protein C1Q-like 1 (C1QL1).

Synaptic pruning in the visual system

Input from the retinal photoreceptors is transmitted to the retinal ganglion that in turn sends projections through the optic nerve to the thalamus. The optic nerve is composed of axons from the retinal ganglion and glial cells. Crossing over of the optic nerve occurs in the optic chiasma, and the fiber then reaches the lateral geniculate nucleus of the thalamus. Fibers extend from the lateral geniculate nucleus of the thalamus to the occipital cortex. Neniskyte and

Gross noted that specific synaptic pruning takes place in these processes and that this is mediated by specific types of glial cells, microglia, and astrocytes. The important observation involved evidence for roles of astrocytes in the geniculate nucleus of the thalamus. There astrocytes were demonstrated to contain engulfed synaptic material.

Neniskyte and Gross noted that complement proteins C1Q and C3 were secreted by glial cells and possibly by neurons. In addition, a complement receptor C3R was expressed on glial cells. There is evidence that in the developing brain, CR3 is expressed exclusively on microglia.

They noted that studies on pruning in the cerebral cortex were restricted because of the complicated synaptic structure there. However, specific chemokine ligands, e.g., CX3C and chemokine receptors including CX3CR, were noted to be present on microglia. They were shown to play roles in pruning of synapses associated with pyramidal cells in specific brain regions.

Brain disorders and synaptic pruning

Children who develop autism spectrum disorders were reported to have increased brain size in early childhood (Courchesne et al., 2003). There is some evidence that this may be related to inadequate synaptic pruning.

Neniskyte and Gross noted that symptoms of schizophrenia occur between 15 and 25 years of age and that this coincides with the later period during which synaptic pruning occurs. They noted that aberrant adolescent synaptic pruning in schizophrenia may lead to excitatory inhibitory imbalance in circuits. Woo et al. (2020) reported that overall complement pathway activity is increased in schizophrenia. They particularly highlighted the C4 complement component.

Neniskyte and Gross reported evidence that some forms of epilepsy may be associated with aberrant synaptic pruning. Autosomal dominant lateral temporal lobe epilepsy (TLE) has been shown to occur in individuals with mutations in LGI1, leucine-rich glioma inactivated 1. Mice with pathogenic mutations in this gene were shown to have decreased synaptic elimination and increased excitatory neurotransmission in the hippocampus.

8. Complement in the central nervous system

Druart and Le Magueresse (2019) reported that a number of studies have drawn attention to the production of complement by neurons and glial cells and that there is evidence that complement is produced during embryonic and postnatal life. Complement receptors have also been shown to be present in brain. There is some evidence that microglial cells are the primary source of C1Q. They noted evidence that complement may play roles in neuronal migration. There is definite evidence that complement plays roles in synaptic pruning and that complement C3 and its receptor C3R are particularly important in synaptic pruning.

Autophagy in the regulation of synaptic plasticity

Liang reviewed evidence for the role of autophagy components and their potential impact on synaptic plasticity and synaptic function. Synaptic plasticity was noted to require protein synthesis but also regulated protein degradation. Liang noted that neurons are long-lived cells; however, remodeling of the synaptic proteome occurs in neurons, in part through activity of the ubiquitin proteasome system and in part through the autosome lysosomal pathway. Autophagy was noted to involve degradation of cytoplasmic proteins, lipids, polysaccharides, nucleic acids, and organelles. Liang noted that autophagy had been shown to target synaptic components including neurotransmitters, receptors, and postsynaptic density components and also presynaptic components.

9. Memory and engram

In 2017, Josselyn et al. published a review of the history of the engram concept, entitled "Heroes of the Engram." They noted that in 1904 Richard Semon introduced the term engram. In 1921, the engram concept was published and was noted to refer to "the enduring though primarily latent modification of the irritable substance produced by a stimulus."

Josselyn et al. (2017) noted that lately the engram definition had been modified and was defined as "the lasting physical changes in the brain state and structure introduced in response to an event, an experience." They noted the later research provided evidence that memory is not located in a particular brain region but that it emerges from activity in many brain regions.

Donald Hebb made considerable contributions to understanding of mechanisms of learning and memory. Josselyn et al. noted that Hebb proposed that during a particular event, groups of neurons were stimulated and that synaptic strengthening occurred between them and that they then served as the basis for future recall.

Wilder Penfield first demonstrated that electrical stimulation at a specific site in the brain could lead to memory recall. Penfield carried out his experiments during the course of brain surgery designed to treat epilepsy. Surgery was not carried out under general anesthesia so that the surgeon retained contact with the patients and the surgeon could avoid resecting areas that would impair function.

Josselyn et al. (2017) reported that Wilder Penfield and Brenda Milner studied patients who had undergone brain surgery and determined that the hippocampus and adjacent temporal lobe were essential for memory retention. In time, concepts changed partly and investigators concluded that although memories were initially dependent on the hippocampus and temporal lobe, later memories were represented in a distributed cortical network.

Vernon McConnell carried out experiments on Planaria worms that were trained to associate light with a shock. He and his coworkers documented that this recognition was associated with RNA production (see Corson et al., 1970).

Thompson et al. (1997) used electrophysiological experiments and demonstrated that neurons in the cerebellum were important in forming memories of specific forms of stimulation.

The six investigators documented before were included in the Josselyn et al. report of heroes of the engram.

Stabilization of memories

Takeuchi et al. (2014) reported that stabilization of connections and synthesis of new proteins and synaptic plasticity were required for memory retention.

Hippocampus and memory

Lisman et al. (2017) expressed viewpoints on how the hippocampus contributes to memory navigations and cognition. They noted that cellular and network studies have recently provided new information on aspects of hippocampal function and on the extent of connections of other brain regions to the hippocampus. They defined specific segments in the hippocampal tail and uncus that in primates contain neurons that gather information on space, time, sound, and memory. There was consensus in the group that the hippocampus is required for spatial information, representation of the environment, and episodic memory.

Specific investigators proposed that the hippocampus provided storage of information that could later be used in an episodic context. Place cells in the hippocampus were noted to be key to space mapping. The hippocampal entorhinal system was noted to have bidirectional neocortex connections. The entorhinal cortex is located in the medial temporal lobe. The entorhinal cortex was reported to connect to the neocortex that lies in the outer layers of the cerebrum.

Specific regions of the hippocampus were noted to include the dentate gyrus, CA3, CA2, and CA1 regions. The CA regions of the hippocampus were reported to contain pyramidal cells. Signaling pathways were noted to pass from the hippocampal dentate gyrus to the CA regions of the hippocampus and then to the entorhinal cortex, located in the medial temporal lobe.

Lisman et al. noted that the hippocampus must be considered to be important beyond the context of space memory. This information came from individuals who had sustained hippocampal damage.

Engram cells, memory, and consolidation

Tonegawa et al. (2018) noted that the theory of systems consolidation of memory (SCM) implies that maintenance of memory over time requires changes in brain circuitry and networks. They also noted that advances had been made in analysis of connection between memory engram cells and brain circuits. Tonegawa et al. noted that in the mammalian brain, the hippocampus is the node of episodic memory formation. They defined engram cells as "neurons activated during experience that have undergone physical or chemical changes and that these changes are again impacted when the experience is repeated."

The SCM also involves processes that alter brain networks in response to a specific stimulus. Engram cells and engram-activated systems have been identified. There is also evidence that synaptic plasticity is involved in memory consolidation.

Tonegawa et al. noted that theories differ as to whether the hippocampus is required to react certain systems when a memory is reactivated. They noted that there are specific brain regions that are involved in memory consolidation. These include the frontal cortex, the temporal cortex, and the cingulate cortex. In humans, the prefrontal cortex was shown to be particularly important in the retrieval of older memories. The basolateral amygdala was noted to be important for storage of memory. The dorsal hippocampal engram cells were noted to deliver input to this region through a specific pathway.

FOS-expressing neurons were noted to form memory cells. Tonegawa et al. noted that epigenetic changes are also important in memory formation.

Neurobiology of memory consolidation

Takehara-Nishiuchi (2020) reviewed the neurobiology of memory consolidation and noted advances made in part due to development of new techniques. These include the ability to genetically tag-specific neurons activating during memory formation and to then specifically determine which specific cells were activated during memory formation and retrieval. In addition, advances in electrophysiological imaging enable detection of neuronal firing patterns.

Takehara-Nishiuchi noted that earlier studies had demonstrated that damage to the medial temporal led to impairment of episodic memory and autographical memory and that the hippocampus and its connection ere important in retrieving original memories. Two processes that were postulated to be important in stabilization of memory were development of connections from hippocampal neurons to the neocortex and neocortical neuron modifications and synthesis of new protein.

Some investigators proposed that once memories became represented in different regions of the cortex, the hippocampus was no longer necessary.

Specific studies were carried out to identify engram cells. In the hippocampus, engram cells were identified in the dentate gyrus, in the CA3 and CA1 regions of the hippocampus. Specific genes differentially expressed in engram cells included the activation transcription factor 3 (ATF3) and cyclic AMP-responsive element–binding protein (CREB).

Takehara-Nishiuchi noted that there is evidence that reactivation of hippocampal neurons also leads to reactivation of neurons in the neocortex that were activated during the initial memory generating exposure. There is now evidence that the hippocampal engram cell expression lasts and hippocampal engram-related expression is induced in reactivation of the original memory. However, some investigators are not persuaded, and there is conflicting evidence regarding the permanence of hippocampal memory.

There is evidence that neocortical memory traces are refined over time. Some investigators have proposed that input from the hippocampus is important in maturation of neocortical memory.

Takehara-Nishiuchi noted that there is evidence for strengthening of memory traces through strengthening of connections between different neocortical regions activated by a memory.

Expression of CFOS in response to reactivation of memories has been identified.

There is consensus that engram cells can be described as subsets of neurons that change firing rates during a specific experience.

10. Role of sleep in memory

Rasch and Born (2013) reviewed concepts and findings related to mechanisms through which sleep impacts memory. They noted that there is abundant evidence that sleep deprivations and disruption lead to cognitive and emotional problems.

Two phases of sleep were particularly studied in mammals: slow-wave sleep (SWS) and rapid eye movements sleep (REM). SWS sleep was reported to occur early in the sleep cycle and to then increase. SWS sleep was found to be characterized by high waves on electroencephalography (EEG). REM sleep occurred later in the sleep cycle and was characterized by fast low-amplitude EEG waves and also by muscle atonia. A specific sleep phase in humans was defined as non-REM stage N2 sleep with waxing and waning spindle EEG waves.

Rasch and Born reported that specific neuromodulators impacted sleep. These included acetylcholine, noradrenaline, serotonin, and cortisol.

In considering memory, Rasch and Born noted three major processes: encoding, consolidation, and retrieval. The encoding phase of memory was reported to give rise to a memory trace that was susceptible to decay. In the consolidation phase, the memory trace was reported to strengthen and to be integrated into previously established networks. In the retrieval phase, memory is reaccessed.

They noted that neuropsychologists distinguish between declarative memory and nondeclarative memory. Declarative memory includes episodic memories, and autographic memory is thought to involve the spatiotemporal cortex. Nondeclarative memory was noted to include motor skills, sensory skills, and specific forms of learning and was thought to involve structures that included motor areas, striatum, cerebellum, and sensory cortices. Acquisition of nondeclarative memory was reported to be slow and to require multiple trials.

Rasch and Born noted that specific theories on the benefits of sleep have been generated. More recent theories propose that sleep promoted the consolidation phase of memories, because during sleep there is less interference from incoming stimuli. Specific studies have been carried out to determine the sleep phase when memory consolidation takes place. Rasch and Born concluded that REM sleep may be most beneficial to memory consolidation. Some studies indicate that different phases of sleep serve optimal for different types of memories.

11. Molecular mechanisms of the memory trace

Asok et al. (2019) reviewed molecular biology of memory. They considered cellular mechanisms of system consolidation of memories. They emphasized that consolidation of long-term memories requires synaptic plasticity and that synaptic plasticity requires signaling cascades that in turn strengthen synaptic connections in a particular region.

They noted that long-term memory consolidation required electrophysiological, genetic, proteomic, and epigenetic processes. They noted further that there is now agreement in the establishment that long-term memory generation requires de novo RNA transcription and protein synthesis.

Earlier studies on the organism Aplasia revealed that short-term memories involved release of glutamate from presynapses and stimulation of postsynaptic receptors. However, long-term memories were reported to require de novo transcription, translation, new protein synthesis, and synaptic growth. Particularly important observations were that activity in the MAP kinase pathway, protein kinase a (PKA), and CREB (cAMP-responsive element binding protein) were required for long-term memory consolidation.

The translation of the catalytic subunit of PKA in the cytoplasm and its transfer to the nucleus led to phosphorylation of CREB and to gene transcription. Included among transcribed genes were genes with cyclic AMP response elements. Asok et al. noted that these studies demonstrated how activation of synaptic signaling was followed by intracellular signaling that altered nuclear function. Specific early genes transcribed included the following:

- CFOS (FOS) forms part of a complex that functions as AP1 transcription factor
  - ZIF268 (EGR1) is a nuclear protein and functions as a transcriptional regulator.
    - ARC activity–regulated cytoskeleton–associated protein
    - CAMKII (CAMK2G) subunit of Ca(2+)/calmodulin-dependent protein kinase
    - PKC family of protein kinases involved in phosphorylation of hydroxyl groups

Activated MAP signaling was reported to also alter distribution of neural cell adhesion molecules, which in turn promoted synaptic growth.

Asok et al. noted that additional questions arose regarding persistence of memories. Important candidate molecules potentially involved in LTP included CAMKII and protein kinase zeta PRKCZ that has broad expression in the brain.

PRKCZ mRNA was noted to be transported to dendrites and to regulate endocytosis of glutamate receptors and to maintain long-term synaptic potentiation. The ARC mRNA was also reported to be transported to synapses and to regulate AMPA glutamate receptor endocytosis.

Asok et al. noted that the Kandel lab also investigated to functional role of prions. These are proteins that are soluble unless they are activated when they form complexes. One prion studied was an RNA-binding protein. The question that arose was whether a self-sustaining mRNA could be transported on a protein between pre- and post-synaptic regions, thereby modifying synapses.

The role of epigenetic factors was also investigated. Epigenetic modification of the gene that encodes calcineurin was reported to impact memory consolidation. Calcineurin encodes a phosphatase that was proposed to impact memory consolidation.

Additional questions arise as to how a particular neuron is selected to house a memory trace. It is interesting to note that there is additional evidence for the deposition of mRNAs into ribonucleoprotein complexes at the synapse. Roy et al. (2020) noted that such complexes can then be readily accessed for translation. They also noted the importance of chemical modifications of mRNA in regulating gene expression at synapses.

12. AMPA glutamate receptors and synaptic plasticity

Diering and Huganir (2018) reported that abundance of postsynaptic AMPA glutamate receptors were the key factors controlling synaptic plasticity including LTP and Ltd. Trafficking of these receptors to the synapse was modulated by auxillary subunits, protein interactions, and protein modifications.

Kennedy (2018) noted that LTP involves increased activity of AMPA glutamate receptors and can also be accompanied by increases in the head size of the synapse and increased branching of synaptic cytoskeleton actin. Kennedy noted that LTD is associated with decreases in the number of AMPA receptors and spine head shrinkage. Kennedy also noted that the postsynaptic density structures also play important roles in LTP and LTD. The postsynaptic density is comprised of four classes of scaffold proteins, MAGUKS (membrane-associated guanylate kinases); SHANKS, scaffold proteins of the postsynaptic density that connect neurotransmitter receptors; ion channels; and other membrane proteins linked to the actin cytoskeleton and G protein–coupled signaling pathways. Also important are SAPAPs (synapse-associated proteins) often referred to as PSDs (e.g., PSD95) and as HOMER, adapter proteins.

13. Neuron and astrocyte energetics in memory and learning

Dienel (2017) reported that several studies had drawn attention to the important roles of glucose, glycogen, and lactate in neuronal gene expression, signaling, and memory consolidation. He also noted evidence for the importance of astrocyte to neuron transfer of specific metabolites. There is evidence that glycogen, localized primarily in astrocytes, serves as the glucose reservoir in the brains and is used to support memory consolidation.

Dienel noted that in the resting brain, glucose is phosphorylated by hexokinase and is subsequently oxidized. During mental activation and sensory stimulation, glucose metabolism leads to the production of lactate. Lactate can also be taken up into the brain during physical exercise when blood levels of lactate are high.

Dienel reported that catecholamine neurotransmitters modulate the oxygen and glucose index in the brain. The locus coeruleus was noted to be the main sources of noradrenaline in the brain. Astrocytes have alpha1, alpha2, beta1, and beta2 adrenergic receptors, and activation of these receptors and downstream signaling were reported to modulate glucose transport, glycolysis, and oxidative metabolism. In addition, adrenergic stimulation modulated glutamate uptake and glutamate hydrolysis.

Astrocytes were reported to have greater lactate uptake than neurons and to also shuttle lactate to neurons.

Dienel noted that glycogenolysis has been shown to play an essential role in memory processing. Glycogen turnover was shown to be regulated in part by noradrenaline produced by the locus coeruleus. Hippocampal glycogenolysis was shown in studies on rats to be important in establishment of short-term working memory.

Pharmacological inhibition of monocarboxylate transporters that transport lactate and pyruvate was shown to impact in long-term memory processing.

Extracellular brain lactate levels rise during exercise. Specific G protein–coupled lactate receptors were shown to be present in the hippocampus. These receptors include HCAR1 (GPR81) that is present in the hippocampus on postsynaptic neuronal membranes and on astrocyte processes. Dienel concluded that glucose is the key source of brain carbohydrate and that it can be derived from glycogen in the brain. Lactate is also an important metabolite in the brain, and it passes into cells and impacts brain energetics and functions.

Alberini et al. (2018) reviewed evidence of the critical roles played by nonneuronal cells, astrocytes, oligodendroglia, and microglia, in providing energy for memory formation in neuronal cells. They emphasized the roles of astrocyte glycogenolysis and lactate formation in memory production and noted the recruitment of beta2 adrenergic receptors in the generation of emotional memories.

Alberini et al. noted potential differences in biological mechanism involved in the generation of short-term memories from those involved in the generation of long-term memories. They noted evidence that short-term memory generation utilized posttranscription processes and protein modifications, while long-term memories required de novo gene expression.

Alberini et al. noted that astrocytes were excitable through calcium fluctuations. Astrocytes were reported to release glia transmitters and to play roles in synaptic plasticity. Astrocytes were reported to have end feet processes that communicate with blood vessels and that played roles in coupling blood flow to the brain dependent on brain activity. Astrocyte end feet were also noted to have contact with neuronal synapses.

Astrocytes were shown to produce the glucose transporter GLUT1. In addition, lactate produced in astrocytes was shown to be transported to neurons via monocarboxylate transporters. Alberini et al. noted evidence from rat studies that astrocyte-derived lactate was important in memory generation.

They emphasized that long-term memory formation required phosphorylation of CREB and expression of immediate-early genes including ARC, and phosphorylation of the actin-binding protein cofilin was also important.

Early studies produced evidence that learning and memory consumed glucose. Alberini et al. noted that neurons do not store glucose as glycogen and that glycogen is present in astrocytes and can be broken down to fuel energy demands in learning memory and memory storage. Astrocyte glycogenolysis was shown to be essential for memory formation.

Lactate was shown to be an energy source in neurons. Lactate was also reported to impact intracellular levels of nicotinamide adenine dinucleotide (NADH). Lactate was shown to signal through the G protein–coupled receptor HCAR1 (GPR81). This receptor is located in neurons on the synaptic membrane and on intracellular organelles. Alberini et al. noted that additional research is required to define all the roles of lactate in neurons. There is some evidence that astrocyte–neuron interactions may differ between neonatal and adult brain.

14. Environmental enrichment and brain plasticity

Kemperman (2019) reported that experience and behavioral activities alter the brain. Specific experiments designed to demonstrate this have been carried in rats and mice in laboratories.

In human studies, Mora (2013) reported that the aging brain retained considerable functional plasticity. Specific factors including environmental enrichment, aerobic exercise, nutrients, and reduction in stress were factors that could reduce age-related defects.

Kondo (2017) noted that specific mechanisms through which the environments and exercise impacted brain structure and plasticity were poorly understood. Specific studies in animals had revealed changes at the cellular level. The size of neuronal synapses and the number and length of dendrites were reported to change on environmental enrichment. Some studies revealed that environmental enrichment increased synaptogenesis.

Van Praag (1999) reported that environmental enrichment increased synaptogenesis in the hippocampus. Specific studies also correlated morphological changes with increased gene expression.

Environmental enrichment was also reported to increase expression of neurotrophins and expression of BDNF (brain-derived neurotrophic factors) and NGF (nerve growth factor). Other studies reveled increased levels of synaptic protein, including synaptophysin and postsynaptic density protein PSD95 on environmental enrichment.

Kondo (2017) focused attention on the kinesin family of proteins (KIFs) described as microtubule-based molecular motor proteins involved in the transport of various cargoes. Kondo and collaborators examined expression of kinesin protein and learning improvement following environmental enrichment. Their studies revealed that environmental enrichment led to increased expression of a number of different KIFs especially KIF1A. Increased levels of KIFs were found to be dependent on increased levels of BDNF. Enhanced production of KIF1A was shown to increase axonal transport. KIF1A increased expression and also promoted synaptogenesis.

Kondo noted that motor stimulation, increased exercise, and physical activity impacted brain structure, e.g., through increased hippocampal neurogenesis. Studies provided evidence that expression of the serotonin receptor HTR3A (5-hydroxytryptamine receptor 3A) was necessary for hippocampal dentate gyrus cell proliferation. Expression of this receptor was also shown to be necessary for fear extinction.

15. Activity-dependent changes in myelin

Mount and Monje (2017) reviewed evidence of activity-dependent neural plasticity associated with changes in myelin. They noted that the production of myelin on axon membranes by oligodendrocytes and activity of clustered ion channel at nodes greatly enhanced electrical impulse conduction. Key factors that influence conduction include the thickness of the myelin sheath and the spacing of the of the internodes.

Myelination is known to occur in late prenatal life and in postnatal life through proliferation of oligodendrocyte precursor cells and their differentiation into myelinating oligodendrocytes. Studies have also revealed that in adult life, specific axons manifest variable degrees of myelination. Mount and Monje considered the possibility that experience might modify the myelin and that myelination could be adaptive.

Studies on mice revealed the impact of social interactions. Juvenile mice individually isolated after weaning were found to have reduced branching of oligodendrocytes in the prefrontal cortex, thinning of myelin sheaths, and reduced numbers of internodes. They noted recent evidence for important roles of myelin in motor learning. Imaging studies of white matter in humans revealed changes in myelination associated with proficiency in learning specific skills, e.g., juggling, training on musical instruments. Changes in myelination were also reported in individuals who learned a second language, indicating that adaptive myelination was not only associated with increasing motor skills.

Mount and Monje considered mechanism through which adaptive myelination occurred. They noted that there is evidence that neuronal activation on axons impacts myelination. This has been observed through studies on the optic nerves in mice. Action potential along an axon was shown to activate adenosine receptors on oligodendrocyte precursor cells. There is also evidence that stimulation could derive from vesicular glutamate release. Results of particular studies supported evidence for the roles of ionotropic and metabotropic glutamate receptors in oligodendrocyte lineage cells. Glutamate receptor activation was shown to trigger increases in calcium levels in oligodendrocyte lineage cells.

Mount and Monje noted that there is also evidence that glutamate signaling impacts the segmentation of myelin sheaths.

Production of specific growth factors, e.g., BDNF and neuregulin, was also shown to impact myelination. However, question still remains regarding mechanism through which neuronal activity stimulates myelination.

Neuroglial interactions and myelin plasticity

In 2018, de Faria et al. reported that myelin plasticity is invoked as a mechanism that contributes to the learning process. They noted that myelination was previously considered static, but there is now evidence that it is plastic and responsive to neuronal activity. They also noted that increases in white matter in specific brain regions have been correlated with learning of specific skills. Reading was reported to lead to corpus callosum white matter increases and to increased interhemispheric connectivity.

Different forms of myelin plasticity were considered by de Faria et al. They proposed that activity-dependent myelination can represent new myelin formation on previously unmyelinated axons. There is evidence that oligodendrocytes detect neurotransmitters. There was evidence that neuronal circuit activity stimulated oligogenesis and de novo myelination. Furthermore, insertion of new nodes and changes in internodal distances likely serve to fine-tune neuronal conduction. It is also possible that existing myelin could be modified, e.g., thickened. There is also the possibility that specific modification may take place at the nodes (nodes of Ranvier).

In conclusion, de Faria et al. noted that there is clear evidence for experience-driven myelination. There is evidence that learning a new sensory skill increased the number of myelinating oligodendrocytes at a particular site.

Monje (2018) reviewed myelin plasticity and nervous system function and development. He noted that certain nerves, including spinal nerves and the optic nerves, were reported to have advanced myelination in early infancy. However, in certain regions of the central nervous system, myelination profiles could be suboptimal, even in adults. These regions included the neocortex, subcortical projections, and the corpus callosum.

Monje noted that myelination occurs in the central nervous system through activity of oligodendrocytes and in the peripheral nervous system through activity of Schwann cells. He emphasized that the speed of electrical conduction is impacted by the thickness of the myelin sheath and the internode distance.

Monje distinguished between innate myelination and adaptive myelination. In considering adaptive myelination, he noted evidence that white matter abnormalities occurred in children who experience severe socioemotional neglect. In mice, enhanced myelination in motor circuits was demonstrated in response to specific activities, e.g., wheel running, maze running. Performance of motor tasks was shown to induce oligodendrogenesis in the motor cortex. The cytokine MIF was shown to be necessary for enhanced myelination in response to activity.

Specific studies in children revealed that improvement in reading coincided with increased myelination in the anterior centrum semiovale, as detected by fraction anisotropy brain imaging. The centrum semiovale is defined as a white matter region superior to the lateral ventricles and corpus callosum; it contains cortical projection fibers and association fibers.

16. Cerebrovascular plasticity

Bogorad et al. (2019) reviewed architecture of brain microvessels and circulatory changes that underlie brain plasticity. They noted that cerebral blood flow has been shown to be directly proportional to the cerebral metabolic rate. The cerebral vasculature in addition not only supplies oxygen and nutrients to the brain, but also, through the blood–brain barrier, it protects the brain to some extent from damaging substances. They noted that perturbation in cerebral vasculature is closely related to alterations in brain architecture.

There is evidence that increases in cerebral blood flow through dilatation of capillaries and arterioles occur during exercise and there is evidence that cerebral blood flow decreases during sensory deprivation.

Bogorad et al. particularly focused on plasticity of cerebral microvasculature and noted that local increases in neural activity were associated with dilatation of capillaries and arterioles and increased cerebral oxygen consumption. Particular factors that modulated these responses included release of vasodilators, in part by endothelial cells and also by glial cells. These include nitrous oxide, vasoactive intestinal factors, and prostaglandins.

Bogorad et al. noted that persistent changes in neural activity led to neurovascular changes. Studies on rodents revealed that when they were raised in stimulating environments, there was evidence of increased capillary density and increased capillary branching. In contrast, rodents raised in a continuously dark environment showed lower density of vessels and lower cortical thickness.

Bogorad et al. noted that increased physical activity in humans has also been shown to lead to increased cerebrovascular plasticity. Aerobic exercise in humans has also been shown to lead to increased neural plasticity, but that mechanisms through which exercise increased neuroplasticity were not clearly defined. Important contributory factors included increased levels of vascular endothelial growth factor (VEGF), insulin-like growth factor (IGF1), and BDNF.

During aging, both microvascular structure and neuronal density decreased. Altered permeability of vessels was also noted.

Stimpson et al. (2018) undertook studies to determine exercise how exercise improved cognition. Their studies provided evidence that exercise promotes angiogenesis and improves cerebral circulation. They noted through studies in mice that exercise promoted proliferation of endothelial progenitor cells and increased production of nitric oxide synthase. Exercise was also shown to increase production of specific neurotrophins. These included BDNF and insulin-like growth factor. Increased levels of CREB and protein kinase B (AKT1) also resulted.

Stimpson et al. noted that increased angiogenesis particularly resulted from regularly repeated exercise.

References

Alberini, C.M., Cruz, E., Descalzi, G., Bessières, B., Gao, V., 2018. Astrocyte glycogen and lactate: new insights into learning and memory mechanisms. Glia 66 (6), 1244–1262. https://doi.org/10.1002/glia.23250.

Asok, A., Leroy, F., Rayman, J.B., Kandel, E.R., 2019. Molecular mechanisms of the memory trace. Trends Neurosci. 42 (1), 14–22. https://doi.org/10.1016/j.tins.2018.10.005.

Benito, E., Barco, A., 2010. CREB's control of intrinsic and synaptic plasticity: implications for CREB-dependent memory models. Trends Neurosci. 33 (5), 230–240. https://doi.org/10.1016/j.tins.2010.02.001.

Bogorad, M.I., DeStefano, J.G., Linville, R.M., Wong, A.D., Searson, P.C., 2019. Cerebrovascular plasticity: processes that lead to changes in the architecture of brain microvessels. J. Cerebr. Blood Flow Metabol. 39 (8), 1413–1432. https://doi.org/10.1177/0271678X19855875.

Corson, Golub, F., Masiarz, T., Villars, J., 1970. McConnell Behavior induction\ or \memory transfer. Science 169 (3952). https://doi.org/10.1126/science.169.3952.134. PMID: 5454149.

Courchesne, E., Carper, R., Akshoomoff, N., 2003. Evidence of brain overgrowth in the first year of life in autism. J. Am. Med. Assoc. 290 (3) https://doi.org/10.1001/jama.290.3.337.

de Faria, O. Jr., Pama, E.A.C., Evans, K., Luzhynskaya, A., Káradóttir, R.T., 2018. Neuroglial interactions underpinning myelin plasticity. Dev. Neurobiol. 78 (2), 93–107. https://doi.org/10.1002/dneu.22539. PMID:28941015.

Dienel, G.A., 2017. The metabolic trinity, glucose-glycogen-lactate, links astrocytes and neurons in brain energetics, signaling, memory, and gene expression. Neurosci. Lett. 637 https://doi.org/10.1016/j.neulet.2015.02.052 (Review.PMID).

Diering, Huganir, R.L., 2018. The AMPA receptor code of synaptic plasticity. Neuron 100 (2). https://doi.org/10.1016/j.neuron.2018.10.018. Review.PMID).

Druart, M., Le Magueresse, C., 2019. Emerging roles of complement in psychiatric disorders. Front. Psychiatr. 10 https://doi.org/10.3389/fpsyt.2019.00573.

Flavell, S.W., Greenberg, M.E., 2008. Signaling mechanisms linking neuronal activity to gene expression and plasticity of the nervous system. Annu. Rev. Neurosci. 31, 563–590. https://doi.org/10.1146/annurev.neuro.31.060407.125631.

Forrest, M.P., Parnell, E., Penzes, P., 2018. Dendritic structural plasticity and neuropsychiatric disease. Nat. Rev. Neurosci. 19 (4), 215–234. https://doi.org/10.1038/nrn.2018.16.

Gallegos, D.A., Chan, U., Chen, L.F., West, A.E., 2018. Chromatin regulation of neuronal maturation and plasticity. Trends Neurosci. 41 (5), 311–324. https://doi.org/10.1016/j.tins.2018.02.009.

Ghandour, K., Ohkawa, N., Fung, C.C.A., Asai, H., Saitoh, Y., et al., 2019. Orchestrated ensemble activities constitute a hippocampal memory engram. Nat. Commun. 10 (1), 2637. https://doi.org/10.1038/s41467-019-10683-2. PMID:31201332.

Hardingham, G.E., Pruunsild, P., Greenberg, M.E., Bading, H., 2018. Lineage divergence of activity-driven transcription and evolution of cognitive ability. Nat. Rev. Neurosci. 19 (1), 9–15. https://doi.org/10.1038/nrn.2017.138.

Hearn, B.R., Jaishankar, P., Sidrauski, C., Tsai, J.C., Vedantham, P., Fontaine, S.D., Walter, P., Renslo, A.R., 2016. Structure-activity studies of bis-O-arylglycolamides: inhibitors of the integrated stress response. ChemMedChem 11 (8), 870–880. https://doi.org/10.1002/cmdc.201500483.

Heinz, S., Romanoski, Benner, C., Glass, C.K., 2015. The selection and function of cell type-specific enhancers. Nat. Rev. Mol. Cell Biol. 16 (3) https://doi.org/10.1038/nrm3949 (Review. PMID).

Hinnebusch, A., Ivanov, I., Sonenberg, N., 2016. Translational control by 5′-untranslated regions of eukaryotic mRNAs. Science 352 (6292), 1413–1416. https://science.sciencemag.org/content/352/6292/1413.

Josselyn, S.A., Frankland, P.W., 2018. Memory allocation: mechanisms and function. Annu. Rev. Neurosci. 41, 389–413. https://doi.org/10.1146/annurev-neuro-080317-061956. Epub 2018 Apr 25. PMID: 29709212.

Josselyn, S.A., Köhler, S., Frankland, P.W., 2017. Heroes of the engram. J. Neurosci. 37 (18), 4647–4657. https://doi.org/10.1523/JNEUROSCI.0056-17.2017.

Kashiwagi, K., Yokoyama, T., Nishimoto, M., Takahashi, M., Sakamoto, A., Yonemochi, M., Shirouzu, M., Ito, T., 2019. Structural basis for eIF2B inhibition in integrated stress response. Science 364 (6439), 495–499. https://doi.org/10.1126/science.aaw4104.

Kempermann, G., 2019. Environmental enrichment, new neurons and the neurobiology of individuality. Nat. Rev. Neurosci. 20 (4), 235–245. https://doi.org/10.1038/s41583-019-0120-x.

Kennedy, M., The, 2018. Protein biochemistry of the postsynaptic density in glutamatergic synapses mediates learning in neural networks. Biochemistry 57 (27). https://doi.org/10.1021/acs.biochem.8b00496. PMID.

Kondo, M., 2017. Molecular mechanisms of experience-dependent structural and functional plasticity in the brain. Anat. Sci. Int. 92 (1) https://doi.org/10.1007/s12565-016-0358-6.

Leonard, H., Cobb, S., Downs, J., 2017. Clinical and biological progress over 50 years in Rett syndrome. Nat. Rev. Neurol. 13 (1), 37–51. https://doi.org/10.1038/nrneurol.2016.186.

Liang, Y., & Emerging. (n.d.). Concepts and Functions of Autophagy as a Regulator of Synaptic Components and Plasticity. Cells (Vol. 8). https://doi.org/10.3390/cells8010034. (PMID).

Li, J., Wilkinson, B., Clementel, V.A., Hou, J., O'Dell, T.J., Coba, M.P., 2016. Long-term potentiation modulates synaptic phosphorylation networks and reshapes the structure of the postsynaptic interactome. Sci. Signal. 9 (440), rs8. https://doi.org/10.1126/scisignal.aaf6716.

Lisman, J., Buzsáki, G., Eichenbaum, H., Nadel, L., Ranganath, C., Redish, A.D., 2017. Viewpoints: how the hippocampus contributes to memory, navigation and cognition. Nat. Neurosci. 20 (11), 1434–1447. https://doi.org/10.1038/nn.4661.

Lister, R., Mukamel, E.A., Nery, J.R., Urich, M., Puddifoot, C.A., et al., 2013. Global epigenomic reconfiguration during mammalian brain development. Science 341 (6146), 1237905. https://doi.org/10.1126/science.1237905. PMID:23828890.

Lüscher, C., Malenka, R.C., 2012. NMDA receptor-dependent long-term potentiation and long-term depression (LTP/LTD). Cold Spring Harbor Perspect. Biol. 4 (6), 1–15. https://doi.org/10.1101/cshperspect.a005710.

Monje, M., 2018. Myelin plasticity and nervous system function. Annu. Rev. Neurosci. 41, 61–76. https://doi.org/10.1146/annurev-neuro-080317-061853.

Mora, F., 2013. Successful brain aging: lasticity, environmental enrichment, and lifestyle. Dialogues Clin. Neurosci. 15 (1), 45–52. http://www.ncbi.nlm.nih.gov/pmc/articles/PMC3622468/pdf/DialoguesClinNeurosci-15-45.pdf.

Mount, C.W., Monje, M., 2017. Wrapped to adapt: experience-dependent myelination. Neuron 95 (4), 743–756. https://doi.org/10.1016/j.neuron.2017.07.009.

Nishiyama, J., 2019. Plasticity of dendritic spines: molecular function and dysfunction in neurodevelopmental disorders. Psychiatr. Clin. Neurosci. 73 (9), 541–550. https://doi.org/10.1111/pcn.12899.

Penfield, W., Rasmussen, T., 1950. The Cerebral Cortex of Man: A Clinical Study of Localization of Function.

Rasch, B., Born, J., 2013. About sleep's role in memory. Physiol. Rev. 93 (2), 681–766. https://doi.org/10.1152/physrev.00032.2012.

Romanoski, C.E., Glass, C.K., Stunnenberg, H.G., Wilson, L., Almouzni, G., 2015. Epigenomics: roadmap for regulation. Nature 518 (7539), 314–316. https://doi.org/10.1038/518314a.

Roy, R., Shiina, N., Wang, D.O., 2020. More dynamic, more quantitative, unexpectedly intricate: advanced understanding on synaptic RNA localization in learning and memory. Neurobiol. Learn. Mem. 168 https://doi.org/10.1016/j.nlm.2019.107149.

Salter, M., Stevens, B., 2017. Microglia emerge as central players in brain disease. Nat. Med. 23 (9), 1018–1027. https://doi.org/10.1038/nm.4397. PMID:28886007.

Schacter, D., Eich, E., Tulving, E., 1978. Richard Semon's theory of memory. J. Verb. Learn. Verb. Behav. 17, 721–743. https://scholar.harvard.edu/files/schacterlab/files/schactereichtulving_semontheory1978.pdf.

Sonenberg, N., Hinnebusch, A.G., 2009. Regulation of translation initiation in eukaryotes: mechanisms and biological targets. Cell 136 (4), 731–745. https://doi.org/10.1016/j.cell.2009.01.042.

Stimpson, Davison, G., Javadi, A., 2018. Joggin' the noggin: towards a physiological understanding of exercise-induced cognitive benefits. Neurosci. Biobehav. Rev. 88 https://doi.org/10.1016/j.neubiorev.2018.03.018 PMID.

Takehara-Nishiuchi, K., 2020. Neurobiology of systems memory consolidation. Eur. J. Neurosci. https://doi.org/10.1111/ejn.14694 (ahead Review.PMID).

Takeuchi, T., Duszkiewicz, A.J., Morris, R.G.M., 2014. The synaptic plasticity and memory hypothesis: encoding, storage and persistence. Phil. Trans. Biol. Sci. 369 (1633) https://doi.org/10.1098/rstb.2013.0288.

Thompson, 1997. Int. Rev. Neurobiol. 41, 151–189 https://doi.org/10.1016/s0074-7742(08)60351-7.

Tonegawa, S., Liu, X., Ramirez, S., Redondo, R., 2015. Memory engram cells have come of age. Neuron 87 (5), 918–931. https://doi.org/10.1016/j.neuron.2015.08.002. Review. PMID:26335640.

Tonegawa, S., Morrissey, M.D., Kitamura, T., 2018. The role of engram cells in the systems consolidation of memory. Nat. Rev. Neurosci. 19 (8), 485–498. https://doi.org/10.1038/s41583-018-0031-2.

Van Praag, H., Kempermann, G., Gage, F.H., 1999. Running increases cell proliferation and neurogenesis in the adult mouse dentate gyrus. Nat. Neurosci. 2 (3), 266–270. https://doi.org/10.1038/6368.

Von Bernhardi, R., Eugenine-von Bernhardi, L., Eugenin, J., 2017. The plastic brain. In: Advances in Experimental Medicine and Biology. Springer. https://link.springer.com/bookseries/5584.

Woo, J., Pouget, J., Zai, C., Kennedy, J.L., 2020. The complement system in schizophrenia: where are we now and what's next? Mol. Psychiatr. 25 https://doi.org/10.1038/s41380-019-0479-0. PMID.

Yap, E.L., Greenberg, M.E., 2018. Activity-regulated transcription: bridging the gap between neural activity and behavior. Neuron 100 (2), 330–348. https://doi.org/10.1016/j.neuron.2018.10.013.

Zylka, M.J., Simon, J.M., Philpot, B.D., 2015. Gene length matters in neurons. Neuron 86 (2), 353–355. https://doi.org/10.1016/j.neuron.2015.03.059.

Zyryanova, A.F., Weis, F., Faille, A., Abo Alard, A., Crespillo-Casado, A., Sekine, Y., Harding, H.P., Allen, F., Parts, L., Fromont, C., Fischer, P.M., Warren, A.J., Ron, D., 2018. Binding of ISRIB reveals a regulatory site in the nucleotide exchange factor eIF2B. Science 359 (6383), 1533–1536. https://doi.org/10.1126/science.aar5129. PMID: 29599245.

CHAPTER

5

Gene expression, regulation, and epigenetics in brain

1. Blueprint for development

In a 2015 review, Bae et al. (2015) summarized aspects of the blueprint for human development as formulated on the bases of modern genetic and genomic studies; it includes 3 billion base pairs of DNA, approximately 21,000 proteins coding genes, RNAs including microRNAs small nucleolar RNAs, long noncoding RNAs, and cis-regulatory genomic elements including promoters, enhancers, repressors, and transposable elements.

Factors involved neural tube closure: insight from neural tube defects

Insights into factors involved in neural tube closure have been obtained in part through determination of gene defects and environmental conditions associated with increased frequencies of neural tube defects. These were reviewed in 2019 by Wolujewicz and Ross (2019). They noted that the most common neural tube defects included anencephaly and myelomeningocele, and they emphasized evidence of multifactorial causation.

Specific signaling pathways relevant to neural tube closure were noted to include the sonic hedgehog pathway SHH, the beta catenin canonical WNT signaling pathway, the noncanonical WNT signaling pathway, and the planar cell polarity pathway.

Wolujewicz and Ross noted that given the reduction in neural tube defects through folate ingestions during and prior to pregnancy, studies on neural tube defect causation also concentrate on aspects of folate metabolism, particularly on the one carbon metabolism.

Core components of the sonic hedgehog pathway include

- SHH, sonic hedgehog signaling molecule
  - PTCH1, patched 1, the receptor for the secreted hedgehog ligands
    - SMO, smoothened, frizzled class receptor
    - GPR161, G protein–coupled receptor 161
    - GLI 1, 3 family zinc finger 1 and 3, transcription factor activated by sonic hedgehog signal transduction SHH target genes

Core components of the planar cell polarity pathway (PCP) include

- CELSR1 cadherin EGF LAG seven-pass G-type receptor 1

https://doi.org/10.1016/B978-0-12-821913-3.00014-7

- FZD3/8 FZD3/8 frizzled class receptors 3 and 8
  - VANGL1/2 VANGL PCP protein members of the tetraspanin pathway
  - PTK7 protein tyrosine kinase 7, plays a role in multiple cellular processes including polarity and adhesion
  - SCRIB scribble planar cell polarity protein, scaffold protein involved in cell polarization processes
  - DVL1,2,3 disheveled segment polarity proteins 1,2,3, cytoplasmic phosphoprotein
  - DAAM1, disheveled associated activator of morphogenesis 1, also promotes the nucleation and elongation of new actin filaments
  - JNK1, JNK1/MAPK8-associated membrane proteins
  - RHOGTPase signaling protein

The activated PCP pathway impacts cytoskeletal organization.

One carbon pathway (Findley et al., 2017)

Findley et al. (2017) carried out sequencing studies on 348 patients with meningomyelocele and identified rare variants in folate transporters SLC19A1 and variants in folate receptors FOLR1, FOLR2, and FOLR3.

Wolujewicz and Ross noted that several publications have implicated mutations in gene products active in the one carbon pathway carbon metabolism in neural tube defect causation. These included the following:

MTHFR, methylenetetrahydrofolate reductase involved in generations of homocysteine t and methionine

BHMT1, betaine—homocysteine S-methyltransferase, involved in methionine generation

SLC25A3, solute carrier mediates transfer into the mitochondria

One carbon metabolism takes place in the cytoplasm and in mitochondria. Methionine and S-adenosyl methionine are generated in the one carbon metabolism, and they influence epigenetic processes. Components generated in the one carbon pathway also make their way to mitochondria and impact metabolic processes.

Wolujewicz and Ross noted that the precise mechanisms through which epigenetic regulation influences neural tube defects require more attention. Regulatory impacts extend beyond methylation and impact histone modification and chromatin modification.

In addition, defects in histone and DNA modifiers and chromatin regulators were implicated, in causation of neural tube defects. These included the following:

CECR2 histone acetyl-lysine reader, bromodomain-containing protein involved in chromatin remodeling

DNMT3B DNA methyltransferase 3 beta

SMARC4 SWI/SNF-related, matrix-associated, actin-dependent regulator of chromatin

2. Evolution and brain development

In a review of the human brain, Miller et al. (2014) emphasized that the human brain cortex has undergone significant evolutionary expansion, likely due to increased rates of expansion of progenitor cell proliferation during neurogenesis. In addition to the cortex, other brain regions that have undergone expansion during evolution include the subventricular region, the transient subplate zone, and the transient subpial granular zone.

The transient subplate zone contains the first generated neurons in the cortex. It was reported to be important in establishing neuronal connectivity. This zone was reported to disappear during postnatal development.

Miller et al. noted that analyses of gene expression revealed differences in gene expression in different major brain regions. They noted that gene expression differences were also found in different cell populations within different brain regions.

They analyzed transcriptomes in two brains from 16 weeks postconception fetuses and two brains from fetuses at 21 weeks postconception. A finding of interest was the expression of the folate receptor FOLR1, which was selectively upregulated in the ventricular zone and in ganglion eminences in the medial lateral and caudal eminences in the subventricular zones.

Miller et al. particularly documented regions of expression of genes associated with human-accelerated conserved noncoding sequences, ncHARs.

In studies on brain evolution, emphases have been placed on specific sequence elements that have undergone sequence changes during evolution and divergence of humans from chimpanzees. These sequences occur primarily in non–protein-coding regions and are referred to as ncHARs.

In 2013, Capra et al. (2013) reported that 2649 ncHAR sequences occurred in humans and there was evidence that 30% of these acted as enhancer elements.

In their 2015 review, Bae et al. noted that much effort has been devoted to defining genetic changes that occurred in brain and cognition during the evolutionary transition from lower primates to humans. They emphasized that these changes particularly impacted regulatory elements. Other important factors include changes in splicing patterns of certain gene transcripts and the use of alternate promoter elements in specific genes. They noted that there were also specific protein coding sequences that occurred in a number of different genes during evolutionary transitions.

Another key fact was the extended period of progenitor proliferation in humans. There was also evidence for cell cycle differences.

Significantly the subventricular zone in humans was reported to have more diverse progenitors than occurred in other species, and studies have been carried out to determine possible interspecies differences in gene expression in the subventricular zone.

Mitchell and Silver (2018) noted that single-cell MRNA sequence studies revealed that human cells were enriched in expression of lncRNAs (long nonprotein coding RNAs) and also in target genes of neurogenin 2 (NEUROG2) and in expression of gene products in the early growth response (EGR1), FOS, and STAT signaling pathways. Other genes highly expressed in human versus other species were genes that have undergone duplication in humans including the following:

- ARHGAP11B Rho GTPase activating protein 11B
  - SRGAP2 SLIT-ROBO stimulates GTPase activity of Rac1, involving cortical neuron development
    - DUFF1220 domain genes
    - TBC1D3 TBC1 domain family member 3
    - BOLA genes in this family important for the assembly of the mitochondrial respiratory chain complexes

Mitchell and Silver noted that higher expression of SRGAP2 in humans was reported to promote increased neuronal migration and denser dendrite spine formation and to impact both inhibitory and excitatory synapses.

Nucleotide coding sequences changes in human evolution involved certain specific genes, e.g., the gene that encodes FOXP2, forkhead P2, evolutionarily conserved transcription factor, which may bind directly to approximately 300–400 gene promoters in the human genome to regulate the expression of a variety of genes.

Mitchell and Silver noted that non–protein-coding regulatory sequences, particularly sequences in enhancer elements, seemed to be important in evolution, 2700 ncHAR genes (noncoding human-accelerated region) occurred in humans, and 150 of these were reported to be involved in brain development and to act as enhancers in brain development. Specific ncHAR genes were shown to impact expression of specific genes, e.g., ncHAR5 impacts expression of FZD8 that encodes a receptor involved in WNT signaling.

DUF1220 domain

Dumas and Sikela (2009) reviewed information on the DUF1220 domain that has undergone duplication and also chromosomal translocation during primate evolution. The DUF1220 domain copy number in gorillas is 99; in humans, the copy number is 272. Specific subgroups of DUFF domains have been identified. The major location for DUF1220 domains in humans is chromosome 1q21.1–1q21.2, and DUF1220 domains also occur on chromosomes 3, 4, and 5.

NOTCH2 gene and NOTCH2 paralogs

Fiddes et al. (2018) and Suzuki et al. (2018) reported that duplication of the NOTCH2 gene and generation of NOTCH2 paralogs were associated with increased brain expansion in humans and were associated with prolonged proliferations of neuronal progenitors.

Genes important in brain that demonstrate significant sequences differences between chimpanzee and human lineages

Sousa et al. (2017) compiled a list of these genes, and they distinguished different groups including genes that were expressed in both inhibitory and excitatory neurons, genes expressed primarily in excitatory neurons, and genes expressed primarily in inhibitory neurons.

Gene products expressed in both excitatory and inhibitory neurons

- NRGN, neurogranin, postsynaptic protein kinase substrate
  - NLGN3, neuroligin 3, neuronal cell membrane protein
    - AHI3, Abelson helper integration site, required for development of cortex and cerebellum
    - TSC2, tuberous sclerosis 2, stimulates GTPase activity
    - UBE3A, ubiquitin ligase 3A AGAP1 ADP ribosylation GTPase activating protein
    - HTH3, histamine receptor H3 involved in regulation of neurotransmitter release
    - GRIN1 *N*-methyl-D-aspartate receptors, glutamate-regulated ion channel
    - CLCN4 voltage-dependent chloride ion channel
    - COMT, catechol-O-methyl transferase plays role in degradation of catecholamine neurotransmitters

Gene products expressed primarily in excitatory neurons

- MET receptor protein tyrosine kinase
  - GLRA3, glycine receptor alpha 1
    - GRM3, metabotropic glutamate receptor
    - KCNN1/2 potassium voltage-gated channel subfamily N members 1 and 2
    - KCNG1, potassium voltage-gated channel modifier subfamily G member 1
    - KCND3, potassium voltage-gated channel subfamily D member 3
    - KCNJ11, potassium voltage-gated channel subfamily J member 11
    - PKD2L1, polycystin-2 like 1, transient receptor potential cation channel

Gene products expressed primarily in inhibitory neurons

- HTR5A, 5-hydroxytryptamine (serotonin) receptor
  - GAD1, glutamic acid decarboxylase
    - CHRNA2, cholinergic receptor nicotinic alpha 2 subunit

- HCN4 hyperpolarization-activated cyclic nucleotide-gated potassium channel 4
- KCNN2, potassium calcium-activated channel subfamily N member 2
- ADRA2A, adrenoceptor alpha 2A
- KCNC1, potassium voltage-gated channel subfamily C member 1
- KCND3, potassium voltage-gated channel subfamily D member 3

Included in the categories of gene products with differences between chimpanzees and humans are gene products known to have mutations that lead to specific neurodevelopmental disorders. These include the following:

- NRGN reported to be implicated in psychiatric disorders (Gurung and Prata, 2015).
  - NLGN3 implicated in autism (Quatier et al., 2019).
    - TSC2, tuberous sclerosis (Sampson and Harris, 1994).
    - UBE3A Angelman syndrome (Buiting et al., 2016).
    - GLRA3 epilepsy (Sobetzko et al., 2001).
    - HRH3, histamine receptor 3 (Sousa et al., 2017) implicated in Tourette syndrome (Pittrenger, 2020).
    - GRIN1 epilepsy (Lemke et al., 2016).
    - GRM3 schizophrenia (SCZ) (Maj et al., 2016).

Additional insights in gene product functions through investigations in disorders of human cognition and behavior

Doan et al. (2018) noted that studies on specific neurological conditions also give insights into roles of specific genes in determining human cognition and behavior. In this regard, they drew attention to the specific genes and functions; some of the genes listed are also known to undergo evolutionary changes.

- FOXP2 DNA-binding protein, specific mutations in FOXP2 protein impact facial movements and speaking
- ASPM microtubule assembly factor, defects in this protein lead to microcephaly
- GPR56 G protein receptor, defects impact radial glial migration lead to polymicrogyria

Doan et al. also drew attention to the HAR genomic elements (human-accelerated region) and their importance in neural processes. They noted that the first HAR discovered, HAR1, was shown to encode RNA expressed in the Cajal–Retzius cells important in organization of the cortex during development.

Specific functions of the HAR elements continue to be discovered. They have been noted to be rich in transcription factor–binding sites. In addition, there is evidence that they bind CTCF (CCCTC-binding factor) and may be involved in chromatin looping that is important in gene expression regulation.

3. Regulation of gene expression and more recently analyzed genomic segments

In a 2020 report, Wang and Goldstein (2020) investigated the size and numbers of enhancer elements associated with specific genes. They noted that transcriptional regulation of a specific gene is often influenced by arrays of enhancer elements. These elements are often located at some distance from the gene they regulate, e.g., the regulatory elements may be located about one megabase away from the gene they regulate.

They noted that there is some evidence for redundancy of enhancer elements and that this redundancy serves to protect gene expression from genomic disruptions, e.g., genomic deletions that delete a region containing the regulatory element for a specific gene.

Wang and Goldstein noted that genetic variants that are associated with increased risk for certain common diseases often occur in genomic regions that harbor enhancer elements. However, currently the specific target genes impacted by specific enhancers are poorly defined.

They carried out analyses to determine if disease pathogenicity was impacted by the degree of redundancy of enhancers that determined regulation of expression of a specific gene. They developed a specific enhancer domain score to reflect both the size and redundancy of enhancers for a specific gene.

Analyses carried out in their study revealed that enhancer size and enhancer redundancy showed correlation with the level of expression of a specific gene and with the capacity of that gene to maintain essential body functions. Their study revealed that specific essential genes were associated with large and redundant enhancer domains.

Long noncoding RNAs

Fang and Fullwood (2016) proposed that lncRNAs play roles in chromatin remodeling, in transcriptional and posttranscriptional regulation. There is some evidence that lncRNAs impact chromatin looping.

Roberts et al. (2014) noted that lncRNAs had been postulated to play roles in transcription regulation, in epigenetic alterations of transcription, in modulation of RNA splicing, and in numerous other functions. They noted that specific motifs in lncRNAs have potential protein-binding properties. One notable example was the lncRNA Hot Air that acts as a negative regulator of expression of the HOXC gene cluster.

They specifically addressed involvement of lncRNAs in neurodevelopment and brain function. Increased expression of lncRNAs was documented in pluripotent stem cells as they underwent neuronal differentiation. Increased lncRNA expression was documented as pluripotent stem cells underwent neuronal differentiation. Specific association of lncRNAs with SUZ12, a component of the polycomb repressive complex, was noted. Association of specific lncRNAs with the neurogenesis associated transcription factor SOX2 was also documented.

Beermann et al. (2016) reviewed studies on non—protein-coding RNAs including microRNAs, lncRNAs, and circular RNAs. They also stressed that the amount of non—protein-coding RNA increases as organismal complexity increases. Small noncoding RNAs are often 200 nucleotides in length or shorter, whereas lncRNAs can be several kilobases in length.

They noted that lncRNAs likely play roles in many cellular processes and that their expression is regulated. lncRNAs can be transcribed from the intergenic region or from introns; however, sometimes, they contain segments that are antisense to coding segments of genes. They can be transcribed by polymerase II or polymerase III, and they can also undergo editing.

Beermann et al., (2016) noted evidence that lncRNAs can regulate transcription and can also regulate splicing. There was also some evidence that lncRNAs in cytoplasm could be involved in posttranscriptional control. One function postulated is that the lncRNAs can act as sponges for microRNAs to prevent interaction of microRNAs with the 3' untranslated region of their specific target genes.

They emphasized the roles of lncRNAs in neural differentiation, in genomic imprinting, and in chromatin modifications. They noted that XIST is one of the best studied lncRNAs. It is involved in blocking transcription from one of the two X chromosomes present in cells of females.

Andersen and Lim (2018) reviewed lncRNA functions in brain. There is some evidence that lncRNAs regulate expression of specific genes important in brain function.

Long non—protein-coding RNA biology

In a 2018 review, Mattick (2018) reviewed noncoding RNA biology. He noted that complex organisms produce non—protein-coding RNAs (lncRNAs) from intergenic and intronic regions. In addition, antisense lncRNAs are produced.

Earlier studies in molecular biology documented mRNA as templates for proteins, and other forms of RNA included ribosomal RNAs, transfer RNAs, and spliceosomal RNAs. Long noncoding RNA was first documented as participating in X chromosome inactivation.

Mattick noted that in studies across species, there is evidence for lineage specificity in lncRNAs. He also noted that the metazoan proteome is static with approximately 20,000 genes in humans and nematodes. However, the non–protein-coding genome has expanded as developmental complexity of organisms has expanded.

Questions arise regarding the functionality of lncRNAs. Mattick noted evidence for expression of specific lncRNAs during cell-type development. There is also evidence that 3' untranslated regions of genes give rise to lncRNA particularly during embryonic development.

One function postulated for lncRNAs is that they act as scaffolds and as epigenetic guide molecules. Mattick noted that most lncRNAs are nuclear and are also associated with chromatin. However, there is evidence that lncRNAs also occur in the cytoplasm.

AIRN is a macroRNA that silences the paternally derived region on chromosome 6q25.2 in humans. This region is the location of the IGF2 receptor and SLC22A2 and SLC22A3 genes that encode polyspecific organic cation transporters.

Hezroni et al. (2020) reviewed lncRNAs and their roles in neural lineages. They noted the great diversity of lncRNAs in the nervous system, the rich repertoire of lncRNAs in brain, and the differential expression of lncRNA types in different brain regions. Databases have been established that list lncRNAs; these include GTEX, ENCODE, and Allen Brain atlas.

Hezroni et al. noted that different brain cell types and cells at different stages of differentiation expressed different lncRNAs. Specific studies have been carried out to derive information on the roles of lncRNAs in neurodevelopment and brain function.

4. Signaling pathways and neural development

Notch signaling

Engler et al. (2018) noted associations of Notch signaling with neural stem cells and radial glial cells in the ventricular zone of the developing brain. Primary stem cells in the brain give rise to intermediate progenitor cells. Progenitor cells then migrate out of the ventricular zone and into the subventricular zone where they give rise to neuroblasts that migrate to form the cortical layers. Neuroblasts were reported to migrate along radial glial fibers.

There is evidence that some neural stem cells persist even in the adult brain and are located in the lateral wall of the subventricular zone and in the dentate gyrus, specifically in the subgranular zone.

Engler et al. noted evidence that adult neural stem cells can amplify and give rise to neuroblasts and neurons. They noted that Notch signaling plays a key role in neural stem cell maintenance. Neural stem cells express a Notch receptor on their surfaces that conduct signaling into the stem cells.

Four Notch genes occur in vertebrates and give rise to four different receptors. In humans, NOTCH1 is widely expressed in different tissues. NOTCH3, in humans also referred to as CADASIL, are encoded on chromosome 19p13.2 and are reported to play a key role in neural development.

Zhang et al. (2018) studied Notch receptors on cell surfaces and noted that they are comprised of an extracellular domain with epidermal growth factor–like repeats. The external domain is involved in ligand binding. The extracellular domain is linked to a transmembrane that links to an intracellular domain that is involved in signaling. A number of different ligands bind to Notch receptors.

Specific proteolytic cleaving is required for Notch receptor proteins to mature. This proteolytic cleave follows ligand binding, and maturation of the receptors occurs within the cell membrane.

In humans, there are five Notch ligands:

- JAG1, Jagged 1 encoded on human chromosome 20p12.1
  - JAG2, Jagged 2 encoded on chromosome 14q32.33
    - DLL1, Delta-like 1 encoded on chromosome 6q27
    - DLL3, Delta-like 3 encoded on chromosome 19q13.2
    - DLL4, Delta-like 4 encoded on chromosome 15q15.1

Different signaling pathways have been described for Notch. The canonical signaling pathway, described by Kopan and Ilagan in 2009 involves the intramembrane proteolysis of the receptor to generate a mature active receptor. Following this proteolysis, the intracellular domain of the Notch receptor passes to the nucleus. Within the nucleus, the intracellular domain binds to a DNA-binding protein that in turn binds to DNA to activate the expression of specific target genes.

Andersen et al. (2012) described a noncanonical Notch signaling pathway. This involves interaction of Notch and the Wnt beta catenin signaling. Notch was reported to antagonize Wnt signaling, which then impaired differentiation of neural stem cells. Loss of Notch signaling leads then to premature differentiation of stem cells.

RAS signaling pathway in neurodevelopment

Nishiyama (2019) reviewed the importance of RAS signaling in neural development and in the proper function of dendritic spines. The RAS superfamily of small GTPases was noted to comprise more than 250 gene products that also play roles in actin reorganization and dendritic spine morphogenesis. Nishiyama noted that HRAS was the most abundant form of RAS in the brain.

The RASGDP RASGTP cycle is regulated by guanine nucleotide exchange factors (GEFs) and GTPase-activating proteins (GAPs). GEFs act by promoting conversion of RASGDP to RASGTP, the active form of RAS. GTPases alter RASGTP to RASGDP and are therefore inactivating.

Nishiyama reported the importance of calcium influx into the dendritic spine and binding of calcium to calmodulin kinase to generate CAMKIIA that promotes long-term potentiation of neural signaling. Activated RAS and CAMKIIA promote actin dynamics and dendritic plasticity.

RAS signaling pathway defects lead to neurodevelopmental abnormalities. These disorders are sometimes referred to as RASopathies and include neurofibromatosis, Noonan syndrome, Costello syndrome, and Legius syndrome (Rauen, 2013).

Transcriptome profiling in the developing brain

Li and others in the Psych ENCODE consortium Li et al. (2018) reported information on the regulatory genome and transcriptional processes in brain development. They carried out studies on brain tissues from different regions and on different cell types, at different life stages from embryonic life on and in adult brain. They studied transcription and methylation of DNA and histone modifications. Increased levels of transcription occurred in specific genes during formation of dendrites and synapses and were coincident with neuronal activity. Their studies also revealed relationships between epigenetic modifications and neurodevelopmental processes.

Li et al. concluded that nervous system development was dependent on spatiotemporal transcriptome regulation. Furthermore, they noted that there was growing evidence that neurodevelopmental factors, including transcriptional regulations and epigenetic modification when disrupted, could impair brain function and increase risk for neuropsychiatric diseases.

In 2019, Mukai et al. (2019) reported that mice with heterozygous loss-of-function mutations in Setd1A, a histone lysine methyltransferase, exhibited alterations in axonal branching and in cortical synaptic functions. They also noted working memory deficits in these mice. They also demonstrated that LSD1 (Kdm1A) lysine demethylase impacts Setd1A and that suppression of Kdm1A may counteract the effects of diminished Setd1A.

In recent years, much information on gene transcription in the brain has been gathered through studies in the ENCODE project and also through studies carried out by the Allen Brain Institute. In a review in 2015, Bae et al. noted that between 80% and 95% of protein genes are expressed in human brain at least during one period of life. The Allen Brain project involves detailed cellular, morphological, and transcription analyses of human brain from conception on. The ENCODE projects assesses long range regulatory elements, enhancers, repressors, silencers insulators, promoters, and transcripts.

Bae et al. emphasized evidence that different neocortical areas differed in their transcription and transcript splicing programs. They also noted transcriptional differences between germinal and postmitotic layers in the cortex. Recent studies have also provided evidence that somatic mutations arise in the brain.

Transcription information has also come from studies carried out as part of the Brain Span project. It has also come from analyses of postmortem brain samples from individuals with behavioral or psychiatric disorders carried out along with analyses of postmortem brain samples from controls.

In February 2018, Gandal et al. reported results of transcriptome profiling of postmortem brain samples from individuals with autism (ASD, 50 cases), SCZ (159 cases), bipolar disorder (BD, 94 cases), major depressive disorder (MDD 87 cases), alcohol misuse disorders (AMD 17 cases), and controls (293 samples).

Their studies revealed differential expression of a large number of genes in cases versus controls. Differential gene expression (either increased gene expression or reduced gene expression) was documented in 1099 genes in ASD, 890 genes in SCZ, and 112 genes in BD. Gandal et al. also reported substantial overlap in these disorders in specific genes that manifested differential expression. Specific modules were identified based on analyses of networks and interactions. Key modules implicated in differential gene expression include neurons, endothelial cell, astrocyte, and microglia modules.

It is important to note that changes in gene expression may also arise due to genomic segmental copy number changes (Fig. 5.1).

Functional genomic analyses in the brain psych ENCODE project

Wang et al. (2018) carried out detailed analyses on 1866 brain samples in the Psych ENCODE project. They emphasized that progress in defining the molecular basis of psychiatric disorders was very slow, despite the fact that genome-wide associated loci had been defined for some psychiatric disorders. The genome-wide association study (GWAS) signals were predominantly located in non–protein-coding genomic regions.

Large-scale studies by the Psych ENCODE consortium include genotyping, single-cell RNA sequencing, and transcriptome studies on 32,000 cells from different brain regions and chromatin conformation studies. These studies led to generation of data on genome functional elements including quantitative trait loci (QTLs) and regulatory network linkages. Activities in this project led to generation of

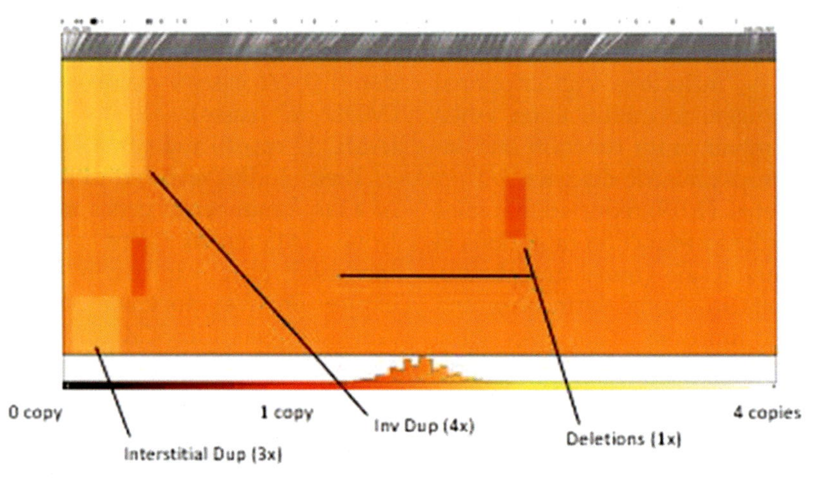

FIGURE 5.1 Microarray image of chromosome 15 in 5 cases of autism. Note presence of duplications (bright yellow increased signal) and deletions (red) indicating decreased signal.

intermediate phenotype data defined as functional information on genomic elements, and data were correlated with clinical variables.

One important finding from the study was that cell type changes accounted for a significant degree of variation in gene expression.

Wang et al. also reported that the level of non–protein-coding gene transcription is much higher in brain than in other tissues.

Importantly, researchers identified brain regulatory networks that linked transcription factors and enhancers with specific target genes.

Prior GWAS studies had identified 142 SCZ-associated genomic loci. The Psych ENCODE activities were able to associate 321 target genes with these GWAS loci. Analyses of transcription of these genes improved prediction of phenotype alterations. Results highlighted genes that functioned in immunological, synaptic, and metabolic pathways as having key roles in pathogenesis of psychiatric disorders.

Enhancers

In the Psych ENCODE project, enhancers were defined as open chromatin regions enriched in specific modifications including histone H3 lysine 27 acetylation (H3K27ac) and depleted in histone3 lysine4 trimethylation (HK4me3). Open chromatin was identified on the basis of DNAse1 hypersensitivity in DNAse sequencing.

Wang et al. noted that variability in enhancer activity was marked in different individuals. However, activity of specific enhancers could be mapped to particular brain regions.

Quantitative trait loci analyses

In addition to studies of QTLs that impacted gene expression, Wang et al. undertook analyses to identify QTLs that impacted the abundance of specific cell types. They designated these loci as fQTLs and determined that abundance of excitatory neurons was most significantly impacted by these loci.

A specific eQTL that impacted expression of the *FZD9* gene also acted as an fQTL and influenced the expression of a specific excitatory neuron-type Ex3.

It is interesting to note that neural progenitor cells derived from induced pluripotent stem cells

(IPS cells) developed from Williams syndrome patients by Chailankarn et al. were reported to have alterations that included aberrant calcium oscillation signals and abnormal numbers of synaptic spines. In further studies, the abnormal cellular phenotype was shown to be due to deletion of the gene that encodes FDZ9, a receptor for Wnt signaling proteins. Wang et al. reported that the specific QTL that altered expression of FZD9 also altered the proportion of excitatory neurons of the E3 type.

CQTLs

CQTLs were defined as specific loci that impact chromatin state and transcription factor binding. Wang et al. reported that there was some overlap in eQTL (expression QTLs) and CQTL (connectivity quantitative trait) loci. They also noted that enhancer promoter interaction occurred within TADs (transcriptionally active domains).

Transcriptionally active domains

These domains are separated from nonactive domains by sequences that bin CTCF (CCCTC-binding factor). CTCF binding blocks communication between enhancers and upstream promoters, thereby regulating expression.

Fetal and adult brain genomes were found to differ with respect to the genes included within TADs. Fetal brains were particularly rich in TADs that impacted expression of genes in neurons.

CTCF molecules bind to the boundaries at each end of a specific genomic segment to define a TAD, and the TAD is then stabilized by a cohesin complex ring and loop extrusion occurs (Rowley et al., 2019).

5. Neuronal activity and epigenetics

Belgrad and Fields (2018) reviewed epigenetic interactions in neuronal activity. They documented steps in neuronal function that began with patterned action potential followed by intracellular network activation and specific phosphorylation reactions of specific proteins followed by dephosphorylation of other proteins. Action potential also impacts cytoplasmic calcium dynamics, activation of transcription factors, and regulation of epigenetic mechanisms. The latter involves alteration in nucleosome position, alterations in chromatin looping, histone, and DNA modifications.

Belgrad and Fields emphasized that action potential firing could lead to calcium-dependent and calcium-independent cytosolic signaling. They emphasized that specialized gene expression, in consequence of neuronal firing and signaling, requires diverse epigenetic modifications. Importantly, unique neuronal activation patterns and unique epigenome interactions lead to transcription that is stimulus specific.

Belgrad and Fields noted that tests have been undertaken using specific forms of stimulation of neurons in model organisms and in cultured neuronal cells. Optogenetic stimulation of channel rhodopsin represents one such study; channel rhodopsins function as light-gated ion channels. Another study utilized heat activation as a stimulus of TRP1 (transient receptor potential) cation channels.

Calcium signaling functions in part through activation of protein kinase C and ERKMAP signaling that ultimately impact transcription factors. These processes include phosphorylation of CREB, a calcium-responsive transcription factor.

NMDA glutamate receptors are excitatory neurotransmitters reported to activate ERK signaling and calcium/calmodulin kinase.

Specific environmental factors were reported to impact nuclear calcium signaling, and calcium was reported to bind nucleosomes and to impact the degree of nucleosome compaction. In addition, different forms of neuronal firing were reported to impact transcription factor binding to genomic sites.

The transcriptional effects of different histone and DNA modifications were related to the specific position modified and the degree of modification.

Qureshi and Mehler (2018) reviewed epigenetic modification with particular reference to the nervous system. Key factors in epigenetic mechanisms include DNA methylation and hydroxymethylation, histone modifications, nucleosome repositioning and remodeling, and chromatin organization. Also important are noncoding RNAs including short and long non–protein-coding RNA and RNA editing.

Key factors in epigenetic modification of DNA

Modifications of cytosine residues in DNA primarily involve use of enzymes in the DNMT family of DNA methyltransferases. Methylated DNA can also undergo demethylation and key enzymes involved in this process including TET enzymes (ten–eleven translocases). These enzymes facilitate oxidative processes, leading to conversion of 5-methylcytosine to 5-formylcytosine and 5-carboxycytosine.

Qureshi and Mehler noted that 5-formylcytosine and 5-carboxycytosine can also undergo base excision through activity of DNA glycosylase.

Other potential mechanisms of DNA demethylation involve ABOBEC/AID cytidine deaminases with removal of methylated cytosine that can then be replaced by unmethylated cytosine.

Earlier epigenetic studies concentrated primarily on methylation of CpG dinucleotides particularly in gene promoter regions and roles of this methylation in silencing gene expression. Qureshi and Mehler noted that there is now evidence for complex patterns of DNA methylation and hydroxymethylation in different regions of the genome, gene bodies, intron, exons, and untranslated gene regions.

Chromatin

Key elements of chromatin include histones, nucleosome, and binding proteins. Nucleosomes are composed of histones H2A, H2B, H3, H3.3, H4, and H2AZ. Key factors relevant to gene expression relate in part to positioning and degree of clustering of nucleosomes. Movement of nucleosomes is impacted by ATP and enzymes. Tightly packed nucleosomes restrict gene expression, and open chromatin structure with wider spaced nucleosomes facilitates gene expression.

Specific modification of histones can occur on nucleosomes and particularly on histone tail that projects from nucleosomes (Fig. 5.2). Modification can also occur on histones that surround DNA between nucleosomes.

Qureshi and Mehler emphasized that chromatin states are very dynamic due to histone modification and due to variation in structural conformation of chromatin.

Key enzymes involved in chromatin modification include methylase and demethylase, histone acetylases, and histone deacetylase. Modifications in the histone tails are of particular importance; key sites of histone modifications include lysine, and mono- and di- or trimethylation can occur. Another form of chromatin modification involves replacement of core histones with different forms of histones.

Macromolecular complexes bind to histones and can impact gene expression. Complexes that bind to histones are sometimes referred to as chromatin modifiers and readers, writers, and erasers.

Relevance of epigenetics in neurodevelopmental defects (Fig. 5.3)

Iwase et al. (2017); Iwase and Martin (2018) noted that a large percentage of genes that have been shown to carry mutations implicated in the causation of intellectual disability encode epigenetic regulators. These include genes involved in posttranslational modification of DNA and histones and genes that encode chromatin regulators.

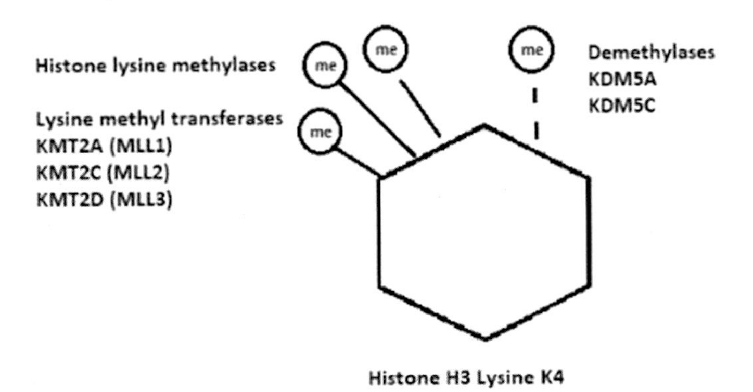

FIGURE 5.2 Methylation and demethylation of histone H3 lysine K4. Shows enzymes involved in methylation and demethylation.

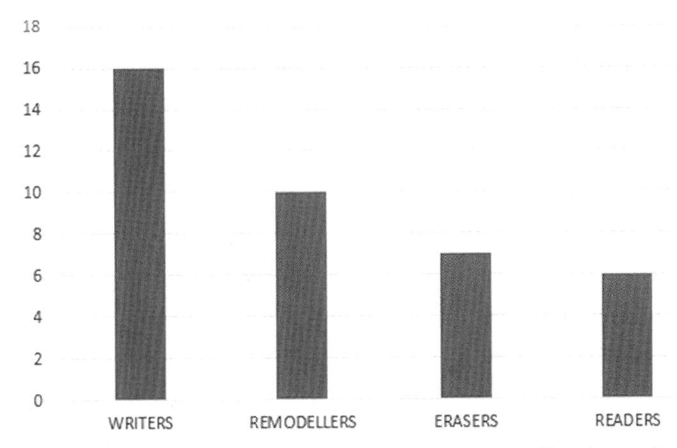

FIGURE 5.3 Major epigenetic mechanisms. Functions of key factors involved.

They noted that development of the brain involved precise temporal and spatial gene regulation. Such regulation was required for regulation of cell fate and cellular processes. They noted too that there was evidence that epigenetic processes impacted changes in neurons and circuits that were required for experience driven neuronal alterations essential to learning and memory.

In 2020, Beck et al. (2020) noted evidence that DNA methylation, histone modification, and chromatin structure play key roles in determining gene expression. They also noted that a significant number of neurodevelopmental defects and growth abnormalities in children were associated with defects in chromatin modification; however, relatively few neurodevelopmental disorders were associated with DNA methylation abnormalities. They noted that functions and disruption of methylation writers, DNMTs, and methylation readers, e.g., MECP2, have been studied but that few studies of intellectual disabilities included analyses of the DNA demethylases and the TET enzymes.

They reported studies on eight families with defective function of the DNA demethylase TET3. Intellectual disability and/or developmental delay were key abnormal phenotypes in these families. Their studies revealed the importance of adequate activity of the DNA demethylase TET3 in neurodevelopmental processes.

6. Additional insights into RNA functions and metabolism relevant to neuronal functions

Non–protein-coding RNAs and roles in impacting gene expression

Long non–protein-coding RNAs were shown to play roles in gene silencing in imprinted regions of the genome. More recently, long non–protein-coding RNAs have been found to impact chromatin looping and remodeling.

Short RNAs including microRNAs play key roles in particularly at posttranscriptional levels.

RNA editing

Modification of nucleotides in RNA particularly involves adenosine to inosine conversion and cytidine to uridine changes. Enzymes involved in RNA editing included specific forms of adenosine deaminase, ADAR1, 2, and 3. Qureshi and Mehler noted that ADAR activity can also impact gene splicing and gene regulatory factors.

Other RNA modifying proteins include members of the APOBEC family (apolipoprotein B mRNA modifying enzymes).

RNA metabolism and RNA-binding proteins

Insights in RNA metabolism and RNA-binding proteins have been derived in part through studies on particular disorders. Nussbacher et al. (2019) reviewed the critical RNA-binding proteins in RNA regulation and functions. RNA-binding proteins play roles in multiple aspects of RNA splicing, expression transfer, and localization.

Insights into unique aspects of RNA function derived from studies in fragile X mental retardation

This disorder results from pathologic expansion in the triplet repeat that occurs in the 5′ regions of the FMRP encoding gene (FMR1) that maps to Xq27.3. Weiler and Greenough (1999) reported that abnormal methylation of the CGG repeat expansion led to impaired gene expression and decreased quantities of the FMRP protein. Studies particularly by Darnell and associates revealed that the FMRP protein played essential roles in the translation of RNA derived from a number of different genes. Normal FMRP protein was found to bind to polyribosomes and to be necessary for MRNA translation of specific genes in neurons. Normal FMRP protein was also found to play roles in mRNA splicing.

RBFOX1 and RNA splicing

Hamada et al. (2013) reported that RBFOX (A2BP1) functions as a mRNA splicing factor. It has been implicated in neurodevelopmental disorders and in autism. They reported that studies in mouse when Rbfox 1 was knocked down revealed defects in radial migration defects of cortical neurons.

7. Environmental stimuli, gene transcription, and neural activity

In 2018, Yap and Greenberg (2018) reviewed activity-dependent gene transcription in brain. They emphasized that neuronal function must adapt to changing environments. In addition, neurons must encode short- and long-term memories. One adaptation method involved changes in synaptic properties; key aspects of this involve coupling of synaptic activity to nuclear function and gene expression.

Yap and Greenberg emphasized the importance of differentiating activity-regulated changes in gene expression from basal gene transcription. They also emphasized the importance of neuronal transcription and epigenome in bringing about behavioral adaptations to specific stimuli.

Early studies by Greenberg and Ziff (1984) documented that an early response to neuronal stimulation was expression of the FOS gene. These studies led to the conclusion that there was rapid communication of stimulated synapses with the nucleus leading to expression of FOS transcripts. This transcription was then followed by transcription other genes, and this became known as the immediate-early gene (IEG) program of response to stimulation.

The effect of the IEG program was to stimulate a subsequent pattern of gene expression referred to as the late response gene (LRG) expression program.

The pathway from early neuronal stimulation to early gene response

Yap and Greenberg (2018) noted that neuronal stimulation is accompanied by calcium influx into the neuron. Calcium influx can occur through voltage-gated ion channels or through specific neurotransmitter receptors AMDA or NMDA glutamate receptors. The calcium influx then stimulates activity of RAS mitogen protein kinases and calmodulin kinase (CAMK) that stimulate preexisting transcription factors that then apparently stimulate CFOS and IER transcription in the nucleus.

FOS expressed in the nucleus then combines with JUN to form a protein complex known as activation protein complex 1. The activation complex 1 (AP1) then triggers expression of the LRGs.

LRGs were reported to encode proteins involved in dendritic growth and spine maturation. Yap and Greenberg (2018) noted that complete identification of the LRGs had not yet been achieved.

Specific studies indicated that the FOS/JUN that forms the activated protein complex binds to specific enhancer elements that lead to activation of gene expression. Yap and Greenberg noted that interesting molecular studies had revealed that the FOS–JUN activation complexes in combination with pioneer transcription factor alter chromatin formation of LRGs. The AP1 complex was shown to interact with SWI/SNF chromatin remodeling complex.

Yap and Greenberg considered additional aspects of neuronal activity and the epigenome. Neuronal activity was shown to impact not only chromatin remodeling but also nuclear histones through activation of histone deacetylase and the NURD complex. The NURD complex is defined as composed of proteins with both ATP-dependent chromatin remodeling and histone deacetylase activities.

Yap and Greenberg noted that questions still arise as to how behavioral information is stored. Possible mechanisms include epigenetic changes and modification induced by neural activity. Indirect evidence has been obtained in that inhibition of DNMT methyltransferase and impaired methylation of DNA may impact memory consolidation.

Methyl cytosine modifications have been studied; however, there is also evidence that methylation occurs at other cytosine nucleotide dimers, e.g., CA, CT, and CC. Lister et al. (2013) reported the occurrence of non-CG methylation in neuronal DNA. These forms of methylation are sometimes referred to as mCA.

Readers of nucleotide methylation, including methyl CG-binding protein MECP3, have been reported to bind to methylated nucleotides and impact transcription.

Yap and Greenberg noted that several studies indicate that genes impacted by MECP2 binding to mCA nucleotides are often long genes.

Sensory input in postnatal brain development

Hubel and Wiesel in 1970 published evidence of important of sensory input to the eye in the development of the visual cortex. West and Greenberg (2011) published evidence for roles of neural-activated gene expression in synapse development and for subsequent synaptic pruning to establish appropriate excitatory–inhibitory balance. They noted the important role of calcium signaling in neuronal-activated gene expression. Particularly important in these processes were expression of activity-regulated genes including BDNF (brain-derived neurotrophic factor), ARC (activity-regulated cytoskeleton-associated protein), HOMER (postsynaptic density scaffold protein), NPTX (neuronal pentraxins), and NRN1 (neuritin). They noted that alteration in the regulation of BDNF impacts neuronal wiring.

Downstream processes after synaptic signaling

ARC protein

ARC is described as an intermediate-early protein. Nikolaienko et al. (2018) reviewed structure, interactions, posttranslational modifications, and functions of ARC protein that are essential for synaptic homeostasis, long-term potentiation, long-term depression, and memory.

They defined synaptic plasticity as the ability of the synapse to change with use and disuse and noted that impaired synaptic plasticity occurred in specific neurocognitive and neurodegenerative disorders.

ARC has been reported to play roles in RNA transfer. HOMER1 expression was reported to be increased in sleep and to likely play roles in memory consolidation. Neuronal pentraxin was reported to be important in the development of GABA interneurons. Interneurons are sometimes referred to as association neurons. Local interneurons are defined as having short axons and as connecting with nearby neurons. Relay interneurons connect specific circuits with other brain regions.

Nikolaienko et al. (2018) emphasized that an additional aspect of synaptic function includes homeoplastic plasticity, also sometimes referred to as synaptic scaling. Specific insights into ARC function had been derived from studies of the downstream effects of genetic defects in ARC protein.

Genes and their products involved in transcription, translation, and synaptic functions

Carmichael and Henley (2018) reviewed signaling pathways and calcium ion passage through channels and through the cytoplasm that follows neuronal excitation. Transmitted signaling molecules and transcription factors pass from the cytoplasm to the nucleus to activate immediate-early gene expression.

Important transcription factors in the nucleus in neurons include CREB and MEF. MEF (ELF4) protein is a transcriptional activator that binds and activates the promoters.

ARC activity–regulated cytoplasmic proteins

ARC is an immediate-early gene product. Immediate-early genes are transcribed shortly after neuronal stimulation, and their transcription is facilitated by transcription factors that are already present in the cytoplasm. Excitation at synaptic receptors triggers signaling pathway activity and passage of factors into the nucleus to trigger expression of immediate-early genes including the ARC encoding gene. Detailed analyses have been carried out by several investigators to identify specific enhancer elements upstream of the ARC gene, and elements within the ARC gene promoter that are immediate are triggering ARC gene expression.

Intense activity in exploring factors involved in ARC gene expression and function of ARC protein has been stimulated by evidence that the ARC protein plays important roles in synaptic plasticity, in learning and memory, and particularly in consolidation of long-term memory. Cyclic AMP was shown to bind to specific response elements upstream of the ARC gene.

ARC protein released into the cytoplasm undergoes a number of different posttranslational modifications that impact its activity.

Nikolaienko et al. (2018) emphasized interactions of ARC protein with proteins in dendritic spines and the impact of these interactions on synaptic plasticity. ARC protein was reported to interact with actin and to promote actin stability. Another important function of ARC is the role that it plays in promoting endocytosis of the AMPAR glutamate receptors, and this activity leads to termination of excitation.

Nikolaienko et al. noted an additional function of ARC in the nucleus, namely the promotion of histone acetylation.

The UniPro database (Q7LC44) information on ARC noted that the ARC mRNA is encapsulated, and these capsules are transferred from the nucleus to the cytoplasm. This RNA can then be translated to protein.

Epstein and Finkbeiner (2018) reported that long-term memory performance is defective in ARC-deficient animals. They also noted that ARC expression is induced during sleep, and this may be related to the role of sleep in mediating long-term memory. They noted the centrality of ARC in mediating conversion of transient experience to long-term adaptive changes in the brain.

Learning and memory and plasticity of circuits

Yap and Greenberg (2018) noted the relationship of dendritic spine dynamics and synaptic remodeling in response to sensory input in the postnatal organism. ARC has been found to play important roles in this process, and also important is activity-dependent expression of MEF2C transcription enhancer factor. HOMER1 expression was reported to be increased in sleep and is likely to play roles in memory consolidation.

DNA methylation and cortical development

Price et al. (2019) reported that DNA methylation changes in neurons occur more prominently during the first 5 years of postnatal life than at any other period in life. These methylation changes involved CpG and CpH methylation changes. They emphasized that during development, experience shaped gene regulation and involved changes in DNA methylation.

Their study involved analyses of neurons from the dorsal prefrontal cortex and focused particularly on neurons. They specifically sorted neuronal cell from glial cells and revealed that age-related methylation differences occurred in neuronal cells.

Stroud et al. (2017) reported evidence that early postnatal experience led to methylation of CA sequences and that these changes impacted gene expression. This methylation particularly involved activity of DNMT3A methyltransferase.

Their studies also revealed that activity-dependent gene expression led to increase expression of DNMT3A expression and to increased methylation CA residues (cytosine adenine). This was then followed by decreased expression from genes that had mCA modifications. The decreased expression of genes was found to be due to MECP2 binding to CA.

The sequence of events then was sensory experience and signaling to increased early gene expression, which in turn led to increased activity of DNMT3A methyl transferase and methylation at CA sites. Binding of MECP2 to methylated CA residues decreased expression of specific genes.

The three-dimensional genome and gene expression

Key factors in initiating gene expression include enhancer–promoter interactions. There is now evidence that chromatin looping brings enhancer sequence elements and promoters into close proximity and that this plays key roles in gene regulation. The enhancer–promoter interaction triggers interaction between promoters and transcription initiating factors.

Yu and Ren (2017) reviewed the three-dimensional organization essential for mammalian gene expression and recently developed molecular tools for investigating these interactions. Chromatin conformation capture techniques involved formalin fixing and cross-linking of DNA stands and bound elements. This was then followed by DNA fragmentation. Specific antibodies to specific bound elements, e.g., transcription factors, could then be used to separate specific fragments and that DNA within those fragments could be sequenced.

Chromatin organization

In recent decades, through advances in gene sequencing, chromatin capture and other techniques clearer insights have been obtained into positions and functions of enhancers, insulators, and promoters and to aspects of chromatin modifications and rearrangements, topologically associated domains (TADs), and the roles of all of these factors in control of gene expression (Perenthaler et al., 2019).

Chromatin remodeling

Chromatin remodeling is defined as the process by which chromatin structure is altered from condensed chromatin to chromatin with open structure that facilitates transcription. One of the processes that take places during chromatin remodeling is the movement of nucleosomes, so they become separated from one another and this facilitates access of transcription factors and other complexes to DNA. Energy for movement of nucleosomes is provided through ATP hydrolysis mediated by specific enzymes.

Staahl and Crabtree (2013) reported important aspects of chromatin remodeling in the nervous system. They reported that 29 different genes encode ATPases that facilitate nucleosome movement and chromatin remodeling. The ATPases are subunits of large complexes referred to as SWI/SNF complexes and BAF complexes.

In humans, protein subunits of BAF complexes are encoded by a number of different genes that include *ARID1A, ARID1B,* and *ARID2*; members of the *SMARC* gene family; and *BCL11 genes, BCL11A,* and *BCL11B*. Significantly the genes that encode BAF subunits have been found to frequently be mutated in neurological diseases.

Chromatin modifying complexes

Kadoch et al. (2015) reviewed PRC2 and SWI/SNF chromatin remodeling complexes. They noted that the PRC2 complex in mammalian cells was comprised of at least five subunits:

- EZH1/EZH2 enhancer of zeste 1/2 polycomb repressive complex 2 subunit
  - EED member of the polycomb-group family of transcriptional repressors
    - SUV subunits act as H1/H2 histone methyltransferase that trimethylates lysine 9 of histone H3
    - RBBP4/7, RB binding protein 4/7, chromatin remodeling factor
    - AEBP2, AE-binding protein 2

They noted that other proteins are required for specific PRC functions and some may act as regulators that include the following:

- JARID2, jumonji and AT-rich interaction domain containing 2
  - PHF19, PHD finger protein 19
    - PHF1, PHD finger protein 1
    - MTF2, metal response element–binding transcription factor 2

Kadoch et al. noted that PRC2 catalyzes methyl transfer from S-adenosyl methionine to specific lysines in histones, e.g., H3K27. They noted that PRC2 is a key enzyme reported to catalyze methylation of H3K27.

SWI/SNF complexes are also sometimes referred to as BAF complexes. These complexes in humans are encoded by genes in the SMARC family. SWI/SNF subunits are encoded in humans by 29 different genes. The subunits assemble into complexes with 10–15 subunits. These complexes utilize energy derived though ATP hydrolysis to impact nucleosome movements and to modulate chromatin accessibility.

SMARCA4 (BRG1) and SMARCA2 (BRM) act as ATPases. Complexes contain either SMARCA4 or SMARCA2.

In addition to ATP-related activities and nucleosome movement, these complexes have other functions. The different functions of a specific complex are influenced by the subunit composition of each complex. There is evidence that at different stages of nervous system development, the complexes of different subunit structure are active.

Setiaputra and Yip (2017) reported on the molecular architecture of chromatin modifying complexes. They noted that these are multisubunit complexes with conformational flexibility. In considering the nucleus of the human cell, they noted reports that 1 meter of genomic DNA is compacted into 1 micron and that the high degree of compaction required is facilitated by proteins that interact with and modify DNA and chromatin. These proteins facilitate the packing of DNA and chromatin and alter the distances between nucleosomes.

Setiaputra and Yip noted the complex and variable composition of chromatin complexes that contain multiple subunits. In addition, dynamic changes occur in the subunit composition of specific chromatin modifying complexes under different conditions, at different stages of differentiation of cells and in different stages of the cell cycle.

Specific chromatin modifying complexes include the polycomb complexes PRC1 and PRC2. These complexes promote transcriptional repression. PRC1 was reported to promote chromatin compaction and to facilitate ubiquitylation of histone H2Alysine 119.

The catalytic domain of PRC1 complex is composed of RING1A and RING1B ubiquitin E3 ligase. These interact with other subunits in PRC1. PRC1 and PRC2 complexes contribute to transcriptional silencing through different mechanisms/

Protein synthesis relevant to neuronal activity and to the integrated strength response

Sonenberg and Hinnebusch (2009) reviewed translation of MRNA to protein. They noted the importance of the AUG translation start codon, methionyl tRNA, and binding of the preinitiation complex and 40S ribosomal subunit. The preinitiation complex includes eukaryotic protein synthesis initiation factors that recognize the M7-guanine cap at the 5′ end of mRNA.

For mRNA translation to proceed, unphosphorylated EIF2 and RNA are transferred to the ribosome, and this transfer is dependent on guanosine triphosphate.

Hinnebusch et al. (2016) reported that the 5' untranslated region of mRNA and the EIF2 translation initiation factors exerted the control of protein synthesis that was implicated in learning and memory.

Hearn et al. (2016) reported that there are a number of different steps that lead to repression of mRNA translation and repression of protein synthesis. A key step in this repression is phosphorylation of serine 51 in the EIF2a subunit of the translation initiation complex. They noted that this phosphorylation can be catalyzed by different kinases including PKR kinase also known as EIF2AK2 and two other kinases that are apparently active under specific conditions including amino acid deficiency and oxidative stress.

Kashiwagi et al. (2019) reported that the binding of translation initiation factor subunits EIF2A and EIF2B was negatively impacted by phosphorylation of EIF2A. The stress-induced kinase EIF2A phosphorylation pathway was designated as the integrated stress response. Following phosphorylation of serine 21, the phosphorylated EIF2A was shown to inhibit a specific function of the EIF2B subunit, namely a guanine nucleotide exchange reaction that is essential for translation to proceed.

Hearn et al. (2016) reported that a small molecule screen led to identification of a class of small molecules that bind to EIF2A and result in impairment of phosphorylation. This impairment of phosphorylation meant that protein synthesis was no longer inhibited. These small molecules were designated as ISRIBs (integrated stress response inhibitors).

Importantly, in rodent studies, ISRIB administration was shown to improve memory consolidation. The target of the small molecule was the interaction between EIF2a and the regulatory subunit encoded by EIB2B3.

Zyryanova et al. (2018) reported that the ISRIB molecule bound at the interface of EIF2B and the EIF2B3 encoded regulatory subunit and that this attenuated the nucleotide exchange activity of EIF2B.

Transcription and memory formation and consolidation

In a 2019 review, Hegde and Smith (2019) noted that substantial progress has been made during the past half century in defining the underpinnings of memory formation and consolidation. Memory formation involves transcription. With respect to transcription, key developments included elucidation of the role of transcription factors including CREB and other factors encoded by immediate-early response genes particularly NR42 and NRA3. Regulation of transcription was also found to be linked to CRTC1 a CREB coactivator. Long-term potentiation of synaptic activity was shown to be linked to NR4A activity. Other important factors in expression of the immediate-early response genes include SRF (serum response factor) that was essential for expression of CFOS and EGR1.

Appropriate translation of mRNA through activity of EIF2A and related factors at the 5′ end of mRNA has also been shown to be important in regulating synaptic plasticity and memory.

Epigenetic factors have also been linked to potentiation and memory formation. Histone acetylation played an important role in memory formation and consolidation.

Posttranscriptional RNA modifications

Widagdo and Anggono (2018) reported that there is evidence that posttranscriptional RNA modification plays roles in brain processes. Their report focused on methylation specifically on methylation at the N6 position in adenosine in RNA. This modification was reported to impact RNA splicing, RNA stability, and RNA translation.

Specific writer proteins were known to induce methylation at N6 in adenosines, and removal of this modification was also induced by specific eraser proteins.

Widagdo and Anggono noted that methylation of adenosine in brain MRNA was first detected using antibodies specific for methylated RNA. There was some evidence that this modification was more prevalent in some neurons than in others.

A specific methyltransferase complex was shown to transfer the methyl group from S-adenosyl methionine to RNA. The complex included METTL3 and METTL14 (methyltrans ferase-like 3 and 14). Studies revealed that methylation occurred cotranscriptionally. The methylated MRNA was recovered from axons, dendrites dendritic spines, and presynaptic nerve terminals. Overlap was detected between adenosine modified RNAs and mRNA known to be targeted by the Fragile X gene product FMRP.

Widagdo and Anggono noted that there was evidence that methylated adenosine in mRNA played an important role in brain development. These conclusions were based on the observation that knockout of specific genes that encode the enzymes involved in this methylation reaction led to developmental defects.

Histone deacetylases and histone acetylases

Schmauss et al. (2017) reviewed the roles of class I histone deacetylases in processes in the hippocampus and prefrontal cortex. Studies on rodent hippocampus revealed that lower levels of HDAC2 were associated with increased associated and learning skills. High levels of histone deacetylases were associated with decreases in association and spatial learning. These studies pointed to evidence that acetylated histone promoted learning.

Their studies indicate that the Nr4a nuclear transcription factor expression is decreased in aging and that this may be in part due to its repression by histone deacetylase Hdac3. These studies may provide insights into possible application of histone deacetylase inhibitors in therapy of aging cognitive impairment.

Questions arise as to which specific amino acids need to be acetylated to promote learning. Lysine residues in histone are frequent targets for acetylation, but impacts may also vary depending on which specific lysine in histone is acetylated. Specific enzymes that acetylate lysine in histone include lysine acetyltransferases (KATs). Enzymes that deacetylate lysine in histone are referred to as KDMs.

Histone acetyl transferase occurs in complexes that are recruited through a specific protein TDP2 (TTRAP) (tyrosyl DNA phosphoe sterase 2). A large collaborative study reported by Cogne et al. (2019) noted that the TRRAP gene was classified as a gene, highly intolerant of mutation. The TRRAP protein was shown to be highly conserved in evolution. Their studies revealed that deletions in TTRAP occurred in two separate groups of patients. One group of patients manifested brain malformations. A second group of patients manifested autistic behaviors, and some had intellectual disabilities. Brain-specific knockout of TRRAP in mice was reported to impact brain development.

MECP2 new insights

Connolly and Zhou (2019) noted evidence that both gain of function and loss of function of MECP2 led to neurological pathologies. They noted further that there was evidence that MECP2 was involved in recruiting corepressor complexes to specific methylated DNA sequences. They proposed that MECP2 functions include organizing chromatin architecture. This organization then impacts gene expression. Gulmez Karaca et al. (2019) noted that MECP2 function was initially described as depression of transcription but that recent evidence indicated that MECP2 had more comprehensive impacts. They reported that two C-terminal domains in MECP2 protein CTD2 and CTD3 interact with proteins that regulate chromatin structure.

They noted further that the transcriptional profiles in neurons play roles not only during development but also later in life. They noted MECP2 plays important roles in chromatin maintenance in mature neurons.

Intact MECP2 was shown to promote formation of chromatin loops and to interaction with other chromatin-binding proteins including CTCF and ATRX. These findings indicate that MECP2 functions as a chromatin organizer.

Gulmez Karaca et al. noted evidence that methylated cytosines in the CpH context (H-A/C/T) occur particularly in the frontal cortex and in the dentate gyrus. CPH methylation was also reported to increase in cortical regions and hippocampus during postnatal development and to remain present there throughout life.

Neuronal cell type–specific promoters and enhancers and gene expression

Nott et al. (2019) carried out a study of noncoding regulatory genomic regions in different cell types in the brain. They isolated nuclei from microglia, oligodendrocytes, astrocytes, and neurons.

They used the ATAC seq method (transposase accessible chromatin sequencing) to identify region of open chromatin in the genome. They also utilized antibodies to H3K27AC and H3K4me to isolate genomic segment known to represent active chromatin and promoter regions.

Applying these techniques to different cells, they identified active promoters and active enhancers in different cell types. These methods led to identification of larger numbers of brain-specific enhancers than were noted to be present in the ENCODE database.

Nott et al. then utilized GWAS that included genomic variants associated with particular neurobehavioral traits. They demonstrated a strong association of reported variants in psychiatric disorders and enhancer elements.

Variants associated with Alzheimer's disease were found to be more enriched in microglia regulatory elements and particularly in enhancers.

Variants associated with psychiatric disease occurred in enhancers and promoter sequences in neurons.

H27Aac was found to frequently occur in promoter regions.

Specific techniques, e.g., PLAC seq revealed chromatin loops between promoters and distal regulatory elements.

In total across different cell types, they identified 2954 superenhancers that interact with specific promoters. In addition, GWAS-associated variants were demonstrated to be in specific superenhancers.

Synaptic function and epigenetics

In a review in 2019, (Campbell and Wood, 2019) noted that gene expression was known to be involved in learning and memory consolidation but that mechanisms underlying these processes had not yet been fully characterized. Studies of the epigenome were being intensely carried out in attempts to shed more light on regulation of learning and memory. They considered the main epigenetic mechanisms involved to include histone modifications, nucleotide modifications, histone variant exchange, chromatin remodeling, and RNA regulation.

Campbell and Wood considered the synaptic activation and signaling processes to be fairly well defined. They noted that following activation postsynaptically, dendritic spines underwent modification in part due to the formation and stabilization of cytoskeletal actin.

Insights into how structural synaptic changes promoted learning in response to experience are being gathered. CREB1 transcription factor is known to be essential for learning and memory. One important discovery was that phosphorylation of serine133 in CREB1 leads to recruiting of CREB-binding protein (CBP) followed by acetyl transferase activity. CBP and histone acetylation were shown to be involved in memory generation. Histone acetylation recruits protein complexes essential for transcription. Deacetylation of histone is carried out by a number of different histone deacetylases.

Campbell and Wood noted that histone methylation has also been shown to impact memory and dependent on the site of the residue in histone that is methylated; methylation can activate or repress transcription. The Ga/Glp complex, an epigenetic silencing complex, influences histone methylation. Histone methylation can also be carried out by a number of different lysine methyl transferases (KMTs). Methyl groups can be removed from histone through the activity of lysine demethylases (KMDs). Defective function of either KMT or KMD enzymes can lead to impaired learning or memory.

DNA modification has different effects depending on the region of the gene that is methylated, e.g., gene body or promoter. DNA methylation is dependent on the activity of DNA methyltransferases (DNMTs). Campbell and Wood noted that DNA hydroxymethylation is particularly important in the brain.

They noted that cytoskeletal protein regulation is important in modifying dendritic spines and promoting long-term potentiation. Important cytoskeletal proteins include actin and cofilin.

Presynaptic cell adhesion molecules were also noted to be important in synaptic function; these include neurexins and neuroligins.

Campbell and Wood noted that many questions remain regarding precise epigenetic modifications that play roles in learning and memory.

MicroRNAs and impact on mRNAs

Bartel (2018) described microRNAs as RNAs with length of approximately 22 nucleotides that function in posttranslational repression of mRNA. He emphasized that together microRNAs impact the expression of many mRNAs. There is also evidence than microRNAs play roles in developmental processes. There is now evidence that microRNAs belong within a broad category of small RNAs. Bartel noted that a hallmark of microRNAs (miRNAs) is that they form hairpins that act as guides in the silencing of MRNAs. In reviewing the biogenesis of miRNAs, he noted that they are transcribed as premiRNAs through activity of RNA polymerase II and that each miRNA transcript has one or more regions that fold back on themselves to produce hairpins. The hairpins then constitute a substrate for a complex referred to as a microprocessor. The microprocessor is composed of DROSHA ribonuclease nuclease and DGCR8 protein. DROSHA cleaves the stem–loop structure to form a pre-miRNA that is 60 nucleotides in length.

The pre-miRNA is then exported from the nucleus to the cytoplasm through activity of XPO5 exportin 5. Within the cytoplasm, DICER1 ribonuclease acts to cleave the pre-miRNA. This cleavage generates a miRNA duplex that is approximately 20bp in length. Subsequently, one of the strands of the duplex is loaded onto Argonaute protein and forms the silencing complex. This silencing complex then interacts with the target mRNA at a site proximal to the AAAAA repeat. Bartel noted that the target site in mRNA usually encompasses approximately 7 nucleotides.

Specific studies have been carried out to document miRNAs and their target genes and in databases. Specific studies in mice led to the documentation of functions of specific through analyses of the effects of knockout of miRNA encoding DNA in the genome.

There is clear evidence that posttranslational control plays roles in neurodevelopment. Rajman and Schratt (2017) reviewed the roles of microRNAs in mRNA translation. They reported that there is evidence that microRNAs are involved early in development and impact neurons and glia. They are also involved later in life particularly in control of neurogenesis in the subgranular zone of the hippocampus. They noted evidence that 116 different microRNAs were enriched in neurons, astrocytes, oligodendrocytes, and microglia, and in the processes of differentiation of progenitor cells to neurons and glia, different microRNAs are expressed.

Sun and Shi (2015) reviewed roles of microRNAs in neurodevelopment and disease. They documented reports of specific microRNAs implicated in altering translation of products of specific target genes. These included miR-9, miR-124, miR-137, miR-219, miR-132, miR-17, miR-200, and miR-195.

The miR-9 was reported to be brain specific, and its important target genes included FOXG1 and FOXP1, members of the forkhead family of transcription factors. The miRNAs miR-124 and miR219 were also noted to be particularly expressed in brain; miR-132 was reported to be involved in synaptic plasticity.

It is important to note that the DGCR8 gene that encodes a subunit of the microprocessor complex involved in generation of primary microRNA transcripts maps to chromosome

22q11.2 a region that is often linked to SCZ. Xu et al. (2010) reported that microRNA dysregulation occurred in patients with 22q11.2 deletions.

8. Additional insights gained into control of gene expression from studies on neurodevelopmental disorders

Rett syndrome and MECP2 defects and regulation of DNA methylation and MECP2

Rajavelu et al. (2018) reported that MECP2 interacts with the DNMT3 and that this binding inhibits DNMT3A activity. They demonstrated that by MECP2, binding is impacted by histone H3 tail modifications. Whether MECP2 activates represses was reported to be dependent on H3 modification. MECP2 was initially reported to be a reader of DNA methylation and to preferentially bind to methylated DNA in CpG dense regions.

MECP2 gene product was subsequently reported to have a protein recruitment domain and to interact with histone deacetylase transcription repressor, with DNMT with ATRX and with SIN3A. Functional studies revealed that MECP2 can bind to methyl groups and sometimes induce expression. It was also shown to interact with chromatin and to impact chromatin looping, and MECP2 binding could impact splice site recognition.

Studies by Rajavelu revealed that MECP2 binds DNMT3A and DNMT3L.

(Gulmez Karaca, 2019) reported that MECP2 has more varied functions than were previously reported. MECP2 was initially reported as a protein that binds to methylated CpG nucleotides in DNA and acts as a transcriptional repressor. MECP2 function was noted to be dependent on two domains, a methyl-binding domain and an NCOR-interacting domain that led to transcriptional repression.

Gulmez Karaca et al. noted that there is now evidence that MECP2 also plays roles in determining the three-dimensional structure of DNA. They emphasized that MECP2 plays important roles throughout life. In the prenatal period, it is involved in neurogenesis and neuronal differentiation. Later in gestation and in the early postnatal period, MECP2 is involved in neuronal maturation. MECP2 was reported to play roles in experience-driven remodeling throughout early postnatal period and later also into adult life. In these time periods, MECP2 impacted chromatin structure and stimulus-dependent gene transcription. These processes facilitated dendritic arborization, synaptogenesis, and circuit formation. During adult life, MECP2 impact on chromatin and gene transcription was reported to facilitate neuronal structural maintenance and activity-dependent neuronal plasticity.

In Mecp2-deficient mice, the excitatory/inhibitory balance of neurotransmission was disrupted. Gulmez Karaca et al. emphasized the important role of MECP2 as a chromatin organizer. MECP2 was shown to bind not only to CpG dinucleotides but also to CpH dinucleotides in DNA (H = A/C/T), and it was reported to bind with higher affinity to CpH dinucleotides than to CpG dinucleotides.

There is also evidence that MECP2 influences activity-related gene expression and expression of neurotrophic factors such as BDNF. Specific reports in the literature indicate that MECP2 protein activity is impacted by phosphorylation.

FMRP protein

This protein, the product of the gene that is impacted in fragile X mental retardation, is considered to be involved in translation of mRNA into protein, particularly in neurons. In 2019, Dockendorff and Labrador (2019) presented evidence that FMRP in the nucleus may impact gene transcription. These authors presented evidence that chromatin structure and gene transcription are influenced by FMRP. Furthermore, they presented evidence that FMRP participates in the accurate repair of DNA damage and therefore promotes genomic stability.

Architectural proteins in the genome: cohesin

Key architectural proteins in the genome include CTCF and the cohesin complex. There is evidence that genes located in segments of the genome between sites that bind CTCF are transcriptionally coregulated. Regions of the genome between sites that bind CTCF and cohesin are referred to as transcriptionally associated domains (TADs).

The cohesin complex was defined as containing structural subunits and regulatory subunits. During the interphase of the cell cycle, cohesin is reported to form rings that facilitate the formation of chromatin–DNA loops. The protein NIPBL was described as being important into loading cohesin onto CTCF-binding sites.

Ouimette et al. (2019) emphasized that specific genetic diseases result from structural genomic changes, including deletions, duplications, and translocations, which disrupt TADs and alter the regulatory environment.

Cohesin complex

The cohesin complex is reported to play roles in sister chromatid cohesion, chromatin structure, gene expression, and DNA repair.

Core subunits of cohesin include the following:

- SMC1A, structural maintenance of chromosomes 1A
  - SMC3, structural maintenance of chromosomes 3
    - RAD21, cohesin complex component
    - STAG1, stromal antigen 1; STAG2, stromal antigen 2

Regulatory cohesin subunits include the following:

- WAPL, cohesin release factor
  - PDS5A/B, PDS5 cohesin-associated factor A and B
    - NIPBL, cohesin loading factor
    - MAU2, sister chromatid cohesion factor

Condensins

Condensins are defined as complexes that compact chromosomes during mitosis. Martin et al. (2016) described two condensin complexes that include key components:

- NCAPD2, non-SMC condensin I complex subunit D2
  - NCAPH, non-SMC condensin I complex subunit H
    - NCAPD3, non-SMC condensin I complex subunit D3

Martin et al. noted that mutations in condensin subunits can lead to microcephaly. These findings implicate mechanisms involved in mitotic chromosome condensation as important in determining cerebral cortex size.

Ouimette et al. noted that the Cornelia de Lange syndrome is a disorder that has provided insights into spatial genomic organization, transcriptional regulation, and the importance of cohesin in generating and enhancing long-range genomic interactions.

When the Barr body was found to represent an inactivated X chromosome. Barr bodies were found to be located at the peripheral of the nucleus and were abundant in heterochromatin, modified DNA that was differentially stained.

Ouimette et al. noted that recent progress in understanding the role of chromatin modification and chromatin interactions has been made through chromatin conformation studies; these studies involve formaldehyde fixation of genomic regions that are closely associated with each other and then isolation and analysis of the DNA sequences and proteins present in specific associated genome regions, referred to as TADs.

The cohesin complex was defined as containing structural subunits and regulatory subunits. During the interphase of the cell cycle, cohesin is reported to form rings that facilitate the formation of chromatin–DNA loops. The protein NIPBL was described as being important into loading cohesin onto CTCF-binding sites.

Ouimette et al. emphasized that specific genetic diseases result from structural genomic changes, including deletions, duplications, and translocations, which disrupt TADs and alter the regulatory environment.

CTCF, cohesin, chromatin

Davis et al. (2018) reviewed CTCF and cohesin in relation to neurodevelopmental disorders. CTCF is a protein that binds to specific DNA elements in the genome; it was first reported by Fillippova in 1996 as a protein that contains 11 zinc fingers. CTCF is encoded by a gene on human chromosome 16q22.1. CTCF was initially reported to bind to promoter proximal regions in the genome, and methylation of CpG in DNA was shown to inhibit CTCF binding.

Detailed studies have now revealed that CTCF can act as transcriptional activator or repressor of promoters and that can also act to insulate enhancers. Marino et al. (2019) noted that the different functions of CTCF are determined by the binding to different proteins. Marino et al. documented the CTCF interactome. CTCF was previously reported to bind to a number of different DNA modules (Nakahashi et al., 2013).

In a 2019 review, (Ouimette et al., 2019) noted that analysis of gene transcription regulation initially focused on the single dimension space of regulatory elements and protein-coding genes. However, recently more evidence had emerged on the importance of long-range interactions and on three-dimensional aspect of the nucleus and genome.

References

Andersen, R.E., Lim, D.A., 2018. Forging our understanding of lncRNAs in the brain. Cell Tissue Res. 371 (1), 55–71. https://doi.org/10.1007/s00441-017-2711-z.

Andersen, P., Uosaki, H., Shenje, L.T., Kwon, C., 2012. Non-canonical Notch signaling: emerging role and mechanism. Trends Cell Biol. 22 (5), 257–265. https://doi.org/10.1016/j.tcb.2012.02.003.

Bae, B.I., Jayaraman, D., Walsh, C.A., 2015. Genetic changes shaping the human brain. Dev. Cell 32 (4), 423–434. https://doi.org/10.1016/j.devcel.2015.01.035.

Bartel, D.P., 2018. Metazoan MicroRNAs. Cell 173 (1), 20–51. https://doi.org/10.1016/j.cell.2018.03.006.

Beck, D.B., Petracovici, A., He, C., Moore, H.W., Louie, R.J., Ansar, M., Douzgou, S., Sithambaram, S., Cottrell, T., Santos-Cortez, R.L.P., Prijoles, E.J., Bend, R., Keren, B., Mignot, C., Nougues, M.C., Õunap, K., Reimand, T., Pajusalu, S., Zahid, M., Fahrner, J.A., 2020. Delineation of a human mendelian disorder of the DNA demethylation machinery: TET3 deficiency. Am. J. Hum. Genet. 106 (2), 234–245. https://doi.org/10.1016/j.ajhg.2019.12.007.

Beermann, J., Piccoli, M.T., Viereck, J., Thum, T., 2016. Non-coding rnas in development and disease: background, mechanisms, and therapeutic approaches. Physiol. Rev. 96 (4), 1297–1325. https://doi.org/10.1152/physrev.00041.2015.

Belgrad, J., Fields, R.D., 2018. Epigenome interactions with patterned neuronal activity. Neuroscientist 24 (5), 471–485. https://doi.org/10.1177/1073858418760744.

Buiting, K., Williams, C., Horsthemke, B., 2016. Angelman syndrome - insights into a rare neurogenetic disorder. Nat. Rev. Neurol. 12 (10) https://doi.org/10.1038/nrneurol.2016.133 (Review. PMID).

Campbell, R.R., Wood, M.A., 2019. How the epigenome integrates information and reshapes the synapse. Nat. Rev. Neurosci. 20 (3), 133–147. https://doi.org/10.1038/s41583-019-0121-9.

Capra, J.A., Erwin, G.D., McKinsey, G., Rubenstein, J.L., Pollard, K.S., et al., 2013. Many human accelerated regions are developmental enhancers. Philos. Trans. R Soc. Lond. B Biol. Sci. 368 (1632), 20130025. https://doi.org/10.1098/rstb.2013.0025.

Carmichael, R.E., Henley, J.M., 2018. Transcriptional and post-translational regulation of Arc in synaptic plasticity. Semin. Cell Dev. Biol. 77, 3–9. https://doi.org/10.1016/j.semcdb.2017.09.007.

Cogne, B, et al., 2019. Missense variants in the histone acetyltransferase complex component gene TRRAP cause autism and syndromic intellectual disability. Am. J. Hum. Genet. 104 (3), 530–541. https://doi.org/10.1016/j.ajhg.2019.01.010.

Connolly, D.R., Zhou, Z., 2019. Genomic insights into MeCP2 function: a role for the maintenance of chromatin architecture. Curr. Opin. Neurobiol. 59, 174–179. https://doi.org/10.1016/j.conb.2019.07.002.

Davis, L., Onn, I., Elliott, E., 2018. The emerging roles for the chromatin structure regulators CTCF and cohesin in neurodevelopment and behavior. Cell. Mol. Life Sci. 75 (7), 1205–1214. https://doi.org/10.1007/s00018-017-2706-7.

Doan, R.N., Shin, T., Walsh, C.A., 2018. Evolutionary changes in transcriptional regulation: insights into human behavior and neurological conditions. Annu. Rev. Neurosci. 41, 185–206. https://doi.org/10.1146/annurev-neuro-080317-062104.

Dockendorff, T.C., Labrador, M., 2019. The fragile X protein and genome function. Mol. Neurobiol. 56 (1), 711–721. https://doi.org/10.1007/s12035-018-1122-9.

Dumas, L., Sikela, J.M., 2009. DUF1220 domains, cognitive disease, and human brain evolution. Cold Spring Harbor Symp. Quant. Biol. 74, 375–382. https://doi.org/10.1101/sqb.2009.74.025.

Engler, A., Zhang, R., Taylor, V., 2018. Notch and neurogenesis. In: Advances in Experimental Medicine and Biology, vol. 1066. Springer New York LLC, pp. 223–234. https://doi.org/10.1007/978-3-319-89512-3_11.

Epstein, I., Finkbeiner, S., 2018. The Arc of cognition: signaling cascades regulating Arc and implications for cognitive function and disease. Semin. Cell Dev. Biol. 77, 63–72. https://doi.org/10.1016/j.semcdb.2017.09.023.

Fang, Y., Fullwood, M.J., 2016. Roles, functions, and mechanisms of long non-coding RNAs in cancer. Dev. Reprod. Biol. 14 (1), 42–54. https://doi.org/10.1016/j.gpb.2015.09.006.

Fiddes, I.T., Lodewijk, G.A., Mooring, M., Bosworth, C.M., Ewing, A.D., Mantalas, G.L., Novak, A.M., van den Bout, A., Bishara, A., Rosenkrantz, J.L., Lorig-Roach, R., Field, A.R., Haeussler, M., Russo, L., Bhaduri, A., Nowakowski, T.J., Pollen, A.A., Dougherty, M.L., Nuttle, X., Haussler, D., 2018. Human-specific NOTCH2NL genes affect Notch signaling and cortical neurogenesis. Cell 173 (6), 1356–1369. https://doi.org/10.1016/j.cell.2018.03.051 e22.

Findley, T.O., Tenpenny, J.C., O'Byrne, M.R., Morrison, A.C., Hixson, J.E., Northrup, H., Au, K.S., 2017. Mutations in folate transporter genes and risk for human myelomeningocele. Am. J. Med. Genet. 173 (11), 2973–2984. https://doi.org/10.1002/ajmg.a.38472.

Gandal, M., Haney, J., Parikshak, N., Leppa, V., Ramaswami, G., & et al. (n.d.). Shared molecular neuropathology across major psychiatric disorders parallels polygenic overlap. Science, 359(6376). https://doi.org/10.1126/science.aad6469.PMID:29439242.

Greenberg, M.E., Ziff, E.B., 1984. Stimulation of 3T3 cells induces transcription of the c-fos proto-oncogene. Nature 311 (5985), 433–438.

Gulmez Karaca, K., 2019. MeCP2: a critical regulator of chromatin in neurodevelopment and adult brain function. Int. J. Mol. Sci. 20 (18) https://doi.org/10.1126/science.aad9868.

Gurung, R., Prata, D.P., 2015. What is the impact of genome-wide supported risk variants for schizophrenia and bipolar disorder on brain structure and function? A systematic review. Psychol. Med. 45 (12), 2461–2480. https://doi.org/10.1017/S0033291715000537.

Hamada, N., Ito, H., Iwamoto, I., Mizuno, M., Morishita, R., Inaguma, Y., Kawamoto, S., Tabata, H., Nagata, K.i., 2013. Biochemical and morphological characterization of A2BP1 in neuronal tissue. J. Neurosci. Res. 91 (10), 1303–1311. https://doi.org/10.1002/jnr.23266.

Hearn, B.R., Jaishankar, P., Sidrauski, C., Tsai, J.C., Vedantham, P., Fontaine, S.D., Walter, P., Renslo, A.R., 2016. Structure-activity studies of bis-O-arylglycolamides: inhibitors of the integrated stress response. ChemMedChem 11 (8), 870–880. https://doi.org/10.1002/cmdc.201500483.

Hegde, A.N., Smith, S.G., 2019. Recent developments in transcriptional and translational regulation underlying long-term synaptic plasticity and memory. Learn. Mem. 26 (9), 307–317. https://doi.org/10.1101/lm.048769.118.

Hezroni, H., Perry, R., Ulitsky, I., 2020. Long noncoding RNAs in development and regeneration of the neural lineage. Cold Spring Harbor Symp. Quant. Biol. https://doi.org/10.1101/sqb.2019.84.039347 PMID.

Hinnebusch, A., 2016. Translational control by 5′-untranslated regions of eukaryotic mRNAs. Science 352 (6292), 1413–1416. https://doi.org/10.1126/science.aad9868.

Hubel, D., Wiesel, T.N., 1970. The period of susceptibility to the physiological effects of unilateral eye closure in kittens. J. Physiol. 206 (2).

Iwase, S., Martin, D.M., 2018. Chromatin in nervous system development and disease. Mol. Cell. Neurosci. 87, 1–3. https://doi.org/10.1016/j.mcn.2017.12.006.

Iwase, S., Bérubé, N.G., Zhou, Z., Kasri, N.N., Battaglioli, E., Scandaglia, M., Barco, A., 2017. Epigenetic etiology of intellectual disability. J. Neurosci. 37 (45), 10773–10782. https://doi.org/10.1523/JNEUROSCI.1840-17.2017.

Kadoch, 2015. Mammalian SWI/SNF Chromatin Remodeling. https://doi.org/10.1126/sciadv.1500447.

Karaca, G., Brito, K., Oliveira, D., AMM, 2019. MeCP2: a critical regulator of chromatin in neurodevelopment and adult brain function. Int. J. Mol. Sci. 20 (18) https://doi.org/10.3390/ijms20184577 (Review.PMID).

Kashiwagi, K., Yokoyama, T., Nishimoto, M., Takahashi, M., Sakamoto, A., Yonemochi, M., Shirouzu, M., Ito, T., 2019. Structural basis for eIF2B inhibition in integrated stress response. Science 364 (6439), 495–499. https://doi.org/10.1126/science.aaw4104.

Kopan, R., Ilagan, M.X.G., 2009. The canonical Notch signaling pathway: unfolding the activation mechanism. Cell 137 (2), 216–233. https://doi.org/10.1016/j.cell.2009.03.045.

Lemke, J.R., Geider, K., Helbig, K.L., Heyne, H.O., Schütz, H., Hentschel, J., Courage, C., Depienne, C., Nava, C., Heron, D., Møller, R.S., Hjalgrim, H., Lal, D., Neubauer, B.A., Nürnberg, P., Thiele, H., Kurlemann, G., Arnold, G.L., Bhambhani, V., Syrbe, S., 2016. Delineating the GRIN1 phenotypic spectrum: a distinct genetic NMDA receptor encephalopathy. Neurology 86 (23), 2171–2178. https://doi.org/10.1212/WNL.0000000000002740.

Li, M., et al., 2018. Integrative functional genomic analysis of human brain development and neuropsychiatric risks. Science 362 (6420), eaat7615. https://doi.org/10.1126/science.aat7615.

Maj, C., Minelli, A., Giacopuzzi, E., Sacchetti, E., Gennarelli, M., 2016. The role of metabotropic glutamate receptor genes in schizophrenia. Curr. Neuropharmacol. 14 (5).

Marino, M.M., Rega, C., Russo, R., Valletta, M., Gentile, M.T., Esposito, S., Baglivo, I., De Feis, I., Angelini, C., Xiao, T., Felsenfeld, G., Chambery, A., Pedone, P.V., 2019. Interactome mapping defines BRG1, a component of the SWI/SNF chromatin remodeling complex, as a new partner of the transcriptional regulator CTCF. J. Biol. Chem. 294 (3), 861–873. https://doi.org/10.1074/jbc.RA118.004882.

Martin, C.A., Murray, J.E., Carroll, P., Leitch, A., Mackenzie, K.J., Halachev, M., Fetit, A.E., Keith, C., Bicknell, L.S., Fluteau, A., Gautier, P., Hall, E.A., Joss, S., Soares, G., Silva, J., Bober, M.B., Duker, A., Wise, C.A., Quigley, A.J., 2016. Mutations in genes encoding condensin complex proteins cause microcephaly through decatenation failure at mitosis. Gene Dev. 30 (19), 2158–2172. https://doi.org/10.1101/gad.286351.116.

Mattick, J.S., 2018. The state of long non-coding RNA biology. Noncoding RNA 4. https://doi.org/10.3390/ncrna4030017 PMID:30103474.

Miller, J.A., Ding, S.L., Sunkin, S.M., Smith, K.A., Ng, L., et al., 2014. Transcriptional landscape of the prenatal human brain. Nature 508 (7495), 199–206. https://doi.org/10.1038/nature13185.

Mitchell, C., Silver, D.L., 2018. Enhancing our brains: Genomic mechanisms underlying cortical evolution. Semin. Cell Dev. Biol. 76, 23–32. https://doi.org/10.1016/j.semcdb.2017.08.045.

Mukai, J., Cannavò, E., Crabtree, G.W., Sun, Z., Diamantopoulou, A., et al., 2019. Recapitulation and reversal of schizophrenia-related phenotypes in setd1a-deficient mice. Neuron. Neuron 104 (3), 471–487. https://doi.org/10.1016/j.neuron.2019.09.014.

Nakahashi, H., Kwon, K.R.K., Resch, W., Vian, L., Dose, M., Stavreva, D., Hakim, O., Pruett, N., Nelson, S., Yamane, A., Qian, J., Dubois, W., Welsh, S., Phair, R.D., Pugh, B.F., Lobanenkov, V., Hager, G.L., Casellas, R., 2013. A genome-wide map of CTCF multivalency redefines the CTCF code. Cell Rep. 3 (5), 1678–1689. https://doi.org/10.1016/j.celrep.2013.04.024.

Nikolaienko, O., Patil, S., Eriksen, M.S., Bramham, C.R., 2018. Arc protein: a flexible hub for synaptic plasticity and cognition. Semin. Cell Dev. Biol. 77, 33–42. https://doi.org/10.1016/j.semcdb.2017.09.006.

Nishiyama, J., 2019. Plasticity of dendritic spines: molecular function and dysfunction in neurodevelopmental disorders. Psychiatr. Clin. Neurosci. 73 (9), 541–550. https://doi.org/10.1111/pcn.12899.

Nott, A., Holtman, I., Coufal, N., Schlachetzki, J., Yu, M., et al., 2019. Brain cell type-specific enhancer-promoter interactome maps and disease-risk association. Science 366 (6469). https://doi.org/10.1126/science.aay 0793 (PMID).

Nussbacher, J.K., Tabet, R., Yeo, G.W., Lagier-Tourenne, C., 2019. Disruption of RNA metabolism in neurological diseases and emerging therapeutic interventions. Neuron 102 (2), 294–320. https://doi.org/10.1016/j.neuron.2019.03.014.

Ouimette, J.F., Rougeulle, C., Veitia, R.A., 2019. Three-dimensional genome architecture in health and disease. Clin. Genet. 95 (2), 189–198. https://doi.org/10.1111/cge.13219.

Perenthaler, E., Yousefi, S., Niggl, E., Barakat, Exome, B., 2019. The non-coding genome and enhancers in neurodevelopmental disorders and malformations of cortical development. Front. Cell. Neurosci. 13 https://doi.org/10.3389/fncel.2019.00352. eCollection 2019 the(PMID).

Pittenger, C., 2020. The histidine decarboxylase model of tic pathophysiology: a new focus on the histamine H3 receptor. Br. J. Pharmacol. 177 (3), 570–579. https://doi.org/10.1111/bph.14606.

Price, A.J., Collado-Torres, L., Ivanov, N.A., Xia, W., Burke, E.E., Shin, J.H., Tao, R., Ma, L., Jia, Y., Hyde, T.M., Kleinman, J.E., Weinberger, D.R., Jaffe, A.E., 2019. Divergent neuronal DNA methylation patterns across human cortical development reveal critical periods and a unique role of CpH methylation. Genome Biol. 20 (1) https://doi.org/10.1186/s13059-019-1805-1.

Quartier, A., Courraud, J., Thi Ha, T., McGillivray, G., Isidor, B., Rose, K., Drouot, N., Savidan, M.A., Feger, C., Jagline, H., Chelly, J., Shaw, M., Laumonnier, F., Gecz, J., Mandel, J.L., Piton, A., 2019. Novel mutations in NLGN3 causing autism spectrum disorder and cognitive impairment. Hum. Mutat. 40 (11), 2021–2032. https://doi.org/10.1002/humu.23836.

Qureshi, I.A., Mehler, M.F., 2018. Epigenetic mechanisms underlying nervous system diseases. In: Handbook of Clinical Neurology, vol. 147. Elsevier B.V, pp. 43–58. https://doi.org/10.1016/B978-0-444-63233-3.00005-1.

Rajavelu, A., Lungu, C., Emperle, M., Dukatz, M., Bröhm, A., Broche, J., Hanelt, I., Parsa, E., Schiffers, S., Karnik, R., Meissner, A., Carell, T., Rathert, P., Jurkowska, R.Z., Jeltsch, A., 2018. Chromatin-dependent allosteric regulation of DNMT3A activity by MeCP2. Nucleic Acids Res. 46 (17), 9044–9056. https://doi.org/10.1093/nar/gky715.

Rajman, M., Schratt, G., 2017. MicroRNAs in neural development: from master regulators to fine-tuners. Development 144 (13), 2310–2322. https://doi.org/10.1242/dev.144337.

Rauen, K.A., 2013. The RASopathies. Annu. Rev. Genom. Hum. Genet. 14, 355–369. https://doi.org/10.1146/annurev-genom-091212-153523.

Roberts, T., 2014. Perspectives on the mechanism of transcriptional regulation by long non-coding RNAs. Epigenetics 9 (1), 13–20. https://doi.org/10.4161/epi.26700.

Rowley, M.J., Lyu, X., Rana, V., Ando-Kuri, M., Karns, R., Bosco, G., Corces, V.G., 2019. Condensin II counteracts cohesin and RNA polymerase II in the establishment of 3D chromatin organization. Cell Rep. 26 (11), 2890–2903. https://doi.org/10.1016/j.celrep.2019.01.116 e3.

Sampson, J., Harris, P.C., 1994. The molecular genetics of tuberous sclerosis. Hum. Mol. Genet. 3 https://doi.org/10.1093/hmg/3.suppl_1.1477 PMID:7849741.

Schmauss, C., 2017. The roles of class I histone deacetylases (HDACs) in memory, learning, and executive cognitive functions: a review. Neurosci. Biobehav. Rev. 83, 63–71. https://doi.org/10.1016/j.neubiorev.2017.10.004.

Setiaputra, D.T., Yip, C.K., 2017. Characterizing the molecular architectures of chromatin-modifying complexes. Biochim. Biophys. Acta Protein Proteonomics 1865 (11), 1613–1622. https://doi.org/10.1016/j.bbapap.2017.06.018.

Sobetzko, D., Sander, T., Becker, C.M., 2001. Genetic variation of the human glycine receptor subunit genes GLRA3 and GLRB and susceptibility to idiopathic generalized epilepsies. Am. J. Med. Genet. Neuropsychiatr. Genet. 105 (6), 534–538. https://doi.org/10.1002/ajmg.1488.

Sonenberg, N., Hinnebusch, A.G., 2009. Regulation of translation initiation in eukaryotes: mechanisms and biological targets. Cell 136 (4), 731–745. https://doi.org/10.1016/j.cell.2009.01.042.

Sousa, A.M.M., Zhu, Y., Raghanti, M.A., Kitchen, R.R., Onorati, M., Tebbenkamp, A.T.N., Stutz, B., Meyer, K.A., Li, M., Kawasawa, Y.I., Liu, F., Perez, R.G., Mele, M., Carvalho, T., Skarica, M., Gulden, F.O., Pletikos, M., Shibata, A., Stephenson, A.R., Sestan, N., 2017. Molecular and cellular reorganization of neural circuits in the human lineage. Science 358 (6366), 1027–1032. https://doi.org/10.1126/science.aan3456.

Staahl, B.T., Crabtree, G.R., 2013. Creating a neural specific chromatin landscape by npBAF and nBAF complexes. Curr. Opin. Neurobiol. 23 (6), 903–913. https://doi.org/10.1016/j.conb.2013.09.003.

Stroud, H., Su, S.C., Hrvatin, S., Greben, A.W., Renthal, W., Boxer, L.D., Nagy, M.A., Hochbaum, D.R., Kinde, B., Gabel, H.W., Greenberg, M.E., 2017. Early-life gene expression in neurons modulates lasting epigenetic states. Cell 171 (5), 1151–1164. https://doi.org/10.1016/j.cell.2017.09.047 e16.

Sun, E., Shi, Y., 2015. MicroRNAs: small molecules with big roles in neurodevelopment and diseases. Exp. Neurol. 268, 46–53. https://doi.org/10.1016/j.expneurol.2014.08.005.

Suzuki, I.K., Gacquer, D., Van Heurck, R., Kumar, D., Wojno, M., Bilheu, A., Herpoel, A., Lambert, N., Cheron, J., Polleux, F., Detours, V., Vanderhaeghen, P., 2018. Human-specific NOTCH2NL genes expand cortical neurogenesis through delta/notch regulation. Cell 173 (6), 1370–1384. https://doi.org/10.1016/j.cell.2018.03.067 e16.

Wang, X., Goldstein, D.B., 2020. Enhancer domains predict gene pathogenicity and inform gene discovery in complex disease. Am. J. Hum. Genet. 106 (2), 215–233. https://doi.org/10.1016/j.ajhg.2020.01.012.

Wang, D., Liu, S., Warrell, J., Won, H., Shi, X., Navarro, F.C.P., Clarke, D., Gu, M., Emani, P., Yang, Y.T., Min, X., Gandal, M.J., Lou, S., Zhang, J., Park, J.J., Yan, C., KyongRhie, S., Manakongtreecheep, K., Zhou, H., Gerstein, M.B., 2018. Comprehensive functional genomic resource and integrative model for the human brain. Science 362 (6420). https://doi.org/10.1126/science.aat8464.

Weiler, I.J., Greenough, W.T., 1999. Synaptic synthesis of the fragile X protein: possible involvement in synapse maturation and elimination. Am. J. Med. Genet. 83 (4), 248–252. https://doi.org/10.1002/(SICI)1096-8628(19990402)83:4<248::AID-AJMG3>3.0.CO;2-1.

West, A.E., Greenberg, M.E., 2011. Neuronal activity-regulated gene transcription in synapse development and cognitive function. Cold Spring Harbor Perspect. Biol. 3 (6), 1–21. https://doi.org/10.1101/cshperspect.a005744.

Widagdo, J., Anggono, V., 2018. The m6A-epitranscriptomic signature in neurobiology: from neurodevelopment to brain plasticity. J. Neurochem. 147 (2), 137–152. https://doi.org/10.1111/jnc.14481.

Wolujewicz, Ross, M.E., 2019. The search for genetic determinants of human neural tube defects. Curr. Opin. Pediatr. 31 (6) https://doi.org/10.1097/MOP.0000000000000817. PMID.

Xu, B., Karayiorgou, M., Gogos, J.A., 2010. MicroRNAs in psychiatric and neurodevelopmental disorders. Brain Res. 1338, 78–88. https://doi.org/10.1016/j.brainres.2010.03.109.

Yap, E.L., Greenberg, M.E., 2018. Activity-regulated transcription: bridging the gap between neural activity and behavior. Neuron 100 (2), 330–348. https://doi.org/10.1016/j.neuron.2018.10.013.

Yu, M., Ren, B., 2017. The three-dimensional organization of mammalian genomes. Annu. Rev. Cell Dev. Biol. 33 https://doi.org/10.1146/annurev-cellbio-100616-060531 (PMID).

Zhang, R., Engler, A., Taylor, V., 2018. Notch: an interactive player in neurogenesis and disease. Cell Tissue Res. 371 (1), 73–89. https://doi.org/10.1007/s00441-017-2641-9.

Zyryanova, A.F., Weis, F., Faille, A., Abo Alard, A., Crespillo-Casado, A., Sekine, Y., Harding, H.P., Allen, F., Parts, L., Fromont, C., Fischer, P.M., Warren, A.J., Ron, D., 2018. Binding of ISRIB reveals a regulatory site in the nucleotide exchange factor eIF2B. Science 359 (6383), 1533–1536. https://doi.org/10.1126/science.aar5129.

CHAPTER

6

Neuroimmunology

1. Neural and hormonal influences and the immune system

Dantzer et al. (2018) noted that studies carried out during the 1970 and 1980s revealed dense sympathetic nervous system activation of the lymphoid tissues. Sympathetic nerves were reported to be located close to blood vessels in lymphoid tissues and that sympathetic nerve terminals came into contact with T lymphocytes and with plasma cells. These terminals were referred to as neuroeffector junctions and were reported to produce substances that include vasoactive intestinal peptide, neuropeptide Y and also neurotensin, and substance P. Parasympathetic nerves were noted to possibly only be activity in lymphoid tissue in the alimentary tract and respiratory system.

Dantzer et al. noted further that cells in the innate immune system have alpha1, alpha2, and beta2 adrenergic receptors. Activation of these receptors triggers a signaling cascade that includes cyclic AMP (adenosine monophosphate) and then either protein kinase activity or guanine nucleotide exchange pathway signaling activity.

The responses to stimulation of adrenergic receptors on immune system cells were reported to be complex but to include production of epinephrine and norepinephrine by T cells and also by macrophages and neutrophils.

Extraadrenal production of glucocorticoids was also discovered, including discovery of these substances by the thymus.

In reviewing neuroimmune reactions, Chavan et al. (2017) noted neuronal cells produced immune modulators including acetylcholine and catecholamine. In addition, Steinberg et al. (2006) discovered evidence for the presence of receptors for neurotransmitters including acetylcholine and evidence for adrenergic neurotransmitter receptors on immune systems cells including macrophages, dendritic cells, and T cells.

Chavan et al. noted that factors that stimulate peripheral somatosensory neurons have also been shown to impact immune cells to produce cytokines.

Somatosensory neurons were also reported to be stimulated by factors released as part of the immune response, e.g., cytokines. Somatosensory neurons were reported to have bidirectional axons with some axonal branches terminating in peripheral tissue and also axonal processes that terminated in the dorsal neurons of the spinal cord. From there, sensations were conveyed to the brain stem, thalamus, and ultimately the somatosensory cortex and anterior cingulate.

In the periphery, pathogens interact with specific molecules on immune cells. These include damage-associated pattern molecules (DAMPs) and pattern-associated pattern molecules (PAMPs),

https://doi.org/10.1016/B978-0-12-821913-3.00012-3

and stimulation of these molecules and receptors subsequently leads to activation of transcription factors and signaling pathways.

Immune cell–derived molecules particularly those generated in the gut were also shown to stimulate the sympathetic and parasympathetic nervous system.

Kamimura et al., (2019) noted that earlier major neuroimmunology research had focused on production of specific hormones by the hypothalamus and pituitary axis influenced the immune system. They noted further that there was evidence for close interactions between circulatory system and the brain despite the existence of the blood–brain barrier. They drew attention to the occurrence of PAMP and DAMP receptors on neuronal cells.

The paravascular pathway and cerebrospinal flow through the brain parenchyma

In 2012, Illiff et al. noted that the brain does not have lymphatic circulation; however, it is clear that a mechanism must exist to remove complex molecules, generated by metabolism and by protein degradation, from the interstitial fluid in the brain. They carried out studies in rodent brain to demonstrate the existence of a perivascular pathway that facilitates clearance of complex substances from the brain. This system was shown to involve activity of glia, and it came to be known as the glia perivascular network or the glymphatic system or as the glial-associated lymphatic system.

In reviewing this pathway in 2018, Plog and Nedergard noted that this system has a pseudolymphatic function. They noted that the interstitial fluid that occurs between brain cells constitutes between 12% of brain fluid and that the majority of fluid is intracellular. They reported that intensive research had been carried out to determine the relationship between cerebrospinal fluid (CSF) and brain interstitial fluid.

CSF is produced by the highly vascular choroid plexus in the ventricles of the brain. This fluid passes through the fenestrated vessels in the choroid flexus vessels into the ventricles; it then passes through the different ventricles in the brain and eventually passes from the fourth ventricle to the subarachnoid space. Experiments with dyed substances were carried out to determine if and how CSF entered the brain tissue. Labeled molecules such as blue dextran introduced into the CSF in the ventricles ended up in perivascular space in the brain. The perivascular pathways in the brain are sometimes referred to as Virchow-Robin spaces.

In a separate approach, experiments were carried out to analyze drainage of interstitial fluid from the brain. Studies carried out in animals involved use of specific dyed substances including blue dextran. Results revealed evidence that substances introduced into interstitial fluid in the brain ended up in the CSF present in the subarachnoid space.

Questions then arose as to how substances moved in the brain. Studies revealed the presence of low-resistance channels through which molecules could move.

Plog and Nedergard (2018) noted that the protein AQ4 (aquaporin 4) is expressed on astrocyte process in the subependymal lining of the ventricles. In the brain interstitial tissue, perivascular astrocyte foot processes occur at the small vessel perivascular brain interstitial tissue interface.

A pathway was therefore shown to exist for CSF flow into the brain in the periarterial space and then into interstitial tissue and for interstitial tissue to enter the perivenular space and perivenous flow system. This system involved astrocytes and became known as the glial-associated lymphatic pathway.

Passage of CSF along the periarterial space of penetrating small arteries and astrocyte end feet led to a convection flow that also propelled interstitial fluid with waste products toward astrocytes end feet that projected to the perivenular space and venules. Clearance into perivenular space ultimately propelled interstitial fluid through a system that led to the subarachnoid space. Arachnoid granulations released fluid to sinuses in the dura and also along the olfactory nerve. There is also evidence of flue of CSF from the subarachnoid space into lymphatic vessels that line the dura. This system will be described further in the following.

In 2013, Illif et al. published evidence that the pathway of flow of CSF within the brain interstitial tissue is impacted by cerebral artery pulsation. There is also evidence that this pathway operates optimally during sleep.

In 2018, Rasmussen et al. noted that studies were needed to determine if this glymphatic pathway was interrupted in specific central nervous system (CNS) disorders.

2. Microglia

Wolf et al. (2017) reviewed the physiology of microglia, the immunocompetent cells of the brain. They also reviewed roles of microglia in brain pathology. A key function of microglia is to maintain brain homeostasis. This is dependent on their production of specific molecules including cytokines, through their function of phagocytosis and through their interactions with immune cells that enter the brain from the periphery during inflammation.

Wolf et al. drew attention to the work of Virchow in 1856 who identified microglia as being distinct from neurons. Ramon y Cajal is reported to have defined three cell types in brain: neurons, astrocytes, and apolar cells (1897). This third population of cells was shown to be activated during brain damage. Microglia have been reported to be distinct from hematopoietic cells, and microglia are derived from neuroepithelium.

Wolf et al. noted that interleukin 34 (IL34) is reported to interact with a receptor on microglia colony-stimulating factor 1 (CSF 1). This interaction was shown to be important for microglial cell proliferation and differentiation. Specific adapter molecules that also bind to receptors on microglia also include DAP12 also known as TYROBP, transmembrane immune signaling adaptor. IRF8 interferon regulatory protein was also shown to be important in microglial development.

Wolf et al. noted that microglia play important roles in brain development and are involved in the programmed cell death that occurs during brain development. This process includes synaptic pruning. Synaptic pruning is also dependent on activity of complement proteins. This will be discussed in the following.

Microglia were also reported to be activated by a specific neuron produced chemokine that interacts with a receptor on microglia chemokine CX3CR1.

Wolf et al. also noted evidence that synaptic pruning also takes place in the adult brain in specific locations. Dendritic spine proteins and proteins in axon terminals have also been identified as binding to microglia.

Microglial heterogeneity

Rio-Hortega in the Cajal school documented different subtypes of microglia that differ in the morphological features. Different types of microglia were documented to predominate in different brain regions (see Sierra et al., 2016).

More recently, microglia in the subventricular region have been found to have low levels of ATP-stimulated purinoreceptors. Different subtypes of microglia also differ in their degree of activation by amines and neurotransmitters.

Wolf et al. (2017) noted that microglial activation occurs as a neuroprotective response to damaging conditions in the brain. There is evidence that they release antiinflammatory factors to protect neurons and nerves. Factors produced by microglia to downregulate inflammation include antiinflammatory agents, interleukin 10 (IL10) and transforming growth factor beta TGF-β).

Analyses of specific marker proteins have revealed that mitochondria are abundant in microglia. One such marker protein is TSPO translocator protein. Investigators have analyzed the transcription profile of microglia and determined that it includes products derived from approximately 100 different genes.

Microglia in pathological conditions

Wolf et al. noted that brain tumors are reported to be infiltrated by microglia and peripheral macrophages. In multiple sclerosis (MS), microglia and macrophage proliferations are more severe in severe forms of the disease. Microglial activation has been shown to cooccur with upregulation products of the major histocompatibility II loci. Wolf et al. also cited evidence of increased microglial response to proinflammatory stimuli in aging.

Microglia have been shown to have pattern receptors that recognize molecular patterns associated with cell damage. Stimulation of pattern receptors DAMPs leads to increase in release of proinflammatory factors.

There is evidence of microglial activation in the neurodegenerative disorders. In the early stages of these diseases, microglial activation is thought to be protective. Later, however, microglial activation may be damaging.

Questions arise as to whether altered microglial regulation occurs in certain neurobehavioral or psychiatric disorders. Wolf et al. noted reports that two networks were proposed to be disrupted in autism, synaptic function was reported to be downregulated, and immunoregulatory genes were reported to be upregulated in autism. Increased levels of microglia and microglial activation have been reported in schizophrenia (Leza et al., 2015).

Questions arise regarding the role of lifestyle factors, stress, and nutrient factors in microglial activation.

3. Central nervous system lymphatic system

In 2015, Louveau et al. and Aspelund et al. reported that functional lymphatic vessels line the sinuses in the dura of the brain.

In 2018, Da Mesquita et al. noted the rediscovery of a meningeal lymphatic system and lymphatic connections between the CNS and periphery. The lymphatic vessels that lined the walls of the dural sinuses were reported to express the same markers as peripheral lymphatic endothelial cells, namely VEGFR3 (also known as FLT3 tyrosine kinase), Prox1 (Prospero homeobox), and Podoplan (PDPN). The meningeal lymphatics were noted to have thinner walls than the peripheral lymphatics.

Da Mesquita et al. emphasized that some of the same fluids that occur in the interstitial brain fluid occur in the CSF and flow into the lymphatic vessels. They concluded that the discovery of direct contact of brain interstitial fluid, CSF, and the lymphatic system has direct relevance to neuroimmunopathology.

Papadopoulos et al. (2020) noted that in prior decades, the CNS was considered to be an immune-privileged system, and the brain was reported not to have a lymphatic system.

In recent years, beginning with studies of tracer dye migrations, connections between the CNS and cervical lymph nodes were discovered. Subsequently, a CNS lymphatic network was demonstrated in meninges of mice. Subsequently a meningeal lymphatic network was demonstrated in humans.

Papadopoulos et al. noted that the brain is surrounded by three layers, the dura mater directly beneath the cranium, the arachnoid mater that is in contact with the dura mater, and the arachnoid mater that is separated from the pia mater of the brain by the subarachnoid space. This space is filled with CSF. Arachnoid trabeculae occur in this space. Beneath the pia mater lies the glia limitans that is directly in contact with the brain.

Papadopoulos et al. noted that the cerebral ventricles and the choroid plexus were found to be lined with distinct epithelial cells that form the CSF from blood plasma and restrict access of large molecules. Studies in recent years have demonstrated that there was contact between the CSF and the interstitial fluid in the brain.

Contact between brain interstitial fluid and CSF was described earlier as taking place through pararterial and paravenular flow tracts and through activity of astrocytic end feet and certain molecules including AQ4.

There is evidence that CSF that has had contact with brain interstitial fluid flows through the paravenular tracts into the subarachnoid space.

Papadopoulos et al. noted that surveillance in the parenchyma of the brain is carried out by microglia. They emphasized that much need to be learned about microglial surveillance and ultimate immune responses.

They noted that DC2 dendritic cells are immune system cells that could ultimately be activated by substances in the CSF. Fluid that has flowed along with interstitial fluid from the parenchyma in the paravenular system flows into the CSF in the subarachnoid space and antigens that occur in the CSF that flows into dural lymphatic system can induce an immune response. They emphasize that lymphatic drainage of the CSF can introduce substances that elicit an immune response and constitute an immune surveillance system.

An extensive lymphatic system was discovered in the dura mater along the transverse and superior sinuses. These lymphatic vessels were shown to serve as conduits for draining of CSF into the cervical lymph nodes and ultimately into the jugular vein.

Additional studies are carried out to determine how CSF components from the subarachnoid space enter the dural lymphatics.

4. Complement

Alexander (2018) reviewed complement in relation to the blood–brain barrier and emphasized the roles of both systems in brain homeostasis. Components of the complement system were noted to be present both systemically in the circulation and in the brain. The complement system was also noted to play a role in brain development.

Alexander noted that complement within the brain acts to protect the brain through surveillance mechanisms and to detect infarcts. Complement was also reported to play roles in repair in the brain.

The blood–brain barrier was noted to be a mechanism to protect the brain from toxins and damaging cells. Compromise of the blood–brain barrier is known to lead to tissue damage. Alexander noted progress in understanding mechanisms that control the blood–brain barrier integrity. These include the structure, products, and functions of the endothelial cells.

Complement proteins were reported to be predominantly synthesized in the liver though there is also evidence that they are synthesized in the brain and in the kidney. Complement includes nine different complement proteins (C1–C9) and also synthesis of factors B (CFB) and D (CFD), CD59,CFH, and DAF (CD59) (regulators of complement activation) and complement receptors C3AR1 and C5AR1.Complement receptors CR1 and CR2 were shown to be synthesized by astrocytes and by brain endothelial cells.

The blood–brain barrier was reported to be formed by endothelial cells in the brain vasculature, in the choroid plexus, and in the epithelium of the arachnoid membrane. There is evidence that the structure of the blood–brain barrier differs in different regions of the brain, particularly in regions of the brain associated with hormonal control, e.g., the pituitary gland and pineal region.

In regions of the brain where the blood–brain barrier is functioning, the endothelial cells have tight junctions and an underlying adherens region. The blood–brain barrier is dependent on connections of functional endothelial cells with the underlying basement membrane. Specific proteins that occur in the tight junctions include claudin, occludin, and vinculin.

Alexander et al. noted evidence that the brain endothelial cells are highly enriched with mitochondria and can therefore be damaged by abnormally high levels of reactive oxygen species.

Alexander documented interaction between the vascular endothelium outer layers and the astrocytes. Proteins involved in these interactions include vimentin, glial fibrillary acidic protein, dystroglycan, dystrobrevin, AQ4, and potassium channel KIR4.1. Gap junction proteins and connexin were also shown to be important in interaction between vessels and astrocyte end feet connections.

Pericytes were reported to occur around microvessels and in proximity to astrocytes and endothelial cells.

Blood–brain barrier dysfunction was reported to occur during aging and during the course of different neurodegenerative disorders and was found to be associated with opening of the tight junction between endothelial cells.

Alexander reported that complement proteins do not normally cross the blood–brain barrier and that complement proteins. However, impaired integrity of the blood–brain barrier can result influx of complement proteins.

Specific cells in the brain can synthesize complement proteins. In addition, certain toxins that cross the blood–brain barrier can serve to activate complement protein synthesis.

Complement and brain development

Coulthard et al. (2018) reported the importance of proteins in the complement system in brain development. They noted that the complement system includes soluble and membrane-bound proteins. Soluble complement components can cause lysis and promote phagocytosis of pathogens. Specific proteolytic products of complement were reported to bind to leukocytes and other cells and to promote their migration to sites of inflammation.

Coulthard et al. noted that inflammatory cells bind complement fragments but that host cells in tissue are protected by complement inhibitor proteins that bind to their surfaces.

Different complement pathways are described: the classical pathway, the lectin pathway, and the alternate pathway. The classical pathway particularly involves C5 complement components. Both C3 and C4 complement components are involved in the lectin pathway. The alternate pathway particularly involves the C3 complement component.

Classical pathway activity was reported to be activated by antigen antibody interactions. The lectin pathway is activated by carbohydrate-rich molecules. The alternate pathway was reported to be activated by inert substances.

A newly described fourth complement pathway was reported to be triggered by serine proteases. The intracellular complement–related system is referred to as the complosome.

Complement in early development

Coulthard et al. (2018) noted that expression of complement components occurs very early in development. In human embryos, the complement receptors in the C5aR category were reported to be expressed in the neuroepithelium and subsequently to be reported to be present in the CSF. Complement was expressed in neuroblasts and glial cells. C1 complement was found to be expressed in neural cell types and microglia postnatally. The lectin pathway complements were reported to play roles in neuronal migration.

Postnatally the classical complement pathway was reported to play roles in synaptic pruning with C1q as the initiating factor. Coulthard et al. noted that the precise mechanism through which complement identified synapses for pruning was not known. However, low activity synapses were shown to be more prone to pruning.

Elevated synaptic pruning has been reported to be a cause of epilepsy in mouse models. Blockade of the complement receptor C5aR1 was reported to reduce epilepsy in mouse models. Coulthard et al. conclude that there is growing evidence for the importance of complement in various aspects of neurodevelopment.

Activity of complement C3 is reported to impact synaptic pruning, and activity of C3 is impacted by activity of complement C4A and C4B (Fig. 6.1).

5. Immune responses in the central nervous system

Waisman et al. (2015) reviewed innate and adaptive immune responses in the CNS in light of information that neural-derived antigens released from the CNS are passed through the glymphatic systems and are detected in the cervical lymph nodes. They noted further that pathological changes in the CNS, e.g., neurodegenerative changes most often result in inflammatory changes. They also presented examples of immune-mediated disorders in the CNS.

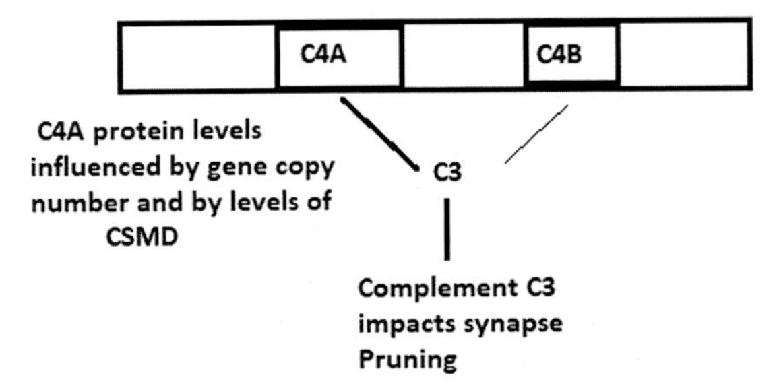

FIGURE 6.1 Complement C4A and C4B impact C3 activity. Complement C3 impacts synaptic pruning.

They noted that the blood–brain barrier is not identical in all brain regions. In specific regions, including the structures that surround the ventricles, blood–brain barrier was noted to be less restrictive.

Waisman et al. noted that the resident immune cells in the brain include astrocytes and microglia. Astrocytes were reported to have several functions that included metabolism, structural support, integrity of the blood–brain barrier, and immune defense.

Monocytes, in the myeloid category of cells, are present in the blood stream and are reported to cross the blood–brain barrier under pathological conditions. Monocytes are reported to produce cytokines and to potentially become phagocytes at sites of inflammation.

Innate immunity that includes phagocytosis by microglia was reported to occur in specific neurodegenerative conditions. Microglia were also be reported to be capable of producing cytokines, chemokines, prostaglandins, complement proteins, and nitric oxide.

Waisman et al. also reviewed aspects of adaptive immunity in the CNS. In certain conditions, e.g., leukodystrophy T cells were reported to occur in the CNS in MS and both B and T cells were reported to be present in the CNS.

They noted further that T cell immune responses were present in the CNS in paraneoplastic conditions. Examples include small cell lung cancer and neuroblastoma where specific proteins act as tumor antigens and act to produce antibody. HuD-antigen is a neuronal RNA-binding protein that can act as an antigen (Ehrlich et al., 2014). Waisman et al. noted that the tumor-derived antibody also activated specific reactive CD8 T cells. The induced antibody and reactive T cells can enter the brain and lead to tissue damage.

Antibody-induced encephalitis

Dalmau et al. (2016) reported evidence for CNS diseases due to the entry into the brain of antibodies generated in the periphery. These antibodies were often found to be induced in response to certain tumors, in some cases by benign ovarian teratomas. Antibodies were induced against specific synaptic proteins including neurotransmitter receptors NMDARs, AMPAR glutamate receptors, GABA receptors, dopamine receptors, or glycine receptors.

Kitley et al. (2014) described patients with antibodies to myelin oligodendrocyte protein (MOG) and patients with antibodies to AQ4. They reported patients with these antibodies who developed a condition in the neuromyelitis optica spectrum. Antibodies to MOG and AQ4 were found in the CSF. Patients usually presented with acute-onset optic neuritis. Some patients presented with manifestations of spinal record damage and myelitis.

T cell responses and infections in the central nervous system

Korn and Kellies (2017) reviewed T cell immunity and the CNS. They reported that viral infections of the CNS may trigger autoimmune disorders. Steps include drainage of antigens from infected and damaged CNS cells and passage through interstitial fluid and lymphatics to the CNS draining lymph nodes.

T cells in these lymph nodes could then be activated by CNS-derived antigen. These activated T cells could then potentially return to the CNS. The activated T cells and/or their products could potentially reach the CNS through the vascular system and the choroid plexus.

Multiple sclerosis

Dobson and Giovannoni (2019) reviewed MS. They noted that incidence of MS is increasing worldwide, in both developed and developing countries. The underlying causes and disease mechanisms have not yet been definitively determined. A number of environmental factors have been implicated. These include prior exposure to Epstein–Barr virus (EBV), low vitamin D levels, smoking, and obesity. Improved methods for diagnosis of MS have been reported.

Dobson and Giovannoni reported that there are two forms of MS, the first characterized by marked inflammatory response and relapsing remitting course. In the second form, there is marked neurodegeneration. Both T and B cell immune responses are implicated, and more recently, immune reconstitution therapies are proving valuable.

Dobson and Giovannoni noted that population studies indicate that the environmental factors influence MS risk. Migration from low-risk countries to high-risk countries such as Northern Europe is associated with an increased disease risk. Negative history of EBV exposure is reported to be a protective factor.

Certain alleles of genetic polymorphic variants also influence disease risk. Presence of the HLA DRB1 15.01 allele increases risk, and risk is more significantly increased in homozygotes than in heterozygotes. The HLA-A*02.01 allele was reported to be protective. Genome-wide association studies have identified a number of other alleles that influence risk.

Dobson and Giovannoni noted that the classical lesion associated with MS is described as an inflammatory lesion that surrounds small vessels and leads to demyelination. The inflammatory lesion was reported to contain CD8 T cells and low numbers of B cells. Demyelination is reported to be related to both the inflammatory response and to oligodendrocyte damage. Later in the course of the disease, axon fibers show damage. The proportion of B cells in the inflammatory lesions were reported to increase during the course of the disease. The severity of disease manifestations increased as the disease progressed from the relapsing remitting form to the more severe progressive form. Clinical symptoms varied depending on the location in the CNS of the most predominant lesions. Optic neuritis, brain stem damage manifestations, and manifestations of spinal cord damage occurred. Dobson and Giovanni noted that MS is a single disease with different spectra of manifestations.

Disease-modifying treatments include immune-suppressor agents and interferon. More recently, immune reconstitution therapy, e.g., with use of hematopoietic stem cells has been initiated in some centers (Karussus and Petrou, 2018).

Calabrese et al. (2015) reported that the pathological features of MS include widespread demyelinating lesions. They noted, however, that there is also evidence of involvement of gray matter in MS. The exact mechanisms involved in gray matter degradation in this disorder were noted to be unclear. One potential mechanism proposed is that axonal damage secondarily leads to neuronal cell damage. Other possible mechanisms include neuronal damage due to products generated during the inflammatory reaction.

Neurotropic viruses and the brain immune system

Miller et al. (2019) published evidence that components of neurotropic viruses remain detectable after the acute infection stage has passed. They noted that neurotropic viruses include West Nile virus, measles virus, and influenza viruses. There is also evidence that corona viruses infect the CNS (Desforges et al., 2019).

Prasad et al. (2019) presented evidence that activated CD8 lymphocytes can enter the brain during viral infections, and the T cells can remain in the brain and can provide protection on reinfection or reactivations of virus. They noted that the majority of T cells that remain in the brain bear the CD103 marker. Increased activity of these adaptive T cells was noted to also have deleterious effects including neurotoxicity.

References

Alexander, J.J., 2018. Blood-brain barrier (BBB) and the complement landscape. Mol. Immunol. 102, 26–31. https://doi.org/10.1016/j.molimm.2018.06.267.

Aspelund, A., Antila, S., Proulx, S.T., Karlsen, T.V., Karaman, S., Detmar, M., Wiig, H., Alitalo, K., 2015. A dural lymphatic vascular system that drains brain interstitial fluid and macromolecules. J. Exp. Med. 212 (7), 991–999. https://doi.org/10.1084/jem.20142290.

Calabrese, M., Magliozzi, R., Ciccarelli, O., Geurts, J.J.G., Reynolds, R., Martin, R., 2015. Exploring the origins of grey matter damage in multiple sclerosis. Nat. Rev. Neurosci. 16 (3), 147–158. https://doi.org/10.1038/nrn3900.

Chavan, S.S., Pavlov, V.A., Tracey, K.J., 2017. Mechanisms and therapeutic relevance of neuro-immune communication. Immunity 46 (6), 927–942. https://doi.org/10.1016/j.immuni.2017.06.008.

Coulthard, L., Hawksworth, O., Woodruff, T.M., 2018. Complement: the emerging architect of the developing brain. Trends Neurosci. 41 (6) https://doi.org/10.1016/j.tins.2018.03.009 (Review. PMID).

Dalmau, J., 2016. NMDA receptor encephalitis and other antibody-mediated disorders of the synapse. Neurology 87 (23), 2471–2482. https://doi.org/10.1212/WNL.0000000000003414.

Dantzer, R., 2018. Neuroimmune interactions: from the brain to the immune system and vice versa. Physiol. Rev. 98 (1) https://doi.org/10.1152/physrev.00039.2016. PMID.

Desforges, M., Le Coupanec, A., Dubeau, P., Bourgouin, A., Lajoie, L., Dubé, M., Talbot, P.J., 2019. Human coronaviruses and other respiratory viruses: underestimated opportunistic pathogens of the central nervous system? Viruses 12 (1). https://doi.org/10.3390/v12010014.

Dobson, R., Giovannoni, G., 2019. Multiple sclerosis - a review. Eur. J. Neurol. 26 (1). https://doi.org/10.1111/ene.13819 (Review.PMID).

Ehrlich, D., Wang, B., Lu, W., Dowling, P., Yuan, R., 2014. Intratumoral anti-HuD immunotoxin therapy for small cell lung cancer and neuroblastoma. J. Hematol. Oncol. 7 (1). https://doi.org/10.1186/s13045-014-0091-3.

Iliff, J., Wang, M., Liao, Y., Plogg, B., Peng, W., et al., 2012. A paravascular pathway facilitates CSF flow through the brain parenchyma and the clearance of interstitial solutes, including amyloid β. Sci. Transl. Med. 4 (147), 147. https://doi.org/10.1126/scitranslmed.3003748.PMID: 22896675.

Iliff, J.J., Wang, M., Zeppenfeld, D.M., Venkataraman, A., Plog, B.A., Liao, Y., Deane, R., Nedergaard, M., 2013. Cerebral arterial pulsation drives paravascular CSF-Interstitial fluid exchange in the murine brain. J. Neurosci. 33 (46), 18190–18199. https://doi.org/10.1523/JNEUROSCI.1592-13.2013.

Kamimura, D., Tanaka, Y., Hasebe, R., Masaki, M., 2019. Bidirectional communication between neural and immune systems. Int Immunol. 32 (11), 693–701. https://doi.org/10.1093/intimm/dxz083. 31875424.

Karussis, D., Petrou, P., 2018. Immune reconstitution therapy (IRT) in multiple sclerosis: the rationale. Immunol. Res. 66 (6), 642–648. https://doi.org/10.1007/s12026-018-9032-5.

Kitley, J., Waters, P., Woodhall, M., Leite, M.I., Murchison, A., George, J., Küker, W., Chandratre, S., Vincent, A., Palace, J., 2014. Neuromyelitis optica spectrum disorders with aquaporin-4 and myelin-oligodendrocyte glycoprotein antibodies a comparative study. JAMA Neurol. 71 (3), 276–283. https://doi.org/10.1001/jamaneurol.2013.5857.

Korn, T., Kallies, A., 2017. T cell responses in the central nervous system. Nat. Rev. Immunol. 17 (3), 179–194. https://doi.org/10.1038/nri.2016.144.

Leza, J., García-Bueno, B., Bioque, M., Arango, C., Parellada, M., et al., 2015. Inflammation in schizophrenia: a question of balance. Nature 523 (7560), 337–341. https://doi.org/10.1038/nature14432. PMID: 26092265.

Louveau, A., Smirnov, I., Keyes, T.J., Eccles, J.D., Rouhani, S.J., Peske, J.D., Derecki, N.C., Castle, D., Mandell, J.W., Lee, K.S., Harris, T.H., Kipnis, J., 2015. Structural and functional features of central nervous system lymphatic vessels. Nature 523 (7560), 337–341. https://doi.org/10.1038/nature14432.

Mesquita, D., Fu, S., Kipnis, Z., 2018. The meningeal lymphatic system: a new player in neurophysiology. Neuron 100 (2). https://doi.org/10.1016/j.neuron.2018.09.022. Review.PMID).

Miller, K.D., Matullo, C.M., Milora, K.A., Williams, R.M., O'Regan, K.J., Ralla, G.F., 2019. Immune-mediated control of a dormant neurotropic RNA virus infection. J. Virol. 93 (18). https://doi.org/10.1128/JVI.00241-19.

Papadopoulos, Z., Herz, J., Kipnis, J., 2020. Meningeal lymphatics: from anatomy to central nervous system immune surveillance. J. Immunol. 204 (2), 286–293. https://doi.org/10.4049/jimmunol.1900838.

Plog, B., Nedergaard, M., 2018. The glymphatic system in central nervous system health and disease: past, present, and future. Annu. Rev. Pathol. 13. https://doi.org/10.1146/annurev-pathol-051217-111018 (Review.PMID).

Prasad, S., Lokensgard, J.R., 2019. Brain-resident T cells following viral infection. Viral Immunol. 32 (1), 48–54. https://doi.org/10.1089/vim.2018.0084.

Rasmussen, M.K., Mestre, H., Nedergaard, M., 2018. The glymphatic pathway in neurological disorders. Lancet Neurol. 17 (11), 1016–1024. https://doi.org/10.1016/S1474-4422(18)30318-1.

Sierra, A., Castro, F., Río-Hortega, De, Iglesias-Rozas, R., Garrosa, J., Kettenmann, M., The, H., 2016. Big-Bang\ for modern glial biology. Trans. Comment. Pío del Río-Hortega 1919 Series Paper. Microg. Glia 64 (11). https://doi.org/10.1002/glia.23046 PMID:27634048.

Sternberg, E.M., 2006. Neural regulation of innate immunity: a coordinated nonspecific host response to pathogens. Nat. Rev. Immunol. 6 (4), 318–328 https://doi.org/doi: 10.1038/nri1810.

Waisman, A., Liblau, R.S., Becher, B., 2015. Innate and adaptive immune responses in the CNS. Lancet Neurol. 14 (9), 945–955. https://doi.org/10.1016/S1474-4422(15)00141-6.

Wolf, S.A., Boddeke, H.W.G.M., Kettenmann, H., 2017. Microglia in physiology and disease. Annu. Rev. Physiol. 79, 619–643. https://doi.org/10.1146/annurev-physiol-022516-034406.

CHAPTER

7

Neurodevelopmental, neurocognitive, and behavioral disorders

1. Introduction

Boivin (2015) defined neurodevelopment as "the dynamic inter-relationship between genetic, brain, cognitive, emotional and behavioral processes across the developmental lifespan. Significant and persistent disruption to this dynamic process through environmental and genetic risk can lead to neurodevelopmental disorders and disability."

It is important to note that new techniques developed in recent decades have expanded possibilities for diagnosis of underlying causes and mechanisms, leading to neurodevelopmental disorders. These techniques include genomic and genetic techniques, advances in metabolomics, lipidomics, and glycomics, and these studies provide insights into disease-specific treatments. However, in many cases, additional in vitro studies and studies in model organisms will be required to identify effective treatments.

Reid (2016) noted that early clinical features of neurometabolic disorders could include feeding difficulties, vomiting, seizures, and abnormal movements. Neurometabolic disorders are characterized by neurologic dysfunction and biochemical abnormalities.

2. Neural tube defects and associated gene defects

Wang (2018) noted that in the mouse, at least 300 different genes had been found to have pathogenic variants that predisposed to neural tube defects and that a number of these pathogenic variants occurred in planar cell polarity (PCP) genes. PCP proteins defective in mouse neural tubes included Vangl1, Vangl2, Celsr1, Dvl 1, 2, 3, and frizzled Fzd3 and Fzd6. PCP proteins frequently localize to specific positions on cell membranes.

Defects in a number of cell polarity pathway genes have been identified in humans and have been found to lead to defects including neural tube defects. Wang (2018) reported studying genes in the PCP pathway in 510 humans with neural tube defects. Variants found on exome sequencing were confirmed using Sanger sequencing. Specific genes that were found to have variants and that encoded products in the PCP pathway included VANGL1 and VANGL2 proteins in the tetraspanin family CELSR1, cadherin EGF LAG seven-pass G-type receptor 1; SCRIB, scribble, a scaffold protein; DVL2 and DVL3, disheveled 2 and 3, which also

Mechanisms and Genetics of Neurodevelopmental Cognitive Disorders
https://doi.org/10.1016/B978-0-12-821913-3.00007-X

function in the WNT signaling pathway; and PTK7, protein tyrosine kinase 7 p, which plays a role in transducing signals across cell membranes. Wang et al. reported finding 74 protein altering variant in these gene products in the individuals studied and noted that these variants all occurred at frequencies <0.01 in the general population. Of particular interest was the finding that digenic variants occurred and CELSR1 +SCRIB variants were found in three spina bifida cases and in one anencephaly case. Digenic variants in CELSR1 and DVL3 were found in an anencephaly case. In spina bifida cases, digenic variants in PTK7 and SCRIB were found. The authors noted that no digenic variants in these genes were found in the 1000 genomes database of control individuals. They concluded that combinatorial PCP variants contributed to the etiology of neural tube defects in humans.

3. Brain growth and cortical expansion defects

Pirozzi (2018) reviewed processes involved in cortical expansion of the human brain. They noted that these take place through neural stem cell proliferation, migration, organization, synaptogenesis, and apoptosis. They noted that alteration in any of these processes could lead to changes in brain size, microcephaly, or macrocephaly. They noted that defects in cell cycle, centrosome formation, spindle orientation, microtubules, and cytokinesis were primarily associated with decreases in cortical volume.

Megalencephaly is defined as brain size more than 2 standard deviations above average at that age in that population. Megalencephaly was found to be associated with defects in the PI3K–mTOR pathways in genes involved in regulation of gene expression and transcription factor activity. RAS-MAP signaling pathways defects were also found to be associated with some cases of megalencephaly. It is important to note that some forms of megalencephaly are associated with somatic overgrowth.

Pirozzi et al. drew attention to new evidence that mutations that decrease activity of the PP2A phosphatase family can also lead to brain overgrowth. These phosphatases, particularly PPP2R5D, normally act to negatively regulate the PI3K–mTOR pathway. When mutations result in decreased PPP2RD activity, there is less inhibition of mTOR and brain overgrowth occurs.

Overgrowth syndromes associated with megalencephaly

Burkardt (2019) identified two classes of somatic overgrowth syndromes associated with megalencephaly: disorders associated with upregulation of levels of growth hormone, insulin, and insulin-like growth factor due to defects in expression of gene in the 11p15.4–11p15.5 region.

It is important to note that megalencephaly and somatic overgrowth also occur in Sotos syndrome that has been found to be due to defects in other gene products, including NSD1, an androgen receptor associated coregulator, and NFIX, a transcription factor.

Growth disorders associated with defects in epigenetic regulators:

- NSD1, nuclear receptor–binding SET domain protein 1, and androgen receptor coactivator (SOTOS syndrome).
- SETD2n overgrowth syndrome SET domain containing 2, histone lysine methyltransferase
- DNMT3A, overgrowth syndrome DNA methyltransferase 3 alpha Weaver syndrome
- EZH2, enhancer of zeste 2 polycomb repressive complex 2 subunit Cohen Gibson syndrome
- EED, embryonic ectoderm development
- SUZ, overgrowth syndrome SUZ12 polycomb repressive complex 2 subunit

4. Specific brain defects due to abnormalities in products of ciliary pathway genes

The structural and functional aspects of cilia and microtubules were described in Chapter 1.

In a 2017 review, Reiter and Leroux reported that 35 different ciliopathy disorders were known due to defects in 187 ciliopathy genes. In addition, 241 ciliopathy genes were potential candidate genes for ciliopathy disorders. They noted that both motile and immotile cilia were involved in ciliopathies. They distinguished first-order and second-order ciliopathies.

First-order ciliopathies included diseases resulting from defects in proteins in the ciliary basal body, or ciliary compartments and disorders in which there was disruption of intraflagellar transport.

Second-order ciliopathies were defined as disorders in which processes involved in cilia formation were disrupted.

It is important to note that in ciliopathies, a number of different tissues may be impacted. In this section, the focus will be on ciliopathies that primarily impact the nervous system. Reiter and Le Roux noted that an extensive array of organ systems can be impaired in ciliopathies. With respect to the brain, they reported that ciliopathies due to defects in nonmotile cilia can result in brain malformation, mental disabilities, epilepsy, and/or ataxia. Defects in the function of motile cilia can lead to hydrocephalus.

They also extensively reviewed the different cilia components and associated pathological defects that resulted from disruptions in genes that encode products in each ciliary component. Joubert syndrome can arise due to defects in products of any one of 15 different genes. In some cases, it results from digenic mutations.

Brain abnormalities that occur in Joubert syndrome include hypoplasia of the cerebellar vermis. A characteristic neuroimaging finding in Joubert syndrome is the so-called molar tooth abnormality. It is caused in part by hypoplasia or aplasia of the cerebellar vermis and reoriented cerebellar peduncles. The clinical neurological defects include ataxia, hypotonia, and oculomotor defects; changes in respiratory rhythm may be observed. Other systems that can be affected in Joubert syndrome include kidney and liver (Valente, 2013).

Stromme syndrome is a ciliopathy due to autosomal recessive mutations in a specific gene CENPF. In this syndrome, multiple systems can be affected, leading to cognitive defects, renal abnormalities, and intestinal atresia. CENPF protein is associated with mitotic spindles.

Several multisystem ciliopathies can also include brain defects. Bardet–Biedl syndrome can result from defects in any one of 19 different genes. This syndrome is reported to be autosomal recessively inherited. It is characterized by behavioral dysfunction and other features that may include obesity, kidney dysfunction, hypogonadism, polydactyly, retinitis pigmentosa, and intellectual impairment (Beales, 1999).

Centrosome and microtubules

Microtubules are involved in transport within cells. Microtubules are composed of different proteins including microtubule-associated proteins (MAPs) and alpha and beta tubulins. The centrosome forms the microtubule-organizing center. Microtubules play critical roles in cells division; they attach to the centromeres of chromosomes as the cell prepares for cell division. Microtubule abnormalities impact cell division and therefore represent a significant cause of microcephaly.

Microtubule centrosome and actin interactions

There is evidence that interactions between actin and microtubules in the cytoskeleton play important roles in cell polarity and in migration (Dogterom and Koenderink, 2019).

Dobyns (2018) noted that two proteins play roles as actin microtubule linkers. The proteins are microtubule and actin cross-linking factor 1

(MACF1) and dystonin (DST). They noted further that many different isoforms of MACF1 occur and that some isoforms are brain specific.

Dobyns et al. described a brain malformation that included lissencephaly, brain stem malformation, defective intercerebral connection, and hippocampal dysplasia. Individuals with these defects manifested intellectual disabilities, seizures, and aberrant muscle tone. Eight individuals with this disorder were found to have deleterious mutation in the MACF1 protein.

Microcephaly can arise due to defects in the following genes that encode microtubule-related products.

- MCPH1, microcephalin may play a role in G2/M checkpoint arrest in the cell cycle
  - ASPM abnormal spindle microtubule assembly involved in mitotic spindle function
    - CEP152, centrosomal protein 152 involved in centrosome function
    - CEP63, centrosomal protein 63 involved in centrosome function
    - NIN, ninein centrosomal protein 125 involved in centrosome function
    - NDE1, nudE neurodevelopment protein 1
    - CENPE, centromere protein E
    - KIF5C, kinesin family member 5A functions as a microtubule motor
    - KIF11, kinesin family member 11 involved in spindle dynamics
    - TUBB2B, tubulin beta 2B class IIb binds GTP and is a major component of microtubules
    - TUBB2A, tubulin beta 2A class IIa
    - TUBG1, tubulin gamma 1 localizes to the centrosome forming ring complex
    - POC1A, POC1 centriolar protein A plays important roles in basal body and cilia formation
    - CENPJ, centromere protein J plays a structural role in the maintenance of centrosome integrity

Microcephaly centrosome and mitotic spindle defects

Alcantara and O'Driscoll (2014) documented clinical forms of microcephaly and specific centrosome or mitotic spindle defects that led to these disorders.

Clinical form of microcephaly, gene product mutated, structure impacted

- MCPH1, microcephalin, centrosome
  - MCPH2 WDR62, mitotic spindle MCPH3 CDK5RAP2 centrosome
    - MCPH3 CEP215, microtubule organization
    - MCPH4 CASC5, spindle assembly kinetochore
    - MCPH5 ASPM, spindle orientation
    - MCPH6 CENJ, centriole biogenesis
    - MCPH7 SCL/TAL1, centrosome centriole biogenesis
    - MCPH8 CEP135, centrosome centriole biogenesis
    - MCPH9 CEP152, centrosome biogenesis genome stability

It is important to note that microcephaly sometimes occurs in combination with defects in other body systems, and sometimes, microcephaly may occur in combination with growth retardation. Mutations in CEP63 (centrosomal protein 63) are reported to lead to Seckel syndrome with reduced growth and microcephaly.

In addition, microtubule and centrosome protein defects may occur in combination with other cortical malformations. Defects in the tubulin gamma complex, TUBG1, lead to microcephaly with complex cortical malformations.

Goncalves (2018), Vandervore (2019) reported that defects in the protein rotatin (RTTN) can lead to polymicrogyria (PMG) and microcephaly. Rotatin was reported to be involved in centriolar organization and ciliogenesis.

Mutations in tubulins and microtubular proteins lead to disorders sometimes referred to as tubulinopathies. Goncalves (2018) reported that brain imaging studies in individuals with tubulinopathies had revealed a range of different structural malformations that could impact basal ganglia, corpus callosum, and cerebellum. In some cases, tubulinopathies were associated with lissencephaly (smooth cortex with reduced gyri and sulci).

Shohayeb et al. (2019) reported that mutations in the WDR62 protein that is required for cilia formation are associated with microcephaly. They reported that WDR62 proteins within the basal body recruit proteins that are required for cilia formation, including centromere-associated J (CENPJ).

5. Defects in DNA replication and congenital microcephaly

Alcantra and Driscoll (2014) reported that mutations in genes that encode products involved in DNA replication led to Meier-Gorlin syndrome, which is associated with microcephaly. Mutations in components of the complex involved in initiation of DNA replication were particularly important in leading to this syndrome.

In individuals with Meier–Gordon syndrome, mutations in proteins in the origin of replication complexes ORC1, ORC4, and ORC6 were identified. There was evidence that this syndrome could also be caused by mutations that led to defects in the prereplication licensing and loading complexes CDT1 and CDC6.

Defects in components of the DNA damage repair response were also noted to lead to microcephaly. Specific proteins that are known to play roles in this response and that are known to have mutations that can lead to microcephaly are listed in the following:

- ATR serine/threonine kinase DNA damage sensor, activating cell cycle checkpoint signaling
- CTIP (RBBP) expressed nuclear protein binds retinoblastoma protein involved in cell proliferation
  - NBS1 serine/threonine kinase and DNA damage sensor
  - RAD1 component of a heterotrimeric cell cycle checkpoint complex
  - PNKP polynucleotide kinase 3′-phosphatase, 5′ phosphorylation and 3′ dephosphorylation
  - XRCC2 involved in the repair of DNA double-strand breaks by homologous recombination
  - XRCC4 involved in the repair of DNA double-strand breaks by homologous recombination
  - NHEJ1 DNA repair factor essential for the nonhomologous end-joining pathway
  - RECQL3 (WRN) plays a role in DNA repair, replication, transcription, and telomere maintenance

Youn and Han (2017) noted that the following brain defects have been reported to result from pathogenic variants in ciliary pathway genes: encephalocele, holoprosencephaly, microcephaly, PMG, heterotopia, intracerebral cysts, hippocampal dysgenesis, corpus callosum agenesis hydrocephalus, cerebellar hypoplasia, and Joubert syndrome. Ciliopathy disorders typically impact a number of different body systems. Cilia are formed from microtubules.

6. Corpus callosum intracerebral connectivity and defects

The corpus callosum contains axonal structures that connect the left and right cerebral hemispheres. Agenesis of the corpus callosum occurs as a birth defect and has sometimes been reported to be associated with cognitive or behavioral problems. In 2014, Palmer and Mowat reviewed outcomes in cases that had presented antenatally or postnatally with agenesis of the corpus callosum. They noted that corpus callosum agenesis led to a range of different outcomes; 25% of patients were found to have intellectual disability. In some patients who were not intellectually impaired, agenesis of the corpus callosum had subtle neurological, social, or learning defects.

Palmer and Mowat noted that some patients with agenesis of the corpus callosum have chromosomal or genetic defects. In addition, antenatal infections, vascular damage, or toxic factors were suspected to play roles in some patients.

Roland (2017) carried out studies on patients on whom surgery had been performed to section the corpus callosum to treat intractable epilepsy. Based on functional magnetic resonance (fMRI) imaging, Roland et al. demonstrated that both callosal and extracallosal connections are responsible for interhemispheric connectivity.

7. Defects in cortex structural differentiation

Guarnieri in a 2018 review emphasized the structure of the cerebral cortex in horizontal layers and cortical columns. They noted that this defined structure was achieved through regulated neurogenesis and neuronal migration that lead to composition of layers with excitatory and inhibitory neurons. Neurons born in the ventricular zone migrated along radial glia toward the surface of the brain and primordial neurons were noted to undergo morphological changes as they migrated. Guarnieri et al. noted that the migration patterns of excitatory and inhibitory neurons differed to some degree.

They emphasized that abnormalities in migration steps could lead to alterations in brain structure. Brain structural abnormalities could also result from inadequate proliferation or excess proliferation of neurons leading to malformations in cortical development. Progenitor proliferation defects leading to excess proliferation could lead to megalencephaly or hemimegalencephaly or to localized malformations in the cortex.

Guarnieri et al. noted that genetic defects, germline or somatic, that led to altered function of the mTOR signaling pathway played roles in structural malformations. Specific associated genes that impacted the mTOR pathway encoded the following protein products:

- PIK3CAm phosphatidylinositol-4,5-bisphosphate 3-kinase catalytic subunit alpha
- PIK3R2m phosphoinositide-3-kinase regulatory subunit 2
  - AKT1, AKT serine/threonine kinase 1, activation occurs through phosphatidylinositol 3-kinase.
    - AKT3, AKT serine/threonine kinase 3
- TSC1, TSC complex subunit 1, hamartin, interacts with GTPase-activating protein tuberin
  - TSC2, TSC complex subunit 2, associates with hamartin in a cytosolic complex
    - NPRL2, NPR2-like, GATOR1 complex subunit
    - DEPDC5, DEP domain containing 5, GATOR1 subcomplex subunit
    - STRADA, STE20-related adaptor alpha

Guarnieri et al. noted that mTOR is essential for cell growth and proliferation through its impacts on protein synthesis and cytoskeletal dynamics. They noted further that the AKT–MTOR pathway played roles in radial glial cell proliferation.

Decreased proliferation of neuronal progenitor cells

This may result in decreased brain size and altered structural differentiation. Guarnieri et al. noted that mutations in gene products that impair centrosome maturation or mitotic spindle formation or function had been shown to lead to microcephaly. They specifically included the following gene products as being implicated in centrosome maturation and spindle formation.

- CDK5RAP2, CDK5 regulatory subunit associated protein 2, localizes to the centrosome
  - ASPM, abnormal spindle microtubule assembly, mitotic spindle regulation
    - CENPJ, centromere protein J, role in centrosome integrity and normal spindle morphology,
    - STIL, STIL centriolar assembly protein, monitors chromosome segregation
    - CEP135, centrosomal protein 135, acts as a scaffolding protein during early centriole biogenesis
    - CEP152, centrosomal protein 152, involved with centrosome function
    - CEP63, involved with centrosome function
    - NDE1, nudE neurodevelopment protein 1, in a multiprotein complex regulates dynein function
    - CDK6, cyclin dependent kinase 6 important for cell cycle G1 phase progression G1/S transition
    - WDR62, WD repeat domain 62

Guarnieri et al. noted that heterotopic location of progenitor cells can arise due to mutation in EML1, an MAP.

Defects in genes that produce proteins active in the nucleus were also noted to cause microcephaly. These included MCPH1 (microcephalin) that may play a role in G2/M checkpoint arrest and ZNF335 that impacts activity of transcription factor REST. NRSF and PHC1 encode components of the polycomb chromatin remodeling complex.

Aberrations in neuronal migration

These were noted to occur as a result of defects in formation of the radial glial scaffolds. Guarnieri et al. specifically noted that periventricular nodular heterotopia could result from defective function of the X-linked filamin A gene product. This protein was noted to regulate the actin cytoskeleton and to be involved in cross-linking of actin filaments.

Other disorders of neuronal migration were reported to result from heterozygous mutation in LIS1 and DCX (doublecortin) proteins. Normal versions of these proteins interact with the protein dynein that stabilizes microtubular motor functions. Defects in DCX function were reported to lead to formation of subcortical heterotopic bands. LIS1 defects can lead to lissencephaly, characterized by defective gyri formation. Other protein defects that can lead to lissencephaly occur in tubulin-related genes TUBA1A, TUBB2B, and TUBG1. Lissencephaly may also arise due to aberrant function in microtubule motor proteins kinesins KIF2A, KIF3C, or DYNC1, a dynein a microtubule-activated ATPase that functions as molecular motor.

They noted that mutations in TBC1D24, a protein that interacts with GTPases, can impact radial neuronal migration. Defects in this protein have been found in cases of epilepsy.

Pial basal membrane and neural migration

Guarnieri et al. noted that the dystrophin glycoprotein complex is a transmembrane complex that connects extracellular matrix with cellular cytoskeleton. Several gene products responsible for glycosylation of dystroglycan have been shown to have undergone mutations that interfere with anchorage of radial glial cells in the pial basement membrane and interfere with radiation of glial cell processes. Defects in this anchoring process were reported to lead to overmigration of neurons and to neuronal heterotopias and cobblestone malformations.

Gene products involved in dystrophin glycosylation include the following:

- ISPD (CRPPA) CDP-L-ribitol pyrophosphorylase A

- POMT1 and POMT2, protein O-mannosyltransferase 1 and 2
  - LARGE1, LARGE xylosyl- and glucuronyltransferase 1
  - FKTN, fukutin, a transmembrane protein
  - FKRP, fukutin-related protein

Guarnieri et al. noted that overmigration of neurons was also shown to occur as a result of impaired function of GPR56, a G protein–related complex receptor, which facilitates anchorage of glial cells in collagen. GPR56 or collagen defects were shown to lead to cortical lamina defects and heterotopias.

Reelin

Reelin was first discovered in mice as a gene product that when mutated led to abnormal movements in mice. Wasser and Herz (2017) reviewed effects of reelin mutations in mice and in humans.

Adequately functioning reelin in prenatal life was initially described as being essential for layering of neurons in the cortex, hippocampus, and cerebellum. It was described as a protein secreted by the Cajal–Retzius cells. Reelin was found to bind to specific receptors including APOER2 designated LRP8 in humans and VLDLR, both are lipoprotein receptors. Reelin was also reported to interact with DAB1 adaptor protein that is also involved in migration of cortical neurons.

Wasser and Herz (2017) reported that in post-natal life, Reelin acts as a neuromodulator in that it modulates axonal and dendritic outgrowth in part through impacting cytoskeletal stability. Following binding to VLDLR and APOER2 (LRP8), reelin enters the cell where it activates DAB1 that in turn activates the phosphatidylinositol (PI3K/AKT) pathway that leads to activation of mTOR. Activation of the PI3K pathway was reported to also lead to activation of LIMK1, to interact with cofilin that impacts actin and the cytoskeleton. Frotscher et al., (2017) reported that through its induction of LIMK1, reelin led to phosphorylation and inactivation of cofilin that led to cytoskeletal stabilization that promoted neuronal migration and corrected orientation in cortical lamina.

Reelin was reported to influence dendritic spines and synaptic contacts. Wasser and Herz emphasized the role of reelin in enhancing synaptic contact at excitatory glutamate receptors. They noted that in mice with reelin defects, the hippocampal neurons were shown to have reduced numbers of dendritic spines.

Long-term synaptic plasticity was reported to be enhanced by reelin. Reelin was shown to increase passage of calcium ions through NMDAR receptors, which then promoted phosphorylation of CREB (cyclic AMP response element–binding protein). There is also evidence that downstream of CREB phosphorylation transcription and translation of ARC protein was increased. Werner and Herz noted that reelin therefore likely plays roles in memory and learning.

The APOER2 (LRP8) receptor mRNA was reported to undergo alternative splicing, and the protein was shown to undergo posttranslational modification. The different forms of this receptor protein were shown to influence reelin binding.

Wasser and Herz noted that roles for impaired reelin function in neuropsychiatric and neurodegenerative diseases have been proposed.

Whittaker (2017) noted that reductions in reelin expression, due to recessive mutations, have been shown to lead to cerebellar hypoplasia in humans. They determined through studies in mice that the ATP-dependent chromatin remodeler Chd7 plays a key role in reelin gene expression.

Lissencephaly syndrome

Fry (2014) reported roles for mutations in three different genes in causation of lissencephaly (smooth cortex with reduced gyri and sulci). They noted that severe lissencephaly is sometimes associated with a specific syndrome

Miller–Dicker syndrome. Individuals with that syndrome manifested lissencephaly, developmental delay, intellectual disability, seizures, motor function impairment, and facial dysmorphology. Individuals with this syndrome were found to have deletion of chromosome 17p13.3. This deletion removed several genes including one gene that was shown to be key to lissencephaly causation. This gene was designated PAFAH1B1 and is also known as the LIS1 gene. It encodes an acetyl hydrolase.

Another gene located in the region of deletion on 17p13.3 is YWHAE that encodes tyrosine 3-monooxygenase/tryptophan 5-monooxygenase activation protein.

Fry et al. also reported that two other gene products when mutated can lead to lissencephaly. These gene products were doublecortin (DCX) encoded on the X chromosome and TUBA1A, tubulin alpha 1a, a constituent of microtubules.

In 2018, Di Donato et al. reviewed data on 811 patients with lissencephaly, and they reported that pathogenic mutations were detected in 81% of patients. In total, the mutations involved products of 17 different genes.

Di Donato et al. noted that a spectrum of malformations occurred in these patients, and they included agyria, pachygyria, and subcortical band heterotopia. They noted that patients had intellectual impairments that ranged from mild in patients with subcortical band heterotopia to severe impairments with limited survival in patients with agyria.

Di Donato et al. also reported on the frequency of specific genetic defects. LIS1 defects including deletion in 17p13.3 or mutations were most frequent and occurred in 40% of cases. DCX (doublecortin) mutations occurred in 23% of cases, TUBA1A mutations were found in 5% of cases, ATRX (chromatin remodeler) mutations occurred 4% of patients, and DYNC1H1 (dynein heavy chain) mutation occurred in 3% of cases.

These investigators also documented that mutations in the following gene products occurred in 1% of patients. These gene products included TUBB2B (tubulin beta 2B class IIb), ACTG1 (actin gamma 1), ACTB (actin beta), TUBG1 (tubulin gamma 1), CADD (caspase recruiting protein), and RELN (reelin).

LIS1, DCX, TUBA1A, and DYNCH1 together accounted for mutations in 70% of individuals.

LIS1 defects were noted to most frequently be associated with diffuse agyria or pachygyria. In DCX cases, a range of abnormalities occurred including agyria, pachygyria, and subcortical heterotopia. DYNCH1 mutations were primarily associated with pachygyria.

With respect to biological functions of gene products impacted by mutations. Di Donato et al. noted that LIS1, YWHAE, and TUBG1 encode MAPs that are also expressed at the centromere. TUBA1A, TUBG, and TUBB2B are MAPs. DYNCH1 is a microtubule motor protein. ACTG1 and ACTB are nonmuscle actions. They emphasized the role of reelin in influencing the actin cytoskeleton.

Doublecortin

Moslehi (2017) reviewed doublecortin (DCX) and its role in neuronal migration and cortical layering. They emphasized the roles of doublecortin defects leading to neurodevelopmental defects.

Two specific domains within the DCX protein were reported to be linked via a flexible middle domain. DCX was reported to bind to tubulin dimers in microtubules. Deletion of the C-terminal domain of DCX was reported to alter the distribution of microtubulin.

8. Intellectual disability

Defining intellectual disability

Iwase (2017) defined intellectual disability "as a prevailing neurodevelopmental condition associated with impaired cognitive and adaptive behavior."

ID is often defined as an IQ below 70; sometimes, it is defined as defects in adaptation to environment and social milieu. Iwase et al. noted that the worldwide incidence of ID is 2%–3% of the population.

Pathways to discovery of causes of intellectual disability

1. Evidence gathered through epidemiologic studies. These may provide insight into causative factors and may also lead to therapeutic or preventive measures.
2. Evidence gathered through metabolic and biochemical studies may sometimes lead to elucidation of underlying causative defects and lead to amelioration in an affected individual or to preventive measures.
3. Genomic studies may provide clues to causative factors. In rare cases, e.g., discovery that a chromosome imbalance in a child is secondary to a balanced translocation in a parent can be valuable information for parents as they consider subsequent pregnancies. Finding of a genomic change in an affected offspring seldom leads to specific therapy directly related to the genomic defect.
4. Insights gained into causative mechanisms of intellectual disability through analysis of DNA sequence may sometimes lead to therapeutic measures related to causative mechanisms.

Epigenetic regulators including chromatin modifiers and roles in intellectual disability

Important epigenetic factors to consider include enzymes and proteins involved in the methylation and demethylation of DNA and histones, proteins that bind to methylated DNA, or methylated histones. Other factors to consider include histone modifiers including histone acetylation and deacetylation and chromatin remodelers.

Disruption of regulated epigenetic processes and chromatin modifications can impact gene expressions and downstream cellular processes. Specific intellectual disability syndromes have been found to be due to disruption of specific epigenetic processes.

RBFOX1 and cortex development

The protein RBFOX1, which was shown to be defective in some cases of autism (Voineagu, 2011) and when it is defective, impacted splicing of different genes. In addition, defective cytoplasmic RBFOX1 protein was shown to impair translation of a number of different genes that play roles in brain cortex development.

ATRX syndrome

Hypomorphic mutations in the Xq21.1 and the gene that encodes ATRX chromatin remodeler protein that impact the function of the encoded protein lead to seizures, autistic behaviors, and alpha thalassemia. Iwase (2017) noted that the ATRX protein has several functions that together influence gene transcription of a number of different genes. Specific domains present in ATRX include a helicase domain that unwinds chromatin and a histone reader domain. ATRX knockout mouse models are being developed to analyze specific effects of ATRX absence on molecular and cellular components in the brain.

MECP2 and Rett syndrome

Extensive studies on the neurological and molecular impact of impaired MECP2 function have been carried out by Bird and coworkers and by Zoghbi Deleterious MECP2 mutations lead to loss of previously acquired skills in young females. In 1992, Lewis et al. of the Bird group described a protein that binds to methylated DNA, and they established that it bound to methylated cytosine, particularly in the context of CpG dinucleotides. This protein was designated MECP2.

Amir in 1999 reported that Rett syndrome results from mutation in the MECP2 gene that maps to Xq28. In some cases, syndrome-unaffected mothers carried the same mutation as their affected daughters but are apparently protected by favorable X inactivation. In 2016, Zoghbi reported that 95% of cases with Rett syndrome have MECP2 mutations. In a 2016 review, Zoghbi noted that Rett originally published features of this syndrome in 1966. The features documented in this original publication included normal development during the first 9 months with decline; then beginning and early manifestations included gait impairments, impaired speech, stereotypic hand movements, and flat affect. Later in the disease, respiratory impairment developed. Pathogenic MECP2 mutations impact gene transcription.

However, the range of transcriptional defects and the precise molecular bases for these defects are complex (Lyst and Bird, 2015). Early studies revealed that the MECP2 gene located on the X chromosome encodes a protein that binds to methylated DNA. Iwase et al. noted that the exact effects of MECP2 protein may differ in different cell types in the brain. There is now also evidence that it interacts with methylated cytosine in CpH dinucleotides (H = AC or T).

In a detailed review of Rett syndrome, Lyst and Bird (2015) reported that classical Rett syndrome occurs in 1 in 10,000 females. They noted that normal development is often restricted to the first 6 months of life.

Features of brain pathology include the presence of densely packed neurons with reduction in a number of dendritic spines and decreased dendritic spine branching.

Studies by several investigators have established that particular domains in the MECP2 protein are important in its functions. These domains include the methyl binding domain, an NCOR-SMRT domain that impacts chromatin and a hook domain. There is evidence that MECP2 is a transcription regulator. MECP2 is primarily involved in suppressing gene expression; however, there is also evidence that it may increase expression of certain genes. MECP2 may alter chromatin architecture, and it may also impact RNA splicing (Lyst and Bird, 2015). Since Rett-causing mutations led to decreased MECP2 expression, therapeutic approaches being investigated in Rett syndrome are aimed at increasing MECP2 expression. One such approach involves use of antisense oligonucleotides (Zoghbi, 2016).

X-linked intellectual disability and KDM5C defects

Iwase (2017) reported that 2% of cases with X-linked intellectual disability have been shown to have mutations in the protein encoded by KMT5C (lysine demethylase 5C). This enzyme erases histone 3 lysine 4 dimethylation and histone 3 lysine 4 trimethylation. Patients with defective function of this enzyme often manifest intellectual disability. Iwase et al. reported that studies on mouse models of KDM5C mutation have revealed alteration in the levels of expression of hundreds of different genes.

Cognitive impairment in Rubinstein–Taybi syndrome

Patients with this syndrome were first reported often have life-threatening malformations that impact heart, lungs, and kidney. Later less severe forms of the syndrome were described in patients that presented with cognitive impairment. Other features of this syndrome include short stature and broad thumbs. Iwase et al. reported that the Rubinstein–Taybi syndrome can result from mutation in any one of two different genes that encode lysine acetyl transferase. These genes including CREBBP (KAT3A) on chromosome 16p13.3 and EP300 (KAT3B) on chromosome 22q13.

Structural genomic changes, dosage effect, and brain development

15q11.1–q12

This genomic region is 0.5 mb in size, and it contains four genes that are highly conserved in vertebrate species NIBP1, NIPA2, CYFIP1, and TUBGCP5. NIPA1 NIPA2, and CYFIP1 are known to be involved in brain development. NIPA1A and NIPA2 encode magnesium transporters. CYFIP1 and NIPA1 were reported to play roles in brain development and plasticity and to impact axon outgrowth and dendritic spine morphology by De Rubeis (2013). The TUBGCP5 gene encodes the tubulin gamma complex associated protein. Tubulins play important roles in neuronal microtubule structure and function. NIPA1 is defined as a magnesium transporter that associates with cell surfaces on neuronal and epithelial cells. NIPA2 is reported to also likely be a magnesium transporter. De Rubeis (2013) reported that CYFIP1 impacts protein synthesis and actin remodeling. They reported that CYFIP1 is incorporated in a complex that inhibits its translation and that it is released from this complex on signaling and by brain-derived neurotrophic factor (BDNF). Following this release, CYFIP1 facilitates translation of target mRNAs.

CYFIP1

CYFIP1 protein interacts with FMRP protein and with translation initiation factor E, and this complex inhibits translation of RNA to protein. DeRubeis et al. noted that CYFIP1 is enriched in the postsynaptic compartment. CYFIP1 together with CYFIP2 also forms part of another complex referred to the WAVE regulatory complex that is involved in actin polymerization and cytoskeletal dynamics (Nakashima, 2018). The WAVE regulatory complex is composed of five subunits and is recruited to sites where actin undergoes polymerization and, for this process, interactions with RAC1 and ARFGTPase are required.

Low CYFIP1 expression was reported to lead to decreased oligodendrocyte count and decreased myelination. In the presence of increased CYFIP1, such as occurs in cases of duplication in the 15q11.2 region, dendritic spine number was reported to be increased. Studies on mice by Pathania (2014) revealed that altered expression levels of Cyfip1 led to defects in dendrites and in neuronal connectivity. Studies on rats carried out by Silva (2019) revealed that haploinsufficiency and reduced levels of Cyfip1 led to white matter thinning, decreased myelin, and behavioral inflexibility. Winkler et al. (2018) carried out brain MRI analyses and reported that 15q11–15q12 deletion carriers had lower cortical surface area and thicker cortical density than individuals without this deletion. They also reported decreased scores on five of seven cognitive measures in these deletion carriers. Davenport (2019) reported that appropriate levels of CYFIP1 and CYFIP2 are required to achieve excitatory inhibitory balance.

CYFIP1 and bilateral brain connectivity

Dominguez-Iturza (2019) noted that defects in communication, social behavior, learning, and motor coordination had been reported in some cases of corpus callosum agenesis. In addition, based on fMRI studies and diffusion tensor imaging (DTI), corpus callosum abnormalities were detected in some patients with autism or schizophrenia. They also noted that structural defects in the chromosome 15q11.2 region occur with higher frequency in patients with autism and schizophrenia patients and patients with intellectual disability than occur in the general population. The 15q11.2 gene that was reported to be of particular interest in this region is CYFIP1, which encodes a protein that interacts with the fragile X mental retardation protein (FMRP). Studies on mice with Cyfip1 defects were reported by Oguro-Ando et al. in 2015 and revealed altered axonal growth and increased dendritic spine numbers. Dominguez-Iturza et al. carried out on mice with heterozygous Cyfip1 and DTI studies revealed microstructural alteration in the corpus callosum. In addition, in Cyfip$^{\pm}$ mice, fMRI studies revealed alterations in functional connectivity between the two hemispheres.

Brain studies in 16p11.2 copy number variant individuals

Sonderby and the Enigma CNV consortium in 2019 reported results of studies carried out in 12 cases with 16p11.2 deletion and 12 cases with 16p11.2 duplications. They compared findings in these cases with findings in 6882 individuals who did not have 16p11.2 changes. The presence of low copy repeat sequence elements in the 16p11.2 region predisposes to structural changes, leading to genomic deletion and duplications. This region lies between 28.3 and 28.9 Mb on chromosome 16. Nine protein-coding genes occur in the region between 28.7 and 28.9 Mb. Specific clinical features associated with 16p11.2 deletion include intellectual disability, obesity, and schizophrenia risk. Duplication of 16p11.2 leads to reduced body mass. Both deletion and duplication of this region increase risk for autism and epilepsy. Sonderby (2020) noted that 16p11.2 copy number variants (CNVs) occur with low frequency in the population. Reported frequencies of the deletion range between 0.012 and 0.019, and duplication frequencies range between 0.030 and 0.038 for the Icelandic and UK populations studied. Results of brain imaging studies revealed alteration of basal ganglia volumes, specifically of the nucleus accumbens, caudate, pallidum, and putamen. In this study, individuals with 16p11.2 duplication had significantly lower mass than other individuals. Full-scale IQ was reported to be lower in 16p11.2 duplication than in 16p11.2 of deletion cases.

It is important to note that other chromosome abnormalities may occur in patients with neurodevelopmental disorders.

Specific proteins and interacting molecules and impaired neurofunctions

Calmodulins and interacting molecules with genetic association with diseases with impaired neurofunctions

Calmodulins act as intracellular receptors for calcium. In addition, calmodulin interacts with numerous proteins in the presynaptic region and in the postsynaptic density and with intracellular enzymes involved in signal propagation (Lipstein, 2017). A list of such proteins that have been found to have defects in some diseases associated with impaired neurofunctions is as follows:

- Neurexins, cell-surface receptors that bind neuroligins to form Ca (2+)-dependent complexes
  - Kainate receptors, GRIK2 glutamate ionotropic receptor kainate-type subunit 2
    - GLUD1 and GLUD2 glutamate dehydrogenases
    - MDGA1 and MDGA2 glycosylphosphatidylinositol (GPI)-anchored cell surface glycoprotein
    - Neuroligins, neuronal cell surface proteins
    - Dystroglycans, central component of dystrophin–glycoprotein complex that links the extracellular matrix and the cytoskeleton
    - LRRTMs, leucine-rich-repeat (LRR) transmembrane neuronal proteins involved in synapse organization
    - GABAA receptors, gamma-aminobutyric acid type A receptor alpha subunits
    - Latrophilins, ADGRL1-4 adhesion G protein–coupled receptors
    - SLITRKs, integral membrane proteins with 2 N-terminal LRR domains
    - ILRAPS, interleukin 1 receptor accessory protein

Synaptic cell adhesion molecules implicated in neurodevelopmental functions

- MDGA1 and MDGA2 MAM domain–containing glycosylphosphatidylinositol anchor 1
  - Dystroglycans
    - Latrophilins, teneurins, transmembrane proteins
    - PTPRD/F/S protein tyrosine phosphatase receptor
    - SYNCAMs, cell adhesion molecules
    - Cadherins calcium-dependent cell–cell adhesion proteins
    - Ephrin/ephrin receptor subfamily of receptor protein–tyrosine kinases

Protein defects and cortical malformations

- ARHGAP1B, Rho GTPase-activating protein 1
  - ASPM, abnormal spindle microtubule assembly
    - CDH1, cadherin 1
    - SMO, G protein–coupled receptor, interacts with the patched protein, hedgehog receptor
    - MEK/ERG mitogen-activated protein kinase
    - PTEN, phosphatidylinositol-3,4,5-trisphosphate 3-phosphatase
  - RASGAP family of GTPase-activating proteins
  - TBC1D3, TBC1 domain family; FGF, fibroblast growth factor
    - FGFR, fibroblast growth factor receptor
    - GPSM1, G protein signaling modulator 1
  - PDGF, platelet-derived growth factor
  - PDGFR, platelet-derived growth factor receptor
  - PAX6, paired box 6
    - SMARCC1, member of the SWI/SNF family of proteins, impacts chromatin

Insights into mechanisms and genes involved into mechanism and genes involved in determining normal development of the cortex and other brain structures also derive from detailed studies documented through clinical investigations. In 2016, Parrini et al. published a review of cortical brain malformations in patients and associated genetic and phenotypic findings. Specific forms of cortical malformation that they analyzed included lissencephaly, subcortical band heterotopia, pachygyria, periventricular nodular heterotopia, PMG, microcephaly, and megalencephaly. They noted that by 2016, 100 different gene defects had been identified on patients with cortical malformations.

Key biological pathways impacted by gene defects associated with cortical malformations

- Cell cycle regulation including mitosis and cell division, apoptosis
- Cell fate specification
- Cytoskeletal structure and function
- Neuronal migration
- Basement membrane function
- Inborn errors of metabolism

Studies have shown that in most cases, gene defects had arisen de novo in patients although parental gonadal mosaicism for damaging mutations could not be excluded. In specific cases of megalencephaly, patients were sometimes found to have postzygotic mutations.

It was also important to note that defects in specific genes were found to be associated with different malformations, e.g., lissencephaly or subcortical band heterotopia.

Lissencephaly and subcortical band heterotopia

Specific disorder defective protein and function of normal protein encoded by that gene:

- ILS/SBH LIS1 (PAFAH1B1), acetylhydrolase 1b regulatory subunit 1
  - ILS/SBH DCX, doublecortin cytoplasmic protein binds to microtubules

- ILS/SBH TUBA1A tubulin alpha 1A microtubule constituent
- ILS TUBB2B tubulin beta 2B class IIb microtubule constituent
- LIS variable gradient KIF5C kinesin family member 5c interacts with protein kinase
- RELN, reelin large secreted extracellular matrix protein
- LIS cerebellar hypoplasia
- VLDLR lipid receptor acts in reelin pathway
- LIS cerebellar hypoplasia CDK5 impacts neuronal migration through phosphorylation of proteins
- Pachygyria ACTB, actin B involved in cell motility, structure, integrity, signaling
- Pachygyria ACTG1, actin gamma 1 component of the cytoskeleton
- Posterior pachygyria TUBG1, tubulin gamma one mediates microtubule nucleation
- PV nodular heterotopia FLNA actin-binding protein cross-links actin filaments that link actin to membrane
- PV nodular heterotopia ERMARD, endoplasmic reticulum membrane–associated RNA degradation
- PVNH and microcephaly
- ARFGEF2, ADP ribosylation factor guanine nucleotide exchange factor 2

Syndromic forms of periventricular nodular heterotopia

- PV nodular heterotopia DCHS1 member of the cadherin superfamily encode calcium adhesion molecules
- PV nodular heterotopia LRP2 low-density lipoprotein-related protein 2 multi-ligand endocytic receptor

Polymicrogyria

- PMG, frontoparietal GPR56 (ADGRG) adhesion G protein–coupled receptor G1
- PMG asymmetric TUBB2B, tubulin beta 2B class IIb major component of microtubules
- PMG aniridia PAX6, paired box 6 two domains that are regulators of gene transcription
- PMG and microcephaly NDE1, nudE neurodevelopment protein 1
- PMG and microcephaly WDR62, WD repeat domain 62
- PMG and microcephaly RTTN, rotatin
- PMG fumaric aciduria FH, fumarate hydratase, catalyzes the formation of L-malate from fumarate.
- PMG, band calcification OCLN, occludin integral membrane proteins
- PMG bilateral perisylvian PIK3R2, phosphatidylinositol regulatory unit
- PMG bilateral temporooccipital FIG4 phosphoinositide 5-phosphatase

Polymicrogyria that can include other cerebral malformations

These that can include corpus callosum agenesis (ACC) and cerebellar hypoplasia (CBLH)

- PMG, ACC DYNC1H1 dynein, microtubule-activated ATPases function as molecular motors
- PMG, ACC EOMES eomesodermin
- PMG, ACC, CBLH TUBA1A, member of TBR1 (T-box brain protein 1) subfamily of T-box genes
- PMG, ACC, CBLH TUBA8, tubulin alpha 8, tubules assemble to form microtubules
- PMG, ACC, CBLH TUBB3, tubulin beta 3 class III
- PMG, ACC, CBLH TUBB, tubulin beta class I

Syndromic forms of polymicrogyria

- PMG and Charge syndrome CHD7, chromodomain helicase DNA-binding protein 7

- PMG and microsyndrome RAB3GAP1, catalytic subunit of an RAB GTPase-activating protein
  - PMG and microsyndrome RAB3GAP2, Rab3 GTPase-activating complex regulatory subunit
  - PMG and microsyndrome RAB18, member of a family of Ras-related small GTPases

Megalencephaly–polymicrogyria

- MPPH2, AKT3, AKT serine/threonine kinase 3, regulator of cell signaling
  - Weaver syndrome EZH2 enhancer of zeste 2 polycomb repressive complex 2 subunit

Megalencephaly–capillary malformation–polymicrogyria syndrome

- PIK3CA protein kinase C alpha can be activated by calcium and diacylglycerol
  - PIK3R2 (MPPH1) phosphoinositide-3-kinase regulatory subunit 2

Dysplastic megalencephaly with focal cortical dysplasia

- FCD DEPDC5 DEP domain containing 5, GATOR1 involved in G protein signaling pathways
  - FCD AKT3 AKT serine/threonine kinase 3, regulator of cell signaling
    - FCD NPRL3, NPR3-like, GATOR1 complex subunit

Holoprosencephaly

The key feature of this disorder is incomplete separation of the cerebral hemispheres due to incomplete cleavage. Different degrees of incomplete cleavage can occur (Hahn et al., 2010). In addition, holoprosencephaly can be associated midline defects of the face, eyes, nose, and mouth. These forms of holoprosencephaly are often described as primary holoprosencephaly. Other bodily abnormalities may occur in association with holoprosencephaly, and conditions with abnormalities are referred to as secondary holoprosencephaly.

Syndromic forms of holoprosencephaly occur in association with trisomy 13, 18, or 22. Other syndromic disorders that may be associated with holoprosencephaly include Smith–Lemli–Opitz syndrome, associated with 7-dehydrocholesterol deficiency and Hartsfield syndrome that occurs due to specific mutations in the fibroblast growth factor gene FGFR1 (Kruszka and Muenke, 2018).

An important feature of primary holoprosencephaly due to specific gene mutations is the great degree of intrafamilial variability, even in family members with the same mutation. An individual with severe manifestations of holoprosencephaly may have a family member with the same mutation who manifests a microfeature of the disorder, e.g., single central incisor or choanal abnormality.

The population frequency of holoprosencephaly in fetal life is estimated as 1 in 300, while the frequency in live-born individuals is 1 in 10,000.

Mutations in single genes leading to holoprosencephaly

In 1996, Roessler et al. reported that dominantly inherited deleterious mutation in sonic hedgehog (SHH) led to holoprosencephaly. SHH is a secreted protein that binds to a specific cellular receptor patched (PTCH). Prior to binding to PTCH, SHH protein is modified and binds to lipids palmitate and cholesterol. Binding of SHH to PTCH releases PTCH from its binding to SMO. This allows SMO to trigger downstream signaling and activation of transcription factors including ZIC2 and SIX3 and to impact TGIF transcription regulator.

Multiple different mutations in SHH, SIX3, and ZIC2 have been found to be associated with holoprosencephaly. At least three different mutations in TGIF are known to lead to holoprosencephaly (Mercier, 2011).

In recent years, mutations in a number of different genes have been found to lead to holoprosencephaly; these include the following:

- STIL, centriolar assembly protein

- DISP1, dispatched RND transporter family member 1, involved in cellular differentiation
    - FAT1, atypical cadherin, involved in cellular communication
    - NDST1, *N*-deacetylase and *N*-sulfotransferase 1
    - CDON, cell adhesion protein

Some mutations in these genes may manifest autosomal dominant effects. However, there is also evidence for mutations that manifest autosomal recessive or digenic inheritance and oligogenic effects.

It is important to note that defects in genes in the ciliary pathway and in the PCP pathway can lead to brain developmental defects.

There is evidence for interactions between the SHH pathway and the ciliary and PCP pathway.

Kruszka et al. (2019) noted that the cohesin pathway gene mutations also led to holoprosencephaly. It is important to note the fetal ultrasound and magnetic resonance imaging can readily detect the abnormalities associated with holoprosencephaly.

There some evidence that environmental factors, including ethanol, may contribute to holoprosencephaly pathogenesis.

Sonic hedgehog pathway in neurodevelopment and circuit formation

Garcia (2018) reviewed updated information on the SHH pathway and information on the roles of SHH in neurodevelopment and in the mature brain.

The SHH gene encodes a secreted protein that was found to undergo significant posttranslational modifications. A number of different gene products act as receptors or coreceptors for SHH; these include BOC and CDON, and both are cell adhesion proteins and glypicans. The expression of specific coreceptors for SHH in the nervous system was shown to differ in different cells and during different stages of neurodevelopment.

Garcia et al. noted that the SHH signaling pathway included PTCH receptor, SMO as coreceptor, and GLI as transcription factor. In the canonical pathway, PTCH1 represses SMO. This repression is negated when SHH binds PTCH, this then allows SMO activation and modulation of the GLI transcription factor.

Garcia et al. noted that SHH signaling regulates neural progenitor cells and ultimately influences the ratio of neurons to glial cells. They noted that in the postnatal brain, neuronal stem cells persist in the subventricular zone and in the dentate gyrus, and at these sites, SHH signaling continues.

There is evidence that secreted SHH plays a role in circuit formation by facilitating growth and guidance of axons and SHH has also been to be expressed in presynaptic terminals. Expression of SHH or of the coreceptor BOC plays roles in synapse formation.

Garcia et al. noted evidence that expression of some synaptic proteins is regulated by SHH. Gli transcription factor was found to be present in a number of different brain regions, including the hypothalamus and cortex.

Hydrocephalus

Verkman (2013) noted the possibility that aquaporins AQ1 and AQ4 play roles in hydrocephalus, associated with the accumulation of excessive fluid in the brain, particularly in the ventricles. Verkman et al. reported that aquaporins assemble in membranes as tetramers and they may aggregate to form large complexes. They transfer water across membranes in response to osmotic gradients. Some aquaporins also transfer glycerol. Aquaporins are expressed in different cell types and tissues.

Trillo-Contreras (2019) reported the presences of aquaporin AQ4 in ependymal cells, in pericapillary astrocyte foot processes, and in the glia limitans, outermost layer of neural tissue. Aquaporin 1 was reported to be present in the epithelial cells of vessels in the choroid plexus. They postulated that aquaporins play a key role in maintenance of cerebrospinal fluid (CSF) homeostasis.

Hydrocephalus and gene mutations

Kousi and Katsanis (2016) reviewed the genetics of hydrocephalus and noted that more than 100 different genes had been implicated as playing a role in this disorder. However, they emphasized that Mendelian forms of the disorder represented only a small fraction of this genetic burden.

In considering the anatomy and physiology of hydrocephalus, these authors noted that CSF is secreted into the lateral ventricles, and it passes through the foramen of Monro into the third ventricle and then through the Sylvian aqueduct into the fourth ventricle. Thereafter the CSH leaves the ventricular system and passes to the corticosubarachnoid space and the spinal subarachnoid space. Ultimately, it drains into venous sinuses.

Hydrocephalus can result from obstruction to flow. However, it may result through other mechanisms. Hydrocephalus is usually associated with increased intracranial pressure except in aged individuals where ventricular spaces are enlarged.

One of the most studied genetic forms of hydrocephalus is due to mutations in the X-linked gene L1CAM. Other genes with mutations that lead to hydrocephalus include the following:

- L1CAM, cell adhesion molecule, a glycoprotein
  - AP1S2, adaptor-related protein complex 1 subunit sigma 2
    - MPD2 (MATR3), matrin 3 stabilizer of mRNA
- CCDC88C, coiled-coil domain containing 88C, a negative regulator of the Wnt signaling

Kousi and Katsanis noted that the L1CAM mutations were reported to lead to stenosis of the aqueduct of Sylvius and may be associated with other brain malformations including agenesis of the corpus callosum.

They also noted that hydrocephalus can occur in association with muscular functional defects and intellectual impairment due to mutations in the gene POMT1 and POMT2, giving rise to Walker Warburg syndrome. This syndrome may also occur as a consequence of mutations in the following gene products:

- FKTN, fukutin, a transmembrane glycosyl transferase
  - FKRP, fukutin-related protein involved in glycosylation
    - POMK, protein O mannose kinase

Hydrocephalus may occur in disorders of glycosylation, e.g., ALG1 or in ciliopathies, e.g., due to mutations in CEP20.

Regulatory gene defects leading to neurodevelopmental impairments

Cornelia de Lange syndrome

Cornelia de Lange syndrome is an example of a genetic disorder with features that include impaired neurodevelopment. It is due to defects in genes that encode products that play key roles in gene transcription. This syndrome was first described by Cornelia De Lange, a pediatrician in 1933 (see Kline, 2018). This syndrome is characterized by distinct facial features, growth disturbance, and limb abnormalities and may also feature global developmental delay. Some patients with this syndrome are reported to manifest autistic behaviors. It is important to note that comprehensive studies by a number of different investigators have provided evidence that this disorder should be considered as a spectrum that includes a classic phenotype, referred to as classic CdLs, a nonclassic phenotype and also disorders that share limited clinical features with CdLs phenotype. Molecular studies have revealed that impaired genes in the CdLs gene spectrum encode products that function in chromatin regulation and particularly in gene products that are related to the formation and function of the cohesin complex. The cohesin complex forms a ring that facilitates positioning and stabilization of chromatin loops that bring together enhancers and promoters to regulate gene expression. The first molecular defect in CdLs was identified in the gene that encodes NIPBL, a protein involved in loading cohesin onto DNA. Studies have also revealed that cohesin has additional functions and plays roles in chromosome segregation, genome stability, genome organization, and regulations of gene expression. Yuan (2019) compiled a comprehensive list of genes that have been identified as having pathogenic mutations or structural defects that lead to the CdLs spectrum. These genes encode proteins important in cohesin loading, structural cohesin proteins, and regulatory cohesin proteins. These include the following:

- NIPBL, cohesin loading factor
  - SMC1A, structural maintenance of chromosomes 1A
    - SMC3, structural maintenance of chromosomes 3
    - RAD21, cohesin complex component
    - BRD4,bromodomain containing 4
    - HDAC8, histone deacetylase 8
- ANKRD11, ankyrin repeat domain 11

In disorders classified as having phenotypic features that overlap with CdLs, additional genes were reported to have pathogenic variants. These included the following:

- EP300, E1A-binding protein p300 that functions as histone acetyltransferase
  - AFF4, AF4/FMR2 family member 4 component of the positive transcription elongation factor

Gene mutations leading to CdLs were reported to most likely be new mutations. It is, however, important to note that mosaicism for CdLS causing mutations has been reported and gonadal mosaicism cannot be ruled out. This needs to be taken into account in genetic counseling.

The non–protein-coding genome and defects leading to cortical malformation and neurodevelopmental disorders

Perenthaler (2019) reviewed the non–protein-coding genome and enhancers and their roles in cortical malformations and neurodevelopmental disorders. They emphasized that regulatory elements in the genome interact over long distances with target genes. An aspect of the genome that needs to be taken into account relates to the organization of chromosomes into different domains within the nucleus, with inactive nonexpressing chromosomes located in the periphery of the nucleus in contact with nuclear lamina and active expressing chromosomes tending to be centrally located in the nucleus.

It is also important to note that enhancer elements for a specific gene promoter may be located at considerable distances from the promoter. Perenthaler et al. reported that a point mutation in a specific brain enhancer element SBE2 that is located 260 kb upstream of the CHH SHH locus was reported to lead to characteristic pathology of SHH defects. There was evidence that the mutation in the enhancer disrupted binding of a transcription factor.

Mendelian disorders of epigenetic machinery and neurodevelopmental disorders

Bjornsson (2015) reviewed Mendelian disorders of epigenetic machinery and characterized them according to which aspects of epigenetic modification were implicated. He described 44 Mendelian diseases that impacted the epigenetic machinery and noted that in 93% of these diseases, neurological dysfunction occurred.

Fahrner and Bjornsson, 2019 reported that although Mendelian disorders of epigenetic machinery are individually rare, collectively they are frequent causes of intellectual disability. These disorders manifest as a result of haploinsufficiency defects, i.e., defects in one member of the gene pair.

It is important to note that a number of different genes in the SMARC family have been implicated in causation of Coffin–Siris syndrome, characterized by mild to severe intellectual impairment, hypotonia, facial dysmorphism, and growth impairment. Although defects in a number of genes in the SMARC family have been associated with rhabdoid tumor, these tumors can sometimes arise in the central nervous system in children.

Chromatin remodelers

In the category of chromatin remodeler, defects were identified in the genes that are included in the following. The gene names are listed followed by gene product, function, and name of the disorder:

- CHD8, chromodomain helicase, DNA-binding protein 8, autism
  - CHD2, chromodomain helicase DNA-binding protein 2, epileptic encephalopathy
- CHD7, chromodomain helicase DNA-binding protein 7, Charge syndrome
- SMARCA2, actin-dependent regulator of chromatin, Nicolaides–Baraitser syndrome
  - SMARCB1, actin-dependent regulator of chromatin B1, Schwannomatosis
    - ARID1A, AT-rich domain, helicase and ATPase activity, Coffin–Siris syndrome
    - ATRX chromatin remodeler, ATRX syndrome
    - SCRP, SNF-related CREBBP activator protein, Floating–Harbor syndrome

Readers of epigenetic modifications

Listed in order of gene product name, function, and disorder:

- RAI1, retinoic acid induced 1 (rich in neurons) Smith–Magenis syndrome
  - MBD3, methyl-CpG-binding domain protein 3, autism intellectual impairment
- PHF6, PHD domain transcriptional regulator, Borjeson–Forssman–Lehmann syndrome (BFLS) cognitive impairment
  - ASXL1, transcription regulator, Bohring–Opitz syndrome
    - BRWD3, bromodomain and WD repeat X-linked intellectual impairment
    - MECP2, methyl CpG-binding protein, Rett syndrome

Erasers of epigenetic modification

Listed in order of gene product name, function, disease:

- SALL1, Spalt-like zinc finger transcriptional repressor, Townes Brock syndrome
  - PHF8, PHF domain histone lysine demethylase, X-linked Siderius intellectual disability
    - HDAC8, histone deacetylase 8, Cornelia de Lange syndrome 5
    - HDAC4, histone deacetylase 4, chromosome 2q37 deletion syndrome

Writers of epigenetic modifications

Listed in order of gene product name, function disease:

- KMT2D, lysine methyltransferase 2D, Kabuki syndrome 1, postnatal dwarfism, long palpebral fissures, intellectual disability
  - DNMT3B, DNA methyltransferase 3 beta, ICF syndrome 1*
    - ZBTB24, zinc finger and BTB domain containing 24, ICF syndrome 2 *ICF syndrome immunodeficiency centromeric instability, and facial anomalies
- KMT2A, lysine aminotransferase 2A, Wiedemann Steiner syndrome, short stature, hairy elbows, facial dysmorphism, developmental delay
  - EHMT1, euchromatic histone lysine methyltransferase 1, Kleefstra syndrome, hypertelorism. epileptic seizures
    - KMT2C, lysine-specific aminotransferase 2C, Kleefstra syndrome spectrum, hypotonia, microcephaly

- KANSL1, KAT8 regulatory NSL complex subunit, 1 Koelen de Vries syndrome, distinct facies, hypotonia, intellectual disability
- KAT6B, lysine acetyltransferase 6B, Ohdo syndrome, intellectual disability, blepharophimosis
- KAT6B, lysine acetyltransferase 6B, genito patellar syndrome, intellectual disability
- NSD1, nuclear receptor–binding SET domain protein 1, Sotos syndrome overgrowth, intellectual disability
- SETD2, SET domain 2 histone lysine methyltransferase, macrocephaly, overgrowth, intellectual disability
- EZH2, enhancer of Zeste 2, polycomb repressive complex 2 Weaver syndrome, overgrowth, distinct facies, developmental delay
- DNMT3A, DNA methyltransferase 3, alpha Tatton–Brown syndrome, tall stature, distinct facies, intellectual disability
- DNMT1, DNA methyltransferase 1, ADCAN syndrome ataxia deafness narcolepsy
- NSD2, nuclear receptor–binding SET domain protein 2, Wolf Hirschhorn syndrome, pre- and postnatal growth deficiency, distinct facies, seizure disorder
    - EP300, E1A-binding protein p300 transcription activator, Rubinstein–Taybi syndrome 2, growth deficiency, broad thumbs, facial dysmorphism intellectual disability
    - CREB-binding protein transcription activator, Rubinstein–Taybi syndrome 1

It is important to note that methyl groups must be available for methylation in specific epigenetic modifications (Fig. 7.1).

S-adenosyl methionine: methyl donor

FIGURE 7.1 **S-adenosyl methionine.** This is an important methyl donor.

Mediator

The main function of this complex is transcription regulation. Deleterious mutations in a number of different Mediator subunits have been identified in neurological and cognitive disorders. MED12 defects are associated with a broad spectrum of developmental defects that include X-linked cognitive impairment.

Polycomb repressive complex

Mutations in components of this complex have been reported in overgrowth syndromes. Cyrus (2019) reported that deleterious mutations in EZH2, a component of Polycomb repressive complex 2 (PRC2), led to Weaver syndrome. Manifestations included prenatal and postnatal overgrowth, advanced bone age, unusual facial features, and intellectual disabilities.

Components of the neuronal activity–dependent transcription pathways and neurodevelopmental disorders

Yap and Greenberg (2018) noted evidence for defects in the following activity-related transcription pathway. Listed in order of gene product function and associated disorder:

- CACNA1C, voltage-gated calcium channel, Timothy syndrome
    - RSK2, ribosomal S6 kinase, Coffin–Lowry syndrome

- CREBBP (CREB-binding protein), Rubinstein–Taybi syndrome
- SMARCA2, SWI/SNF-related chromatin modifier, Nicolaides–Baraitser syndrome
- EHMT1, euchromatic histone methyltransferase, Kleefstra syndrome
- KMT2C, lysine methyltransferase 2C, Kleefstra syndrome
- MECP2, methyl CpG-binding protein, Rett syndrome

Other important gene products in activity response gene expression include FOS (oncogene) AP transcription factor subunit, EGR1 (early growth response protein), and NPAS4 (neuronal protein transcriptional regulator).

Prenatal detection of fetal brain abnormalities

In 2019, Griffiths et al. published results of the MERIDIAN study that was designed to establish the accuracy of MRI in detecting fetal brain abnormalities. They noted that ultrasonography was primarily used in prenatal diagnosis in the United Kingdom, but that in utero magnetic resonance imaging had the potential to improve diagnostic detection of fetal brain abnormalities. Sixteen fetal medical centers in the United Kingdom participated in the collaborative MERIDIAN, and the primary analysis involved analyses of 570 fetuses. Results of the study led investigators to conclude that in utero MRI significantly improved accuracy of diagnosis of fetal brain abnormalities. Diagnostic accuracy was reported to be 68% using ultrasonography and 93% when in utero MI was used.

Hart et al. (2020) carried out follow-up postnatal studies on a subset of the MERIDIAN cohort. Postnatal imaging was carried out at 6 months and in some cases at 3 years, and developmental assessments were carried out. The diagnostic validity of the in utero MRI remained high. They noted, however, that it still lacked accuracy with respect to prediction of developmental outcome.

Tuberous sclerosis brain lesions

Hasbani and Crino (2018) in a review of tuberous sclerosis noted that it remains a clinical diagnosis. They noted that TSC1 or TSC2 gene mutations were not identified in 10%–15% of clinical cases of tuberous sclerosis. They documented the following brain lesions in this disorder: cortical dysplasias including tubers and white matter radial migration lines, subependymal nodules, and subependymal giant cell astrocytomas. Neurological and neurobehavioral features included epilepsy, infantile spasms and intellectual disability, autism anxiety, impaired control of behavior attention deficit hyperactivity disorder (ADHD), and sleep disorders.

In 2013, van Eeghen et al. reported results of evaluation of 30 patients with tuberous sclerosis. They determined that the presence of radial migration lines on diffusion tensor imaging (DTI) was the most consistent white matter abnormality in these patients. Furthermore, they determined that presence of radial migration lines was shown to be strongly correlated with age of onset of seizure, with intelligence quotients, and with levels of autistic behavior.

Marcotte (2012) described the cytoarchitectural alterations in tuberous sclerosis, including cytoarchitectural changes in areas outside the cortical tubers. They noted tubers manifested abnormal cortical lamination, dysmorphic neurons, and giant cells. Other cortical abnormalities found include focal dyslamination, heterotopic neurons, small collections of giant cells, and individual giant cells.

In 2018, de Vries described results from the TOSCA international natural history study of tuberous sclerosis. Their study gathered data from 2216 participants from 31 countries. Categories of manifestations analyzed included behavior, neuropsychological defects, and psychiatric disorders. Behavioral disorders were reported in 10% of patients and included sleep disorders, impulsivity, and mood swings; attention deficit disorders were diagnosed in 19.1% of cases, anxiety in 9.7%, and depression in 6.1%. Intelligence quotient studies in 835 individuals revealed normal IQ in 44%; intellectual disabilities were classified as mild in 28.1% of cases, moderately severe in 13.1%, severe in 9.3%, and profound in 3.1%. Academic difficulties were reported in 58.8%. In TSC2 mutation carriers, academic difficulties and neuropsychologic defects showed a higher frequency. Neural tuberous sclerosis manifestations also occurred in individuals in whom mutations were not found in TSC1 or TSC2 genes.

Inborn errors of metabolism and brain defects

Schiller et al. (2020) reviewed inborn errors of metabolism and neurometabolic diseases that led to neuronal migration disorders. They noted that manifestations of these disorders can include psychomotor developmental delay, intellectual disability, and epilepsy. It is also important to note that in these disorders, body systems beyond the brain may also be involved. Schiller et al. particularly emphasized the roles of peroxisomal biogenesis defect, peroxisomal oxidation defects, and congenital disorders of glycosylation in neurometabolic causes of impaired neuronal migration. They also noted specific organic acidurias that are associated with neuronal migration disorders. They documented the specific forms of neuronal migration disorders associated with the neurometabolic disorders. They include lissencephaly, agyria, and cobble stone forms. Schiller et al. noted that the cobblestone form of lissencephaly was characterized by diminished sulci and neuronal overmigration through the basal membrane leading to collections of neurons in the subarachnoid space. Importantly the cobblestone form of lissencephaly was noted to occur in cases with O-glycosylation metabolic defects. Schiller et al. noted that PMG could result from gene defects from hypoxia and infection and was sometimes associated with peroxisomal defects. Nodular heterotopia and subcortical band heterotopia were noted to often occur as a result of mutations that impacted cytoskeletal structures.

Schizencephaly was described as a malformation in the form of a cleft that extended from the pia to the ependyma. The molecular etiology was not known.

Schiller et al. summarized key metabolic defects, leading to cortical migration defects. They included Zellweger syndrome and peroxisomal oxidation defects that were noted due to impaired peroxisomal biogenesis that can arise as a result of pathological mutations in any one of 13 PEX genes or due to pathogenic mutations in ACOX1 acyl-CoA-oxidase, the first enzyme of the fatty acid beta-oxidation pathway. These disorders lead to increased biochemical levels of very-long-chain fatty acids, decreased levels of plasmalogens, increased levels of phytanic acid and pristanic acid, and increased levels of dehydrocholestanoic acid zellweger syndrome.

O-glycosylation defects that negatively impact neuronal migration were noted to occur in the following forms of disorders that include Walker Warburg syndrome, Fukuyama syndrome, and Muscle brain eye disease; note there are 14 different forms of that disorder.

Other O-glycosylation defects that may be associated with neuronal migrations disorders are due to defect in acetylglucosaminyl transferase–like proteins.

Other biochemical and organic acid disorders associated with neuronal migration defects include the following:

- Pyruvate dehydrogenase deficiency
  - Glutaric aciduria due to defects in ETFDH1, ETFA, and ETFB
    - Nonketotic hyperglycemia defects in aminomethyl transferase (AMT)
    - Defects in glycine decarboxylase (GLDC) and the glycine cleavage system
    - D2HGDH D2 hydroxyglutaric aciduria
    - Fumaric aciduria fumarase deficiency (FMRD)

Neuronal migration disorders may also occur in Smith–Lemli–Opitz syndrome due to 7-dehydrocholesterol dehydrogenase (DHCR7) deficiency.

Neuronal migration defects may also occur in Menkes syndrome due to aberrant copper metabolism due to copper transporting protein ATP7A defects.

Creatine deficiency syndromes

These syndromes arise as a result of impaired creatine synthesis or impaired creatine transport into the brain. Fons and Campistol (2016) reviewed effects of creatine deficiency on the central nervous system. They noted that creatine (alpha methyl guanidine acetic acid) occurs primarily in muscle. Creatine can be derived from the diet or from synthesis in the human body. Key enzymes involved in creatine synthesis include arginine glycine aminotransferase (AGAT) and guanidino-acetate methyl transferase (GAMT). The first step in creatine synthesis involves interaction between arginine and glycine that in the presence of AGAT generates guanidinoacetate and ornithine. Guanidinoacetate in the presence of S-adenosine–methionine undergoes methylation through activity of GAMT to generate creatine. Creatine can undergo phosphorylation and be converted to creatine kinase. Fons and Campistol noted that creatine can be synthesized in muscle and in a number of different tissues including brain. In brain, creatine can be synthesized in neurons and in oligodendrocytes. There is evidence that limited transfer of creatine across the blood brain barrier can occur, which is dependent on activity of a creatine transporter CRTR, also known as SLC6A8 solute carrier. The gene that encodes AGAT maps to chromosome 15q11.2; the GAMT encoding gene maps to chromosome 19p13.3; the SLC6A8 solute carrier is encoded by a gene on Xq28. They noted that in the brain, creatine is involved in processes related to membrane potential and ion gradient homeostasis and it impacts neuronal transmission and signaling and that it impacts GABA receptor function postsynaptically. There is also evidence that SLC6A8 plays roles in release of creatine from the synapse.

Fons and Campistol noted that creatine deficiency syndromes can result from defects in creatine synthesis or defects in creatine transport. Patients do not manifest muscle problems and neurologic symptoms primarily occur. GAMT (guanidinoacetate *N*-methyltransferase)-deficient patients were found to have epilepsy, psychomotor retardation, and movement disorders. Manifestations of AGAT (GATM) (glycine amidinotransferase) deficiency include moderate intellectual disability and language difficulties. SLC6A8 deficiency was associated with intellectual disability, language problems, and epilepsy. Biochemical testing revealed that in AGAT, deficiency levels of guanidinoacetate are decreased in urine. In GAMT deficiency, urinary levels of guanidinoacetate are increased, and levels of this molecule are normal in SLC6A8 deficiency. Brain proton magnetic resonance imaging reveals decreased or absent creatine peaks in key areas, notably the periventricular white matter, cerebellum, parietal, and occipital cortex. Fons and Campistol reported that there is evidence that oral supplementation with creatine is helpful in cases with AGAT or GAMT deficiency but not in cases with brain creatine transport deficiency due to SLC6A8 defects.

Lysosomal dysfunction leading to neurodevelopmental and cognitive disorders

For a number of years, studies on children with neurodevelopmental and neurocognitive disorders revealed that in a significant number of cases, lysosomal function was abnormal, leading particularly to lysosomal storage diseases. Studies in prior decades provided evidence for the important roles of lysosomal acid hydrolase enzymes in degradation of material taken up by endosomes and autophagosomes. In recent years, information on lysosomal function has expanded. New evidence has emerged regarding important functions of lysosomal membrane proteins, including LAMPs (lysosomal-associated membrane proteins) and LIMPs (lysosomal inner membrane proteins) that are often composed of tetraspanins. LAMPs play roles in delivery of phagosomes into lysosomes. LIMPs form a coating on the inside of lysosomes. Lysosomal membrane proteins are reported to be highly glycosylated, and this protects the lysosomal membranes from digestion by the hydrolytic enzymes present in lysosomes (Schwake, 2013). Protons need to be transferred into lysosomes and other organelles to ensure the acid pH and the vacuolar H(+)-ATPase (v-ATPase) complex plays an important role in this transfer. In humans, different subunits compose this complex. In 2018, Fassio et al. reported that de novo mutations of the ATP6V1A gene cause developmental encephalopathy with epilepsy. In 2019, Zhao et al. reported that A subunit of V-ATPases, ATP6V1B2, underlies the pathology of intellectual disability. Lysosomal membrane proteins include cystinosin that is defective in a specific disease. The lysosomal membrane protein LMBRD1 transfers vitamin B12 across lysosomal membranes and is defective in specific diseases (Fettelschoss, 2017). A lysosomal membrane protein, defined as an LIMP, is SCARB2 that transports the enzyme glucocerebrosidase into lysosomes, and defects in this protein can lead to disease with neurological manifestations (Gonzalez et al., 2014). Sialic acid needs to be transported out of the lysosomes, and defects of this transport lead to Salla disease (Gahl, 1987).

There is evidence for interactions of the lysosomal membrane with MTORC1. This interaction is thought to be important in nutrient sensing and in determination as to whether the cell is in an anabolic or catabolic state. MTORC1 function responds to the cellular nutrition state in initiating processes that lead to cell proliferation (Carroll and Dunlop, 2017).

A key transcription factor that influences endosome lysosome and lysosome autophagosome processes is TFEB. When cells are in a stress situation, lysosomes release TFEB from their surfaces. TFEB then makes its way to the nucleus and activates transcription of enzymes that enhance lysosome biogenesis and lysosome function. TFEB release from lysosomes under low nutrient conditions was also reported to stimulate mitochondrial function and particularly fatty acid oxidation (Settembre et al., 2013).

H(+)-ATPases(V-ATPase) proton pump

Cotter (2015) reviewed the structure, function, and regulation of the H+ ATPase proton pumps. These structures play important roles in intracellular organelles and at the outer cell membrane in maintenance of appropriate pH. They are particularly important in acidification of endosomes and lysosomes. In endosomes, acidification brought about by the H+ ATPase proton pumps enables release of components of membranes and enables fusions and subsequent digestions in lysosomes. Under low pH conditions, lysosomal enzymes are released from their bound mannose-6-phosphate, and enzymes are then ready to act in the degradation of complex molecules in the lysosomes. Vesicles active in secretory pathways were reported to harbor V-ATPase proton pumps that alter conditions and promote proteolysis that can act to release active hormones from hormonal precursors. Cotter et al. noted that V-ATPase proton pumps are also present on the cell membranes of particular cell types, e.g., osteoclasts. In those cells, low pH conditions are essential for release of calcium. Key components of the H(+)-ATPases(V-ATPase) proton pump are V1 and V0 domains. The V1 domain is composed of seven subunits and is required for ATP hydrolysis. The Vo domain is composed of five subunits and is involved in the translocation of protons. In addition, there is a central stalk composed of three subunits. The central stalk rotates a proteolipid ring, and this promotes proton transport. ATP hydrolysis drives rotation of the central stalk ring as protons enter the lumen of the structure. Cotter et al. noted that disassociation and reassociation of the V1 and VO domains play key roles in regulation of the proton pump. Dissociation and reassociation are influenced in part by glucose concentration. The assembly of the domains is influenced by levels of growth factors and by presence of mTORC1, phosphatidyl inositol kinase (PI3K) and glucose. Low glucose concentrations promote disassembly of the two domains of the proton pump. There is evidence that regulation and availability of pump components is partly controlled by gene expression under influence of the TFEB transcription factor. Cotter et al. noted that H(+)-ATPases(V-ATPase) also play roles in cell signaling and in cell membrane fusion processes. It is interesting to note that the 17 subunits that form the Vo and V1 domains are encoded in humans by 17 different genes, each located on a different human chromosome.

There is evidence that mutations that impact specific V-ATPase subunits play roles in specific lysosomal storage disease and in neurodegenerative diseases (Colacurcio and Nixon, 2016; Bagh et al., 2017). Colacurcio and Nixon (2016) reviewed the roles of V-ATPase defects in neurodegeneration. They noted that defects in lysosomal functions had been found to be causative of more than 40 neurodevelopmental and neurodegenerative diseases in childhood and that there was also evidence that defective lysosomal function also contributed to late-onset neurodegenerative diseases.

Colacurcio and Nixon suggested that suboptimal lysosomal function tipped toward more serious functional defects by aging-related impairments. They particularly stressed that reduced capacity for lysosomal acidification that led to impaired degradation of macromolecular complexes likely played important roles in late-onset neurodegeneration.

They noted four pathways for delivery of macromolecular complexes to lysosomes, macroautophagy, microautophagy, chaperone-mediated autophagy, and the endocytic pathway. Macroautophagy was particularly involved in delivery of degraded organelles and large macromolecular aggregates to lysosomes while endocytosis delivered smaller complexes to lysosomes. They stressed that maintenance of lysosomal pH between 4.2 and 5.3 was necessary for degradation of most material transported to lysosomes. However, some complexes could be degraded at pH 6.0. Key structures essential fo maintenance of low lysosomal pH were found to be the V-type ATPases. They emphasized the key roles of V-ATPase in nutrient sensing and calcium signaling. Nutrients sensed by V-ATPases included amino acids. When cellular concentrations of amino acids were low, assembly of V-ATPase was accelerated to expedite macromolecular degradation in lysosomes to release amino acids from proteins.

They noted the importance of RAG GTPase and Ragulator in nutrient sensing. Currently, Ragulator is defined as LAMTOR components late endosomal/lysosomal adaptor, MAPK, and mTOR activators. V-ATPase functions were also noted to be influenced by calcium levels. A number of disorders have been associated with defective functions of specific V-ATPase subunits; among these were diseases characterized by neurodegeneration.

V ATPase subunit diseases

- Osteopetrosis with neurodegeneration ATP6AP2 ATPase H+ transporting accessory protein 2
  - X-linked Parkinsonian disease ATP13A2 ATPase cation transporting 13A2
    - Kufor–Rakeb Parkinsonian syndrome ATP6V1A ATPase H+ transporting V1 subunit A

- Epileptic childhood encephalopathy ATP6V1B1 ATPase H+ transporting V1 subunit B1
- Renal tubular acidosis with deafness

Colacurcio and Nixon noted that it is important to consider mutations in the VATPase subunits and also mutations in gene products that impact the regulation of the V-ATPases.

Ceroid lipofuscinoses

Mukherjee (2019) emphasized the two pathways through which materials enter lysosomes: the endosome–lysosome pathway and the autophagosome–lysosome pathway. They noted that key regulatory factors in the vesicular trafficking processes include Ras GTPases. Importantly damaged mitochondria can also fuse with endosomes and Ras GTPase, which are also important in these processes. Mukherjee et al. noted that impaired membrane fusion of endosomes with lysosomes and autophagosomes with lysosomes contribute to the pathogenesis of a number of neurodegenerative diseases including Batten disease and ceroid lipofuscinoses.

These disorders are characterized by the accumulation of autofluorescent material in lysosomes in neurons and also by neuronal apoptosis leading to neuronal degeneration. Thirteen different genes have been shown to have undergone pathogenic mutations leading to this disorder. These disorders are associated with autophagy dysregulation. They noted that functional defects resulting from these mutations can be grouped into four different categories.

Group 1 defects involve abnormal proteins and enzymes within lysosomes including the following:

- CLN1 PPT1, palmitoyl-protein thio-esterase 1, is involved in the catabolism of lipid-modified proteins
  - CLN2, tripeptidyl peptidase 1, cleaves N-terminal tripeptides from substrates
- CLN 5 intracellular trafficking protein involved in degradation of posttranslationally modified proteins
- CLN10, cathepsin D, member of the A1 family of peptidases
- CLN13, cathepsin F cysteine proteinase, major components of the lysosomal proteolytic system

Group 2 defects occur in genes that encode proteins include in membranes and include the following:

- CLN3 batenin, lysosomal/endosomal transmembrane protein
  - CLN7 MFSD8, major facilitator superfamily domain containing 8, has a transporter domain
    - CLN12 ATPase cation transporting 13, transports inorganic cations as well as other substrates
    - CLN6 transmembrane endoplasmic reticulum protein
    - CLN8 localizes to the endoplasmic reticulum (ER), may recycle between the ER and ER–golgi

Group 3 defects occur in genes that encode soluble proteins:

- CLN4 DNAJ heat shock protein family, plays a role in membrane trafficking and protein folding,
  - CLN14 KCTD7 plays a role in membrane trafficking and protein folding

Group 4 genes encode secretory proteins:

- CLN11 GRN precursor is cleaved and glycosylated

The protein CLN1, palmitoyl-protein thiosterase 1, cleaves thioester linkages in the nervous system. Children with defects in this gene product were noted to be normal before 11–18 months of age. They then began to have manifestation of psychomotor retardation. Subsequently, blindness developed due to retinal degeneration. Different degrees of severity of this disorder have been reported. In severely affected cases, death occurs by 4 years of age. Mukherjee et al. noted that there is evidence that altered thioester cleavage impacts a subunit of the V-ATPase proton pump that is involved in lysosomal acidification. Defective thioester cleavage was also reported to impair protein folding. This subsequently leads to the accumulation of excess unfolded proteins and endoplasmic stress.

CLN2 deficiency is described as late infantile form of ceroid lipofuscinosis; retinal degeneration was noted to be prominent in this disorder.

CLN3 is sometimes referred to as juvenile Batten disease or Vogt–Spielmeyer disease. Onset may occur between 6 and 10 years of age; the first manifestation is often visual impairment due to retinitis pigmentosa.

CLN4 disease may lead to a disease in adults referred to as KUFS disease that is associated with myoclonic epilepsy and dementia.

CLN7 was reported to lead to late infantile disease that impacts the midbrain and hippocampus.

CLN8 was reported to be associated with two phenotypes: cognitive impairment in young children and also with juvenile onset epilepsy.

CLN10 was reported to be associated with congenital neuronal damage, leading to respiratory insufficiency and severe seizures in young infants and early death.

CLN11 homozygous mutations that result in complete loss of granulin precursor expression (GRN) lead to the neurodegenerative lysosomal storage disorder neuronal ceroid lipofuscinosis (NCL). This disorder is characterized by rapid visual loss, seizures, and cerebellar ataxia.

Mole (2019) described neuronal ceroid lipofuscinoses as monogenic diseases leading to neurodegeneration that presented with seizures visual failure and progressive cognitive decline. They noted, however, that different gene defects lead to disorders that differed in rate of disease progression.

Brain neuroimaging in ceroid lipofuscinosis

Mole et al. noted that early in the course of ceroid lipofuscinoses, changes could not be detected on MRI. However, in subsequent stages, diffusion tensor imaging revealed white matter tract disorganization and atrophy. In later stages of the disease, gray matter changes could be detected. They noted that in some cases, EEG studies were shown to reveal changes even in early stages, and characteristic spike wave patterns were described. Histopathological changes, in addition to ceroid lipofuscin deposition, included widespread neuronal loss and microglial activation. In the retina, photoreceptor cell damage was evident. Mole et al. noted that in addition to clinical trials of enzyme replacement therapy, a limited number of gene therapy trials had been initiated.

Adult neurodegeneration related to ceroid lipofuscinoses

Individuals heterozygous for specific progranulin (GRN) mutations have been reported to be at increased risk for frontotemporal dementia that is also characterized by the presence of TDP inclusion. Almeida (2016) reported that some individuals heterozygous for pathogenic GRN mutations presented with manifestations of corticobasal syndrome.

Lysosomal storage diseases

These are defined as diseases that arise due to failure of lysosomes to break down complex macromolecules, leading to abnormal levels of accumulation of the macromolecules and eventually to breakdown of lysosomes and accumulation of undegraded or partly degraded material in the cells followed by impaired cellular function and inflammatory reactions.

Symptoms that occur in these disorders impact a number of different organs; hepatomegaly and skeletal dysplasia occur commonly in a number of lysosomal storage diseases. Here, concentration will be primarily on neuropathic lysosomal storage diseases that impact neurodevelopment, cognition, and behavior. Most of these disorders follow autosomal recessive forms of inheritance. It is important to note that there are examples of diseases that fit into the clinical pathology phenotype of lysosomal storage diseases but where specific gene mutations have not yet been identified.

Children with lysosomal storage diseases may first present with manifestation of developmental delay, behavior defects, and motor weakness. Pastores and Maegawa (2013) reviewed neuropathic lysosomal storage diseases. They noted that molecular pathogenesis of these disorders may result from loss-of-function mutations that inactivate specific catalytic sites of enzymes or predispose the gene transcript to non—sense-mediated decay. In addition, mutations can lead to protein misfolding and subsequent accumulation of the misfolded protein products in the endoplasmic reticulum, or disposal of unfolded proteins in the ubiquitin proteasome system.

Vitner (2010) reported that unprocessed macromolecules may also lead to downstream pathologic events. These can include accumulation of secondary metabolites, oxidative stress, and inflammatory reactions. Interference with cell signaling events may also occur.

Pastores and Maegawa (2013) documented specific clinical neurologic manifestations in lysosomal storage diseases. These included visual defects, deafness, myoclonic seizures, peripheral neuropathy, ataxia, and stroke-like episodes. Visual defects included optic atrophy, retinitis pigmentosa, and cataracts and could occur in galactosialidoses, mucopolysaccharidoses, mucolipidoses, and oligosaccharidoses.

Neuroimaging findings in lysosomal storage diseases

These were reviewed by Fagan in 2017. They noted that imaging may reveal enhancement of cranial nerves. Leukodystrophy may be detected in certain lysosomal storage diseases including metachromatic leukodystrophy. In Krabbe disease due to deficiency of galactocerebroside, beta-galactosidase demyelination and gliosis were found. They noted that in Fabry disease, vascular events occur and infarcts associated with small vessels may occur in some patients. In neuronal ceroid lipofuscinoses, MRI may demonstrate diffuse cerebral and cerebellar atrophy, and white matter hyperintensities may also occur.

In mucopolysaccharidoses, MRI findings correlated with disease severity. Fagan noted that white matter lesions occur that may be related to delayed myelination of demyelination. Meningeal thickening and hydrocephalus could be noted in later stages. In gangliosidoses that include Tay Sachs disease, cerebellar atrophy may be noted.

Neurocognitive and behavioral functions in mucopolysaccharidoses

Shapiro (2017) presented aspects of neurocognitive and behavioral functions in individuals with mucopolysaccharidosis. They noted that progressive deterioration of cognition is a sensitive indicator of disease progression. Disease manifestations included cognitive difficulties, language and speech difficulties, behavior abnormalities, and sleep problems. They also noted that some patients may develop seizures.

Aspects of therapy in lysosomal storage disorders

Platt (2018) reviewed lysosomal storage diseases, noting that 70 diseases in this category have now been described and considered aspects of therapy. They noted that some forms of these diseases are treated with enzyme replacement and small molecule therapies that included substrate reduction agents and chaperone agents. They noted that some progress was also being made toward gene therapy.

Toledano-Zaragoza and Ledesma (2019) emphasized that in lysosomal storage diseases, cellular compartments beyond the lysosomes are compromised. They particularly drew attention to oxidative stress and calcium imbalance. They also noted the importance of chronic inflammatory response in these disorders.

Organic acid disorders

Organic acids are derived primarily but not exclusively from metabolism of amino acids, particularly branched-chain amino acids. Detection of organic acids in body fluids has been greatly facilitated by development of mass spectroscopy techniques, particularly gas chromatography combined with mass spectroscopy (GC/MS).

Villani (2017) reported that more than 65 different inborn errors of metabolism have been associated with abnormalities in concentrations of specific organic acids. Specific organic acidurias that occur relatively frequently and are well studied include methylmalonic aciduria, propionic aciduria, maple syrup urine disease, and isovaleric aciduria. Organic acidurias and organic acidemias can also occur in consequence of defects in metabolism of pyruvate and ketones.

Schillaci (2018) noted that children with these disorders may first present with the clinical condition acidosis or anion gap. They noted that a number of these disorders may first be detected in newborn screening programs. However, in particular locations, the numbers of disorders and precise types of disorders screened for may differ. Five conditions commonly screened for in newborns include methylmalonic aciduria, propionic aciduria, maple syrup urine disease, and isovaleric aciduria; other conditions screened for include glutaric aciduria, type 1 beta ketothiolase deficiency, aceto-acetyl-CoA thiolase, mitochondrial (ACAT1) deficiency, biotinidase deficiency, holocarboxylase deficiency, 3-methylcrotonyl CoA carboxylase, and 3-MCC deficiency.

Glutaric aciduria type 1 is an abnormality of lysine metabolism; clinical abnormalities may not present in infants but may present later in childhood and include complex movement disorders due to striatal damage. Schillaci reported that children with glutaric aciduria may be at risk for subdural hematomas.

The primary clinical manifestation in ACAT1 deficiency is ketoacidosis. Neurologic manifestations occur in untreated biotinidase deficiency seizures, hypotonia, ataxia, developmental delay, vision problems, and hearing loss.

Holocarboxylase deficiency is also due to biotin defects; patients can present with ketoacidosis, hyperammonemia, dermatologic manifestations, and developmental delay.

3-Methylcrotonyl-CoA carboxylase (3MCC) deficiency can lead to defects in leucine metabolism, may present with feeding difficulties, and lead to hypotonia, episodes of vomiting, and seizures.

Schillaci noted that in specific forms of organic acidemias, mitochondrial metabolism may also be disrupted. In each organic acidemia, the particular types of organic acids that are altered have been identified, and these data can aid in diagnosis. In addition, levels of specific forms of carnitine derivatives may be altered.

Inborn errors of carnitine metabolism may present early in life or may occur following a period of fasting or acute illness. Almannai (2019) reported that carnitine inborn errors of metabolism may arise due to disorders of carnitine biosynthesis, carnitine transport, or as a result of defects mitochondrial carnitine—acylcarnitine cycle. Carnitine palmitoyltransferase I (CPT1A) is a mitochondrial enzyme, and deficiency can lead to form of disrupted carnitine metabolism, which may lead to encephalopathy characterized by seizures and coma.

CPT1C deficiency was reported to be associated with spastic paraplegia. Carnitine—acylcarnitine translocase (CACT) deficiency is another form of mitochondrial-related disruption of carnitine metabolism. It can present with hypoketotic hypoglycemia and may be associated with cardiomyopathy, seizures, and coma. A number of these disorders respond to carnitine supplementation. Carnitine deficiency may also occur secondary to other inborn errors of metabolism or secondary to administration of certain medications, e.g., valproic acid.

Treatments of organic acidemias

Schillaci (2018) noted that for a number of organic acidemias due to defects in amino acid metabolism, specific dietary formulas have been developed that eliminate the amino acid that cannot be metabolized. Specific vitamins act as cofactors in organic acid metabolism and can also be administered to reduce manifestations of disorders. These include vitamin B12 that may be advantageous in some cases of methylmalonic acidemia. Biotin administration is important in biotinidase deficiency and in multiple carboxylase deficiency. Dietary changes and carnitine administration have proved useful in some cases with glutaric aciduria.

Inborn errors in pyruvate metabolism and in ketone metabolism

Pathogenic mutations in components of the pyruvate dehydrogenase complex or pyruvate carboxylase lead to defects in pyruvate metabolism. Schillaci reported that children with these disorders present with lactic acidosis.

Neurologic manifestations include developmental delay, hypotonia, seizures, and encephalopathy. Specific brain imaging changes that can occur in these disorders include cystic changes, corpus callosum defects, and ventriculomegaly. They noted that ketogenic diet may be useful for treatment of seizures in these cases.

Defects in ketone metabolism may occur due to inborn errors of metabolism. Key enzymes in this category include the following:

- SCOT (OXCT1) 3-oxoacid CoA-transferase 1, a mitochondrial matrix enzyme that transfers coenzyme A from succinyl-CoA to acetoacetate
  - MCT1 (SLC16A1), a monocarboxylate transporter that catalyzes the movement of many monocarboxylates, such as lactate and pyruvate, across the plasma membrane
    - GLUT1 (SLC2A1), a major glucose transporter in the mammalian blood–brain barrier
    - ACAT1, acetyl-CoA-acetyltransferase 1 and acetoacetyl-CoA-thiolase

Schillaci noted that pathogenic GLU1 (SLC2A1) mutations lead to disorder characterized by drug-resistant epilepsy, abnormal movements, and developmental delay. They noted that ketogenic diet is useful in treatment of this disorder.

Fukao et al. (2019) noted that a mitochondrially localized enzyme that catalyzes the reversible formation of acetoacetyl-CoA from two molecules of acetyl-CoA undergoes pathogenic mutations that lead to defects in isoleucine metabolism and ketone body formation. Patients can undergo metabolic crises; however, neurological symptoms including extrapyramidal-related manifestation can occur even in the absence of metabolic crises.

Human glycosylation disorders

These disorders were reviewed by Freeze in 2015. Glycosylation defects lead to more than 100 rare genetic disorders. The key manifestations of these disorders include intellectual disability, developmental delay, movement disorders, alteration in muscle tone, and seizures. Freeze documented specific gene defects leading to these disorders and phenotype variations that were associated with defects in specific glycosylation gene products.

They noted that specific pathogenic mutation in DPAGT1 (dolichyl-phosphate *N*-acetylglucosamine phosphotransferase 1) led to aggressive behavior, seizures, and abnormal muscle tone. It is interesting to note that specific brain abnormalities often occur in this disorder. They include cortical atrophy, hydrocephalus, and corpus callosum defects.

Hyperammonemia and impact on the brain

Hyperammonemia can occur during metabolic crises in a number of inborn errors of metabolism and as a consequence of liver disease. It is a particular feature of inborn errors of the urea cycle. In adults, hyperammonemia was reported to result primarily from liver disease.

Braissant (2013) reported that in childhood, the brain is particularly prone to damage by significantly elevated ammonia levels. They noted that high ammonia levels impact cerebral energy metabolism, neurotransmission, and signal transduction. High ammonia levels were noted to sometimes lead to cerebral edema. Specific structural brain changes induced by hyperammonemia included cortical atrophy and demyelination that could lead to cognitive impairment and movement disorders. Summar and Ah Mew (2018) reviewed inborn errors leading to hyperammonemia. In humans, these diseases were primarily due to defects in enzymes involved in the urea cycle or in specific transporters including any one of the following:

- AAG, arginase; ASL arginosuccinate lyase
  - ASS1, arginosuccinate synthetase
    - CPS1, carbamoyl phosphate synthetase 1
    - NAGS, *N*-acetyl glutamate synthase
    - OTC, ornithine transcarbamylase
    - SLC2A15 (ORNT1), mitochondrial ornithine transporter
    - SLC25A13 (citrin), mitochondrial aspartate glutamate transporter

Summar and Ah Mew particularly stressed the importance of defects in CPS1, OTC, ASS, ASL, and NAGS. They noted that infants with deleterious mutations in these gene products may sometimes not present with difficulties until they encounter health crises, e.g., infection and exposure to certain medications. They noted that hyperammonemia could induce brain edema and glial cell death. There is also evidence that hyperammonemia leads to release of glutamate, leading to impaired function of glutamate receptors on neurons. Acute neurological signs of hyperammonemia including sleepiness and hyperventilation may advance to seizures and coma.

Ornithine decarboxylation

Ornithine that plays roles in the urea cycle also undergoes decarboxylation to give rise to polyamines. Rodan (2018) reported that ornithine decarboxylation is rate limiting in the generation of polyamines including spermine and putrescine. They reported that defects in ornithine decarboxylation lead to a defined neurometabolic syndrome. They reported four cases with defects in ornithine decarboxylase 1 (ODC1) function who manifested developmental delay, behavioral difficulties macrocephaly, and facial dysmorphology. Brain MRI abnormalities included white matter loss and abnormalities of the corpus callosum.

Transporter defects developmental and cognitive impairments

Riboflavin transporter defects

Riboflavin transporter defects can lead to a number of different neurologic manifestations of impaired functions in children. These include sensory or motor impairments, ataxia, hearing loss, vision changes, and bulbar dysfunction. Cases of this disorder have also been reported to first present with a history of developmental regression (Woodcock et al., 2018). Woodcock et al. described a case with regression in behavior and motor abilities who also had episodes of tonic stiffening. Investigations revealed auditory neuropathy and sensory and motor neuropathy in the upper and lower limbs. DNA sequence analysis revealed homozygosity for a mutation in the SLC52A2 that led to an Arg169Cys amino acid mutation. Normal forms of this protein function as a riboflavin transporter. This child's neurologic deficits improved following riboflavin supplementation. Pathogenic mutations in the riboflavin transporter SLC52A3 have been reported in Brown–Vialetto–Van Laere syndrome. Manifestations of this syndrome include optic atrophy, hearing loss, sensory motor neuropathy, and bulbar manifestations including respiratory insufficiency. Woodcock et al. reported that inadequate levels of riboflavin impacted activity of many enzymes. Riboflavin-containing flavin adenine dinucleotide is an important cofactor for acyl-CoA dehydrogenases and enzymes in the electron transport pathway.

Global developmental delay: diagnostic evaluations

Mithiyantha (2017) noted evidence that global developmental delay is identified in between 1% and 3% of children younger than 5 years of age. They reviewed current recommendation from different countries on investigation to attempt to establish causes of developmental delay. Specific manifestations used to identify developmental delay include delay in two or more of the following areas: speech/language, cognition, social interactions, gross or fine motor movements, and daily living activities.

Specific degrees of delay are defined as mild, moderate or severe. In mild cases, scores are less than 33% below scores for chronological age; scores 34%–66% below age appropriate scores are considered moderately delayed; scores 66% or more below age appropriate scores are defined as indicating severe delay. First-line tests include microarray chromosome analysis and fragile X testing; blood tests include analysis of levels of lead, creatine kinase, thyroid functions, homocysteine analysis carnitine levels, amino acids, glucose, and lactate. Urine studies are carried out to analyze organic acids, oligosaccharides, glycosaminoglycans, creatine, purines, and pyrimidines. Tests ordered take into account prior testing in newborn screening programs including thyroid function tests.

Important information to be gathered on clinical examination include history of regression, muscle pain alterations in gait, and alterations in sensory skills including vision. Neurological examination should be designed to test for ataxia, dystonia, cranial nerve abnormalities, ocular skills, and hearing. Patients should also be examined for dysmorphic features and for neurocutaneous manifestations.

Inborn errors of metabolism cognitive impairment and newborn screening

Sklirou and Lichter-Konecki (2018) noted evidence that inborn errors of metabolism (IEMs) are involved in 5% of cases of intellectual disability. In addition, these disorders may be also associated with neurological defects, including epileptic seizures, altered muscle tone, and ataxia. Many inborn errors of metabolism are multisystem disorders. They reviewed four IEM associated with cognitive impairments, phenylketonuria, homocystinuria, and creatine deficiency that are screened for in newborn screening programs in some states and countries.

Phenylketonuria (PKU) when detected in the newborn period and adequately treated with diet manipulation that ensures low intake of phenylalanine can have a favorable outcome (Fig. 7.2). Untreated PKU is associated with impaired cognition, behavioral problems, and dermatologic problems and may be associated with seizure. PKU develops as a result of deficiency of functional phenylalanine hydroxylase; it can also be associated with deficiency of

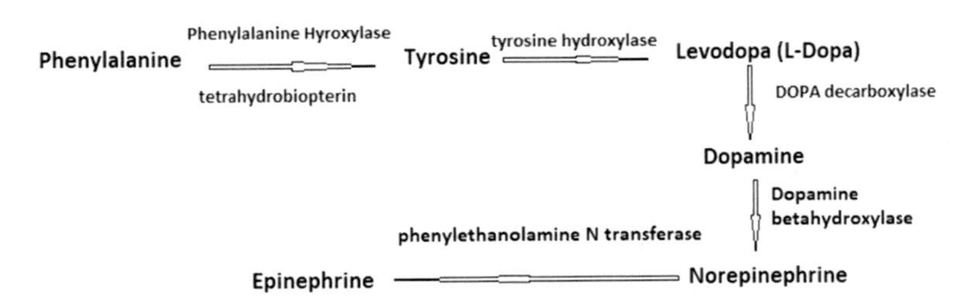

FIGURE 7.2 **Metabolism of phenylalanine and catecholamine biosynthesis.** Note importance of tetrahydrobiopterin.

tetrahydrobiopterin that acts as a cofactor for phenylalanine hydroxylase. It is important to note that deficiency of functional tetrahydrobiopterin can also impact activities of additional enzymes that utilized tetrahydrobiopterin as cofactor; these include enzymes that synthesize serotonin, dopamine, and noradrenaline. It is particularly important that women with phenylketonuria maintain low phenylalanine levels during pregnancy since high levels of phenylalanine in maternal body fluids can enter the fetus and lead to brain damage. It has also become clear that in individuals with PKU, dietary control to ensure low phenylalanine levels be continued into adult life. Pilotto et al. (2019) reported that early-treated PKU patients can exhibit normal IQ levels yet there is some evidence that abnormal dopamine and serotonin can occur in these patients and impair neuropsychiatric function. Serotonin and dopamine levels in CSF were shown to be reduced particularly when phenylalanine levels were not well controlled. Reduced levels of these neurotransmitters can lead to brain atrophy. There was also some evidence reported that even when levels of phenylalanine were well controlled in childhood, later some patients manifested impairments in execution function and attention. Deficits in the latter functions manifest if phenylalanine levels are not well controlled in adult life.

Methionine levels are assayed in many newborn screening programs. Sklirou and Lichter-Konecki (2018) noted that altered methionine levels may result from defects in any one of several enzymes including the following:

- Cystathionine beta synthase (CBS)
- Methionine adenosyltransferase I or III MATI/MATIII
- Glycine *N*-methyltransferase (GNMT)
- S-adenosylhomocysteine hydroxylase (SAAH)

CBS deficiency results in elevation of both methionine and homocystine. Some forms of CBS deficiency are responsive to treatments with vitamin B6 pyridoxine. Low methionine formulas and diets are useful in treating these disorders.

Low levels of methionine and elevated homocystine levels may occur due to defects in levels or function of the enzyme methylenetetrahydrofolate reductase or impairment within the vitamin B metabolic pathway.

Sklirou and Lichter-Konecki reported that brain creatine deficiency can lead to developmental delay and seizures. These can result from disorders that disrupt creatine biosynthesis and defects in creatine transport. There is evidence that creatine plays a critical role in brain and muscle energy provision.

Enzymes essential in generation of creatine include AGAT that catalyzes the synthesis of guanidino acetate (GAA) from arginine and glycine. Defects in function or levels of AGAT were reported to be associated with intellectual disability, muscle weakness, and failure to thrive. Low levels of GAA are present in plasma and urine, and creatine depletion can be demonstrated on brain magnetic resonance spectroscopy. Creatine supplementation can be beneficial.

Males with deficiency of the creatine transporter SLC6A8 that is encoded on the X chromosome can present with mild to severe intellectual disability, speech delay, neurological deficits including altered muscle tone, and ataxia, and they may develop seizures. In these cases, levels of guanidino acetate (GAA) and creatine in plasma and urine are normal. Diagnosis is made through gene studies that detect pathogenic mutations in SLC6A8. Treatments with creatine have been initiated in some cases but may not be successful.

Thurm (2016) reported that creatine transporter defect was diagnosed in a patient following the finding of an abnormal urine creatine/creatinine ratio. Normal conversion of creatine to creatinine requires that creatine be transported into cells, e.g., in muscle and brain. There it undergoes phosphorylation aided by creatine kinase and ATP. Phosphocreatine is then converted to creatinine that is excreted in urine.

Glycine metabolic disorders: nonketotic hyperglycinemia

Glycine in the body can be derived from serine through the activity of serine hydroxy methyltransferase SHMT1 in the presence of tetrahydrofolate. Glycine is metabolized in the body through activity of the glycine cleavage system. Impaired glycine metabolism in the body occurs in nonketotic hyperglycinemia (Stence, 2019).

The glycine cleavage complex is composed of three enzyme proteins encode by four different genes. The specific genes and the proteins that encode are listed in order in the following:

- *GLDC* PLP, pyridoxal phosphate-dependent glycine decarboxylase (P protein)
- *GCSH*, glycine cleavage system protein H (H protein)
- *AMT*, aminomethyltransferase, a tetrahydrofolate-containing protein (T proteins)
- *DLD*, dihydrolipoamide dehydrogenase, NAD-dependent flavin adenine protein (L protein)

Stence et al. noted that different degrees of severity occur in nonketotic hyperglycinemia. Severely affected individuals can manifest epilepsy and profound psychomotor delay. Patients with severe forms of hyperglycinemia manifest in the neonatal period. Some patients have less severe forms of this disorder. They noted in the brain, the glycine/creatine ration can be determined by MRI spectroscopy. Patients with severe disease were noted to have high glycine to creatine ration and alteration in the corpus callosum.

The *DLD*-encoded protein forms a subunit of several multienzyme complexes involved in metabolism.

In addition to defects in the enzymes in the glycine cleavage system, glycine encephalopathy can occur due to defects in either one of two glycine transporters. Kurolap et al. (2016) reported that glycine concentrations in the body are controlled by two glycine transporters: GLYT1 (SLC6A9) and GLYT2 (SLC6A5). These transporters are reported to clear glycine from synapses.

Swanson (2015) reviewed findings in 124 patients with nonketotic hyperglycinemia. They noted that diagnosis can be made on the basis of elevated glycine levels in the CSF. In the majority of patients in their series, disorder arose from defects in the *GLDC* or *AMT* genes. Of their 124 patients, 26 died in the neonatal period (approximately 20%), 45% had severe disease with poor development, and 34% of patients had an attenuated form of the disease. Higher CSF glycine levels occurred in patients with severe disease.

Maple syrup urine disease

This is a disorder in which metabolism of branched chain amino acids is impacted, leading to an accumulation of alpha keto acids in the urine that lead to an odor similar to that of maple syrup. Blackburn (2017) reported that there is a wide spectrum of severity in this disorder. The defective enzymes are part of the branched-chain alpha keto-acid dehydrogenase complex (BCKD). The BCKD complex is thought to be composed of a core of 24 transacylase (E2) subunits and associated decarboxylase (E1), dehydrogenase (E3), and regulatory subunits. The specific gene products impacted in maple syrup urine disease include the following:

- BCKDHA, branched-chain keto acid dehydrogenase E1 subunit alpha
  - BCKDHB, branched-chain keto acid dehydrogenase E1 subunit beta
    - DBT, dihydrolipoamide branched chain transacylase E2
    - DLD, dihydrolipoamide dehydrogenase, sometimes referred to as E3, also occurs in different metabolism-related multienzyme complexes

Defective metabolism of branched-chain amino acids, leucine, isoleucine, and valine can lead to developmental delay and encephalopathy.

Newborns with this disorder may present with abnormal posturing (opisthotonos severe hyperextension) and lethargy and coma. In severe forms of the disease, brain edema can occur. Metabolic crises with severe ketoacidosis can also occur in later life. These disorders can be successfully treated with dietary manipulation including restriction of intake of branched chain amino acids. In addition, thiamine supplementation is important particularly in cases with DBT mutations. The DLD gene product, sometimes referred to as the E3 component of the branched-chain keto acid dehydrogenase complex, forms a subunit in the alpha ketoglutarate dehydrogenase complex and in the pyruvate dehydrogenase complex. Patients with DLD defects may then present with altered levels of alpha keto acids and abnormal levels of lactate alanine and alpha ketoglutarate. Blackburn et al. noted that 190 different pathogenic sequence variants have been identified in patients with maple syrup urine disease. Disease-causing variants are usually homozygous or present in compound heterozygous forms.

9. Mitochondrial defects and impaired neurodevelopment

Mitochondrial homeostasis

Khacho and Slack (2018) noted that impairments in mitochondrial fission and fusion led to impaired neurodevelopment. Specific homozygous mutations in OPA1 mitochondrial dynamin–like GTPase were reported to lead to infantile onset encephalopathy. Mutation in dynamin 1 like (DRP1L), a member of the dynamin superfamily of GTPases, was reported to impact brain development and to also lead to optic atrophy. They noted evidence that aberrant mitochondrial fragmentation occurred in a number of disorders with impacted neurodevelopment, including Wolfram Syndrome and tuberous sclerosis. There is evidence that TSC1 TSC deficiency impairs mitochondrial homeostasis (Ebrahimi-Fakhari et al., 2016). Downregulation of the WFS1 wolframin transmembrane proteins was reported to involve proteins encoded by nuclear DNA and by mitochondrial DNA.

In a review of mitochondrial disease, Molnar and Kovacs (2017) emphasized that the majority of proteins and enzymes active in mitochondria are encoded in the nuclear genome and that mitochondrial DNA encodes relatively few of these (Fig. 7.3). They also emphasized that impaired mitochondrial function most often impairs function of multiple organs; brain and muscle are often involved. They reported that histopathologic findings of mitochondrial functional defects in brain include vacuolation in white and gray matter, loss of oligodendrocytes that provide support to axons, microglial proliferation, capillary proliferation, and mineralization in vessels.

Laboratory investigations in putative mitochondrial diseases

Analysis of fluids: Plasma, Urine Cerebrospinal fluid: Lactate, pyruvate acylcarnitine amino acids

Cell analysis: Electron transfer complexes, analysis of activity
DNA sequencing nuclear DNA, Mitochondrial DNA, search for deletions, mutations
Analyze DNA for specific known pathogenic mutations

Tissue analysis e.g. on muscles: Analyze mitochondrial numbers, mitochondrial histology
Utilize special stains and antibodies directed against
specific mitochondrial components COX, SDH, Coenzyme Q

FIGURE 7.3 Strategy to establish clinical diagnosis in putative mitochondrial disorders.

Mitochondrial DNA mutations frequently impacting neurological functions

Specific mutations in mitochondrial DNA have been reported to lead to encephalopathy, stroke-like episodes, migraine, and cognitive dysfunction. Specific syndromes due to mitochondrial DNA mutations include the following:

- MELAS, mitochondrial encephalopathy, lactic acidosis, strokelike episodes
- NARP, neuropathy, ataxia, retinitis pigmentosa
- MERRF, myoclonic epilepsy with ragged-red fibers in muscle; Leigh syndrome, lactic acidosis, basal—ganglia damage, psychomotor retardation, seizures, movement disorders
- Kearns—Sayre syndrome ophthalmoplegia, retinal pigment degeneration, cardiomyopathy, myopathy
- Leigh syndrome can result from mitochondrial DNA mutations and can also result from specific nuclear DNA mutations that impair mitochondrial functions

Nuclear DNA mutations leading to impaired mitochondrial functions

Molnar and Kovacs separated these into five categories:

- Respiratory complex diseases
- Disorders of mitochondrial protein synthesis
- Disorders of mitochondrial maintenance leading to mitochondrial depletion and mitochondrial DNA deletions
- Defects in mitochondrial dynamics
- Defects in mitochondrial lipid milieu

Other nuclear-encoded gene defects impair a number of structures or organelles including mitochondria and may be associated with epilepsy, ataxia, encephalopathy, and myopathy.

Examples of nuclear gene products when defective impair mitochondrial function include the following:

Friedreich ataxia is due to defect in FXN (frataxin) that functions in regulating mitochondrial iron transport and respiration. There are also extramitochondria forms of frataxin.

Impaired carnitine metabolism can disrupt transfer of long-chain fatty acids across the inner mitochondrial membrane for fatty acid oxidation.

Occipital Horn syndrome can result from defects in ATP7A, a transmembrane copper transporter reported to be essential for redox balance.

Wolfram syndrome WFS1 resulted from defects in a transmembrane glycoprotein and impacts mitochondria and endoplasmic reticulum.

SLC25A19 functions as the mitochondrial thiamine pyrophosphate carrier; deficiency leads to severe 2-ketoglutaric aciduria.

Nuclear and mitochondrial gene interactions in production of proteins relevant to neurodevelopmental defects

Electron transport complexes and oxidative phosphorylation processes and defects

Disorders due to defects in electron transport complexes can result from mutations in nuclear or mitochondrial genes. In a review of defects in electron transport and oxphos system, Ghezzi and Zeviani (2018) considered five multiheteromeric complexes that are embedded in the inner mitochondrial membrane and complexes I, II, III, and IV that are sometimes referred to as the respiratory complexes. They also include two electron shuttles, ubiquinone coenzyme Q, and cytochrome C. Activities of these complexes lead to transfer of protons across the inner mitochondrial membrane to the intermembrane space. This transfer provides the proton motive force for complex V ATP synthase activity.

Specific genes in the mitochondrial genome contribute subunits to complexes Cl, CIII, CIV, and CV. Complex l contains seven mitochondrial encoded subunits MTND1, 2, 3, 4, 4L, 5, and 6. Complex 3 contains three mitochondrial encoded components MTCO1, MTCO2, and MTCO3. The inner core of complex IV is composed of mitochondrial genome-encoded subunits. ATP synthase complex V contains two subunits encoded by the mitochondrial genome MT-ATP6 and MT-ATP8.

Signes and Fernandez Vizaria (2018) reviewed proteins and components involved in these assembly processes. They noted that complex 1 has a total of 45 subunits. Both nuclear and mitochondrial subunits are brought together by assembly factors, and in complex 1, iron sulfur particles are also incorporated. Complex 1 was reported by Guerrero-Castillo (2017) and required 14 assembly factors.

Complex II is formed from nuclear-encoded proteins SDHA, SDHB, SDHC, and SDHD, and it also incorporates iron sulfur clusters.

Ghezzi and Zeviani documents the nuclear-encoded assembly factors for respiratory complexes. These included 11 for complex I, 5 for complex III, 9 for complex IV, and 2 for complex V.

It is also important to note that there are also nuclear encoded products involved in copper and iron sulfur complex incorporation into mitochondrial electron transfer complexes.

Phenotypic abnormalities described in some cases due to defects in the function of mitochondrial complexes include encephalopathy, leukodystrophy, epilepsy, myopathy, and ataxia.

Cytochrome C oxidase (COX) complex IV is the terminal component of the respiratory transfer complex. COX has multiple subunits. It is described as having 11 nuclear-encoded subunits and 3 mitochondrial-encoded subunits. In addition, multiple genes encode assembly factors for COX. A nuclear gene–encoded enzyme farnesyl transferase was reported to be required for expression of function. Manifestations of COX deficiency can include muscle weakness and psychomotor regression. Some patients present with Leigh syndrome.

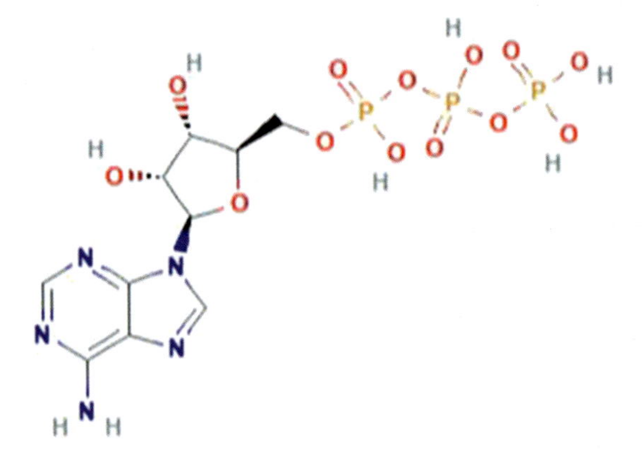

FIGURE 7.4 Structure of ATP (adenosine triphosphate). *From https://pubchem.ncbi.nlm.nih.gov/compound/Adenosine-5_-triphosphate.*

Mitochondrial (Fig. 7.4) complex V (ATP synthase) F1Fo ATPase has as its main function and the utilization of inorganic phosphate and ADP to synthesize ATP in reactions driven by the proton gradient and membrane potential generated by the electron transfer complexes that activate the Fo structure of complex V and this then drives the rotation of the F1 structure. The Fo structure is anchored to the inner mitochondrial membrane, and the F1 structure projects into the intermembrane space. F1 and Fo are linked by inner and outer stalk structures. Fifteen nuclear genes and two mitochondrial genes were reported to give rise to the protein subunits that comprise F1Fo ATPase. The two mitochondrial genes are designated mtATP6 and mtATP8.

Patients with defects in nuclear genes that encode ATP synthase of complex V were reported to present with hypotonia and impaired cardiac function. Biochemically, patient frequently manifested lactic acidosis, hyperammonemia, and 3-methylglutaconic acidemia. Most cases present with neonatal-onset hypotonia, lactic acidosis, hyperammonemia, hypertrophic cardiomyopathy, and 3-methylglutaconic aciduria. Many patients die within a few months or years. Patients with defects in mitochondrial-encoded subunits of ATP synthase may have similar manifestations. MTATPase 6 defects can give rise to a specific syndrome referred to NARP (neuropathy, ataxia, and retinitis pigmentosa). Learning disabilities may also be present in these patients.

More recent studies have revealed that mitochondrial complex V exists in two forms: a monomeric form that is involved in ATP synthesis and a dimeric form. The dimeric form is reported to be important for mitochondrial cristae structure. Siegmund et al. (2018) reported that a small supernumerary protein subunit of ATPsynthase USMG5 plays an important role in dimerization of ATP synthase and formation of appropriate cristae structure. Pathogenic mutation of USMG5 was reported in a case of Leigh syndrome. Barca (2018) reported the occurrence of a founder mutation in USMG5 in the Ashkenazi population that leads to abnormal mRNA splicing. This founder mutation was found in homozygous state in patients with Leigh syndrome and encephalopathy. It is important to note that this protein is also referred to as ATP5MD and DAPIT and is mapped to chromosome 10q24.33.

Transport of proteins into mitochondria

Wiedemann and Pfanner (2017) reviewed machineries for transport of proteins into mitochondria. They noted the five different transport pathways exist. Translocases of outer mitochondrial membrane TOMM and translocases of the inner mitochondrial membrane TIMM play important roles.

Optic neuropathy, deafness, and dystonia have been reported in cases with defects in a translocase of the inner mitochondrial membrane TIMM8A.

Maintenance of high-quality mitochondria

Chan (2020) reviewed the importance of fission and fusion to maintain high-quality mitochondria to segregate damaged mitochondria for degradation. The dynamin family of GTPases were noted to be key components of the fission fusion machinery.

Whitley (2018) reported cases with mutations of the DNM1L gene, also known as DRP1 that is required for mitochondrial division. They noted previous reports of infant with defects in the gene who presented encephalopathy. However, a range of disease severity can be encountered, and some individuals present primarily with epilepsy.

It is important to note that other genes that encode dynamins are also present in the human genome. These include DNM1. Defects in DNM2 were noted to be associated with neuromuscular abnormalities, e.g., Charcot–Marie–Tooth disorder.

It is important to know that mitochondria make contact with peroxisomes and this is also an important function of dynamin-related proteins. Defects in function of DNML1 (also known as DLP1) can lead to a disorder with some features related to peroxisomal functional defects including seizures and encephalopathy.

Mitochondrial genome transcription and translation

In 2018, D′ Souza and Mincuk presented an overview of mitochondrial transcription and translation. In a related review, Boczanadi et al. reviewed clinical syndromes resulting from impairments of mitochondrial DNA transcription and translation.

The mitochondrial genome was noted to encode 2 ribosomal RNAs, 22 transfer RNAs, and messenger RNA that encoded 13 polypeptides that contribute to the generation and function of the oxidative phosphorylation system.

The mitochondrial genome is known to be transcribed bidirectionally from two promoters, one for the heavy chain and another for the light chain strand. Transcripts thus generated cover the entire mitochondrial genome. The polycistronic transcripts were noted to undergo endonucleolytic processing and posttranslational modifications.

Boczanadi et al. noted that mitochondrial (mt) DNA transcription is regulated by three nuclear-encoded transcription factors, and this transcription also requires a specific RNA polymerase. These proteins and the chromosome positions of the genes that encode them are listed in the following:

- TFAM, transcription factor A, mitochondrial, 10q21.1
- TEFM, transcription elongation factor, mitochondrial, 17q11.2
- TFB2M, transcription factor B2, mitochondrial 1q44
- POLRMT RNA polymerase mitochondrial, 19p13.3

Endolytic cleavage of the polycistronic transcript gives rise to mt mRNA, mt ribosomal RNA, and mt TRNAs, and these cleavage products then undergo modifications and the mt RNAs are polyadenylated. D'Souza and Minczuk stress that modification of the transcripts is essential for transcripts to interact with the mitoribosome.

Proteins involved in processing of ribosomal RNAs include the following:

1. MRM1, mitochondrial rRNA methyltransferase 1, 17q12
2. MRM2, mitochondrial rRNA methyltransferase 2, 7p22.2
3. MRM3, mitochondrial rRNA methyltransferase, 17p13.3
4. RPUSD4, RNA pseudouridine synthase D4, 11q24.2
5. TFB1M, transcription factor B1, methylates the conserved stem loop of mitochondrial 12S rRNA, 6p25.3
6. NSUN4 methyltransferase, 1p33
7. Proteins required for assembly and maturation of the mitoribosome include the following:
8. ERAL1 12S RNA chaperone1, binds to the 3′ terminal stem loop of 12S mitochondrial rRNA and is required for proper assembly of the 28S small mitochondrial ribosomal subunit, maps to chromosome 17q11.2
9. RMND1 protein localized in the mitochondria, thought to be involved in translation, gene maps to 6q25.1

The mitoRNA transcripts must be charged with amino acids. It is important to note that mitochondrial TRNA synthetases required for the amino acid loading are encoded by nuclear genes.

The mitoribosome was noted to consist of a 39S large subunit and a 28S small subunit. Mitoribosome assembly was reported by D'Souza and Minczuk to require ERAL1 chaperone.

Proteins that are involved in posttranscription maturation of the primary transcript processing occur in granules in mitochondria; these include ELAC1 and ELAC2 ribonucleases and FASTKD1 and FASTK kinases.

Nuclear-encoded proteins involved in TRNA maturation include the following:

1. NSUN3 RNA methyltransferase 3q11,2
2. ABH1 (ALKBH1) histone dioxygenases 14q24.3
3. MTO1 mitochondrial tRNA translation optimization 1 6q1
4. MTU1 (TRMU) tRNA mitochondrial 2-thiouridylase, 22q13.31
5. TRI1 tRNA isopentenyltransferase 1, 1p34.2
6. TRMT5 tRNA methyltransferase 5 PUS1 pseudouridine synthase 1, 12q24.33
7. TRMT1 tRNA methyltransferase 1, 19p13.13
8. TRMT5 tRNA methyltransferase 5, 14q23.1
9. PDE12 phosphodiesterase 12, 3p14.3
10. MTFM1 mitochondrial methionyl-tRNA formyltransferase15q22.

Translation of proteins from mitochondrial transcripts

Specific nuclear–encoded proteins participate in translation of specific mitochondrial RNAs; these proteins and the chromosomal location of the encoding gene include th following:

1. mtF2 and mtf3 mitochondrial proteins
2. TACO1 translational activator of cytochrome *c* oxidase I, 17q23.2
3. MITRAC (COA3) cytochrome *c* oxidase assembly factor 17q21.2
4. C12ORF62 cytochrome *c* oxidase assembly factor 12q13.12

Translation elongation factors include the following:

1. EFTu (EFTUD2) elongation factor Tu GTP binding domain containing 2 17q21.31
2. EFTs (TSFM) Ts translation elongation factor, mitochondrial, 12q14.1
3. EFGM (GFB1) G elongation factor mitochondrial 1, 3q25.32

Translation termination factors include the following:

1. MTRF1, mitochondrial translation release factor 1, 13q14.11
2. MTRF1A (MTRF1L), mitochondrial translational release factor 1 like, 6q25.2
3. C12ORF65 encodes a mitochondrial matrix protein contributing to peptide chain termination, 12q24.31
4. ICT1 (MROL58)-encoded protein serves as a ribosome release factor 17q25.1
5. mtRRF (MRRF), ribosome disassembly and recycling at termination of mitochondrial translation 9q33.2
6. EFG2 (GFM2), GTP-dependent ribosome recycling factor mitochondrial 2, 5q13.3

Disorders associated with defective transcription and translation in mitochondria

Boczonadi (2018) documented disorders associated with defective mitochondrial transcription and translation. It is important to note that a number of different organs and function are potentially impacted, and liver failure, cardiomyopathy, and skeletal myopathy are often key features. This chapter concentrates particularly on mitochondrial transcription and translation defects that lead to nervous system impairment.

Defects in FASTKD2 that is important in mitoribosome assembly and in 16SRNA stability, can lead to mitochondrial encephalopathy, developmental delay, epilepsy, and hemiplegia and are associated with low levels of COX.

MTPAP was reported to be a nuclear encoded protein a polyA RNA polymerase active in mitochondrial DNA transcription. Defects in function of MTPAP were reported to lead to spastic ataxia and optic atrophy in the Old Amish population. It is encoded by a gene on chromosome 10q11.23.

LRPPRC a leucine-rich protein that has multiple pentatricopeptide repeats (PPR. It is a nuclear encoded protein (2p21) reported to facilitate polyadenylation and translation of mitochondrial mRNAs PRPPRC homozygous mutation were discovered in French Canadian patients with Leigh syndrome, This syndrome is characterized by neurodegeneration in specific brain regions, and by developmental dealy, hypotonia and facial dysmorphisms. These mutations lead to decrease in complex IV and to reduced levels of COX and to reduced levels of several OXPHOS subunits. Pathogenic LRPPRC mutations have subsequently been discovered in other populations.

Specific nuclear gene encode products that play roles in mitochondrial tRNA modification and defects in some of these genes lead to liver failure cardiomyopathy and lactic acidosis. Specific nuclear genes that encode protein involved in mt RNA modification have been reported to lead to developmental, delay seizures and encephalopathy. These mutations lead to defective function of the products listed below.

1. GTPB3 GTP binding protein 3, mitochondrial 19p13.11
2. NSUN3 SUN methyltransferase 3, 2q11.2
3. TRMT5 T RNA methyltransferase 5, 14q23,1
4. TRNT1 TRNA nucleotidyl transferase, 3p26.2
5. MTFMT1 mitochondrial methionyl-tRNA formyltransferase 15q22.31

Genes involved in the aminoacetylation of mitochondrial tRNAs are encoded in the nucleus. Boczanadi et al. noted that defects in these genes can lead to different phenotypes that include encephalopathy with epileptic seizures, and intellectual disability.

Aminoacyl tRNA synthetases with defects that lead to encephalopathy with white matter defects include DARS2 aspartyl tRNA synthetase 1q25.1, EARS2 glutamyl tRNA synthetase 16p12.2, MARS2 methionyl tRNA synthetase 2q33.1 AARS2 alanyl tRNA synthetase 6p21.1.

Aminoacyl TRNA synthetases with defects that lead to epileptic encephalopathy include CARS2 cystinyl TRNA synthetase 13q34 PARS2 prolyl TRNA synthetase 1p32.3 TARS2 threonyl TRNA synthetase 1q21.2 VARS2 valyl tRNA synthetase 6p21.33.

Defects that lead to pontocerebellar hypoplasia occur in RARS2 arginyl TRNA synthetase 6q15.

Defects that lead to intellectual disability occur in RARS2 arginyl TRNA synthetase 6q15 WARS tryptophanyl tRNA synthetase 1p12.

Defects in mitoribosome assembly proteins and in proteins that interact with mitoribosomes can lead to psych motor retardation and neurological deterioration.

These proteins include MRPL and MRPS proteins that are nuclear encoded.

Specific nuclear genes that encode proteins that are involved in mitochondrial translation activation, translation termination, or release of synthesized and that are noted to have defects that cause defects in nervous system function are listed in the following:

1. GFM1, mammalian mitochondrial ribosomal proteins are encoded by nuclear genes and help in protein synthesis within the mitochondrion, encoded on3q25.32. Mutations in this gene have been associated with encephalopathy.
2. TUFM, Tu translation elongation factor, mitochondrial, encoded on 16p11.2, defects associated with lactic acidosis, leukoencephalopathy, PMG
3. TSFM, Ts translation elongation factor, mitochondrial, 12q14.1 C12orf65 encodes a protein that appears to contribute to peptide chain termination in the mitochondrial translation machinery; defects in function of this protein have been associated with Leigh syndrome and spastic paraplegia
4. TACO1 encodes a mitochondrial protein that functions as a translational activator of mitochondrially encoded cytochrome *c* oxidase 1; mutations in this gene are associated with Leigh syndrome
5. Boczonadi et al. also noted that reports of encephalomyopathy in patients with defects in PNPT1, polyribonucleotide nucleotidyltransferase 1. It is involved in import of RNA to mitochondria. Mutations in this gene have also been associated with combined oxidative phosphorylation deficiency and autosomal recessive nonsyndromic deafness

Mitochondrial aminoacyl tRNA synthetase and developmental disorders of myelination and leukodystrophies

Fine et al. (2019) emphasized that defects in mitochondrial aminoacyl TRNA synthetase result in leukodystrophies. They characterized specific mitochondrial TRNA aminoacyl tRNA synthetases involved in specific disorders in which leukodystrophies occur. Leukodystrophies were reported as a result of defects in nuclear genes that encode the mitochondrially active proteins DARS2, aspartyl tRNA synthetase 1q25.1; EARS2, glutamyl tRNA synthetase 16p12.2; MARS2, methionyl tRNA synthetase 2q33.1; and AARS2, alanyl tRNA synthetase 6p21.1.

Encephalopathy leading to psychomotor delay and epilepsy occurred in defects that involved RARS2, arginyl TRNA synthetase 6q15; WARS2, tryptophanyl tRNA synthetase

2, mitochondria 1p12; TARS2, threonyl TRNA synthetase 1q21.2; VARS2, valyl tRNA synthetase 6p21.33; and FARS2, tryptophanyl tRNA synthetase 2, mitochondrial 6p25.1.

Perrault syndrome characterized by sensorineural hearing loss, hyporeflexia, and in some cases by intellectual disability occurred with defects and nuclear changes in genes that encode LARS2, leucyl-tRNA synthetase 2, mitochondrial, 3p21.31; HARS2, histidyl-tRNA synthetase 2, mitochondrial 5q31.3; PARS2, prolyl-tRNA synthetase 2, mitochondria 1p32.3; and NARS2, asparaginyl-tRNA synthetase 2, mitochondrial, 11q14.1. Fine et al. emphasized that clinical variability occurred in cases with this syndrome.

Patient care and treatment in mitochondrial disorders

Muraresku (2018) reported that standards of care for patients with disorders due to mitochondrial functional impairments have been developed and will potentially be complemented by emerging therapies. They noted that preimplantation genetic diagnostic testing has become available to families with a history of mitochondrial diseases.

Care measures need to address specific manifestations that develop in particular organ systems. It is also important in affected individuals to avoid febrile illnesses since fever and food deprivation can trigger decompensation. Muraresku et al. noted that exercise is usually well tolerated and has been shown to be beneficial. There is evidence that exercise increases mitochondrial copy number and activity of electron transfer complexes.

Specific dietary supplements recommended in patients with mitochondrial disease include B vitamins, creatine, and L carnitine (Fig. 7.5). They noted that clinical trials of new therapies are being implemented.

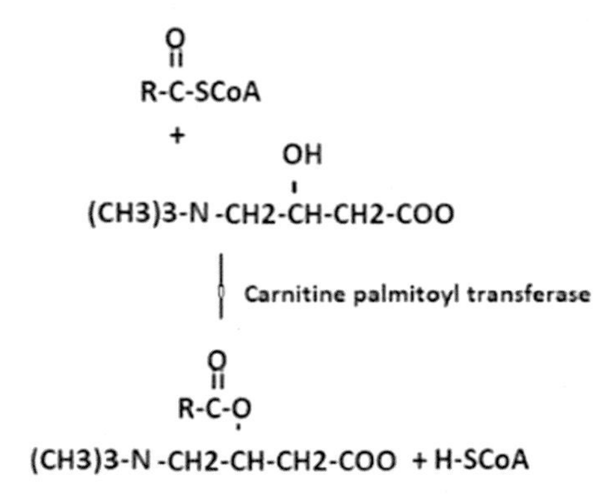

FIGURE 7.5 **Carnitine in fatty acid metabolism.** Carnitine is used in therapy of specific disorders.

Delivery of care to individuals with intellectual and developmental disorders

Ervin (2014) reviewed models of healthcare for individuals with intellectual and developmental disabilities (IDDs). They emphasized the need for training of healthcare providers regarding the needs of these individuals and proposed the development of Intellectual Disability Medicine. Deinstitutionalization of individuals with IDD took place progressively during the 1960s and 1970s. Ervin et al. emphasized that people with IDD can more fully participate in community life if attention is paid to their health status. They noted the difficulties in communication between patients and care providers and complications related to financing systems.

Specific steps to improving care for IDD patients were proposed by David Satcher, the United States Surgeon General in 2011. Specific organizations have been established in the United States to promote care of patients with IDD, which include the Arc alliance and the American Association on Intellectual and Developmental Disabilities (AAIDD). Community healthcare delivery systems for delivery of services to individuals with IDD have been established in the United States.

Ervin et al. also reviewed approaches to healthcare for individuals with IDD in other countries. In the United Kingdom, Australia, New Zealand, and Canada, healthcare delivery to people with IDD was noted to be provided by general physicians and family physicians. Importantly, however, length time for appointments often needs to be longer for patients with IDD.

The Dutch Society of Physicians in 2003 proposed that care for people with ADD could be provided by mainstream health services that were IDD competent and with backup services of specialists with expertise in special needs required by IDD patients.

Ervin et al. emphasized the importance of integration of behavioral healthcare into IDD patient services. They noted that some medical schools have started to include IDD content into their curricula.

Genetic testing in neurodevelopmental disorders: diagnostic yield

A number of studies have investigated the use of exome sequencing as a first-tier test in neurodevelopmental disorders. Srivastava (2019) carried out a literature survey and reported that diagnosis was made on the basis of exome sequencing in 36% of cases of neurodevelopmental disorders defined as disorders with intellectual disability, global developmental delay, and/or autism. In cases where neurodevelopmental disorders were associated with congenital defects that also impacted other body systems, the diagnostic yield was reported to be 53%.

Lindstrand (2019) reported on utility of whole genome sequencing in cases of intellectual disability. They compared diagnostic yield of whole genome sequencing to diagnostic yield obtained with clinical microarray studies. In their analyses, microarray studies led to diagnosis in 12% of cases, while whole-genome sequencing yielded diagnosis in 27% of cases. Lindstrand et al. noted that additional advantages of whole-genome sequencing included ability to detect nucleotide repeat expansions and to detect uniparental disomy.

Other modes of diagnostic testing include use of gene panels that analyze sequences of genes known to be implicated in specific disorders.

Structural and genomic changes relevant to neurodevelopmental and behavioral disorders

Deshpande and Weiss (2018) reviewed the roles of genomic CNVs in neurodevelopmental disorders with emphasis on regions of the genome where pathogenic deletion occurred in some patients while other patients had duplication of the same genomic region.

CNVs in the genome do not always lead to defects; however, CNVs in specific regions have been reported in many studies to have deleterious effects on phenotype. Deshpande and Weiss emphasized the importance of CNV in 1q21.1, 7q11.23, 15q13.3, 16p11.2, and 22q11.2. They noted further that study of gene expression patterns may offer clues to genotype–phenotype correlation.

There is evidence for that deletions and duplications in a specific chromosome region have opposing phenotypes. Deletions of chromosome 1q21.1 were reported to be associated with microcephaly while duplications of 1q21.1 were associated with macrocephaly. However, both of these CNVs were associated with intellectual disability, autism, and ADHD. Deletions of 7q11.23 were associated with gregarious friendly personality and normal speech, while duplications of 7q11.23 were associated with autism spectrum disorder and speech delay.

Deletions of 16p11.2 were reported to be associated with macrocephaly and obesity, speech delay, intellectual disability, and attentional deficit disorder. Duplications of 16p11.2 were associated with microcephaly and low birthweight, autism spectrum disorder, and developmental delay. Deletion of chromosome 22q11.2 was reported to be associated autism and with schizophrenia, while duplication was reported to protect against schizophrenia risk.

Zarrei (2019) reported establishment of a large genomic CNV database relevant to neurodevelopmental and neurobehavioral disorders. Data included information from studies on autism spectrum disorder, ADHDs, obsessive compulsive disorder, and schizophrenia. They also documented rare CNVs found in <0.1% of the population of controls. They documented that the most commonly occurring CNVs associated with neurodevelopmental disorders occurred at 15q11.2 and 16p11.2. They documented that the most frequently deleted gene associated with neurodevelopmental disorders occurred in NRXN1 that maps to chromosome 2p16.3.

10. Autism

Clinical utility of microarray studies in autism

Use of microarray technology has greatly improved the capacity for detection of CNVs. Such variants are frequently detected in individuals diagnosed with autism. However, definitive implications of CNVs as being causative of the autism phenotype are not always possible.

Devlin and Scherer in 2012 reviewed the genetic architecture of autism. They reported that rare de novo or inherited CNVs were observed in 5%–10% of cases with idiopathic autism. They observed CNVs were frequently found to impact genes that encode proteins involved in synaptic functions.

Velinov (2019) considered clinical utility of microarray analyses in autism. They noted difficulties in determining clinical significance of microarray findings when unaffected members of the same family manifested the same CNV as the affected individual and also given the frequency of specific microarray variants in unaffected members of the general population. Furthermore, there is evidence of heterogeneity of the clinical effects of a specific CNV. There is evidence that the CNVs found in autism also occur in other neurobehavioral disorders.

D'Abate (2019) published results of studies on siblings of autism-affected individuals who had specific CNVs. They studied 288 siblings of autism cases and noted that by 3 years of age, 38.8% of these siblings manifested autism features, 18% had atypical development, and 56.8 manifested normal development. In the infant siblings with autism, 13 of 288 shared the same CNV as the first diagnosed autism individual; 4 of these siblings manifested autism, 3 had atypical development, and 1 sibling had normal development.

Rare and common variants contribute to autism, and it is clear that polygenic inheritance also occurs. A number of different projects have revealed evidence of polygenic risk in autism. Weiner (2017) used the transmission equilibrium test in studies of autism. This test determined the frequency with which a specific parental allele is transmitted to an affected child. If a specific variant is transmitted more frequently than half the time to an affected offspring from an unaffected parent, the overtransmitted allele is considered a risk allele.

Weiner et al. utilized genotype data from 493 families in the Simons Simplex c collection where each family included parents, an affected child, and an unaffected child. They also included data on 3870 trios from a separate data set. Their transmission disequilibrium analyses revealed that polygenic factors contributed additively to autism spectrum disorders risk.

Chaste (2017) reviewed genetic research in autism spectrum that has provided evidence that is a complex polygenic disorder in which rare variants, de novo and inherited, act together with common variants and together contribute to liability. They developed a formula for autism liability that includes contributions from genome

G environment E and de novo changes D that together contribute to load L, $L = G + E + D$.

The also defined T that represents the threshold value for autism development. If the value of L exceeds a certain threshold, the individual is affected.

Chaste also noted recent research that indicated a continuum between autism diagnosis and the presence of autism traits in healthy individuals.

Chaste noted the importance of certain genomic CNVs in autism; they included eight genomic regions 1q21.1, 2p16, 3q29, 7q11.21, 15q11.2-q23, 16p11.2, 22q11.2, and 22q13.33.

Specific single genes that they considered to be important in determining autism risk included MBD5 (methyl-CpG-binding domain protein 5), NRXN1 (neurexin), and SHANK genes SH1 and SH3.

In considering exome-sequencing findings, Chaste noted a consistent finding that loss-of-function sequence variants occurred in autism with a frequency twice as great as that observed in autism-unaffected individuals. The frequency of LOF genomic changes is also higher than would be expected based on mutation rates determined in control individuals. There is also evidence that the risk of autism is increased in children fathered by older men.

In their 2017 paper, Chaste noted that clear association of autism with specific genome-wide association study variants had not been obtained. They highlighted the CHD8 chromodomain helicase locus as an example of a gene locus that undergoes rare and common variations that influence autism risk. Furthermore, the CHD8 gene product was shown to impact function of many genes that had been shown to impact autism spectrum disorder.

Heterogeneity in autism

Autism is diagnosed on the basis of specific behaviors including primarily impaired social interactions and repetitive behaviors. Clinical and genetic studies have revealed great genetic heterogeneity. There is also evidence for heterogeneity on neuroimaging studies and on neuropathology defined in limited numbers of postmortem studies (Donovan and Bassson, 2017).

Lombardo (2019) have noted that "Big Data" approaches will be necessary to define autism subgroups and to define underlying molecular etiology in different subtypes of autism.

Autism studies: genetic risk variants in families with multiple autism affected children

Ruzzo (2019) carried out whole-genome sequencing on 2308 individuals. They identified 69 autism risk genes. Their study also included identification of structural variants in non–protein-coding genomic regions.

The most significant variants with the highest level of genome significance and with a false discovery rate <0.01 occurred in the gene products listed in the following. The functions of these gene products are indicated:

- CHD8, chromodomain helicase DNA-binding protein 8, transcriptional regulation, remodeling
  - SCN2A, sodium voltage-gated channel alpha subunit 2, generation and propagation of action potentials ARIDB AT rich domain, a component of the SWI/SNF chromatin remodeling complex
  - SYNCAM1 (CADM1) cell adhesion molecule 1
  - DYRK1A, dual specificity tyrosine phosphorylation regulated kinase 1A, role in a signaling pathway
  - CHD2, chromodomain helicase DNA-binding protein 2, modifies chromatin structure and expression
  - ANK2, ankyrin 2, links integrated membrane proteins to the actin and spectrin in cytoskeleton

- KDM5B, lysine demethylase, demethylates tri-, di-, and monomethylated lysine 4 of histone H3
- ADNP, activity-dependent neuroprotector homeobox, likely functions as a transcription factor

It is interesting that many of these proteins are involved in chromatin modification, and gene transcription and some are involved in synaptic function.

Other genes with significant mutations included the following:

- NRXN1, neurexin 1; neurexins are cell-surface receptors that bind neuroligins and facilitate interactions between the presynaptic and the postsynaptic structures
- SHANK2 and SHANK3, synaptic proteins function as scaffolds in the synapse postsynaptic density; PTEN, phosphatase and tensin homolog, preferentially dephosphorylates phosphoinositide substrates
- SETD5, SET domain—containing 5, may function as a histone methyltransferase

Autism, neuronal networks, and gene expression regulation

In 2019, Sullivan et al. noted that heterogeneous phenotypic manifestations occur in autism and that different degrees of impairment of neuronal networks involved in control of social interactions and behavior occur. They also emphasized that deleterious variants in a large number of different genes have been implicated in autism etiology based on results of DNA sequencing studies in very large cohorts of affected individuals and controls. Earlier studies of gene defects in autism provided evidence of roles of deleterious variants in genes and/or structural genomic variants in synaptic genes in autism etiology.

More recent studies, e.g., those published by De Rubeis (2014) and Satterstrom (2020) provided evidence for variants that increased autism risk including not only those that likely impaired synaptic function but also variants that impacted chromatin modifications, transcription, and regulation of gene expression.

Studies by Williams (2019) led to identification in autistic individuals of de novo mutations in non—protein-coding regions of the genome and particularly mutations in distal promoter regions of genes.

Evidence for transcriptional dysregulation and splicing defects in autism

Studies by Raj and Blencowe (2015) that brain mRNA transcripts are enriched in a specific class of genes that contain short exons. This category of gene transcripts was reported to be particularly dysregulated in mRNA in brain samples from autistic individuals (Irimia et al., 2014). Other transcription alterations were reported in results of postmortem brain mRNA sequencing studies reported by Gandal (2018).

Perturbed epigenetic and transcriptional processes in autism

With respect to genes that encode proteins involved in epigenetic processes, studies in large numbers of autistic individuals have revealed deleterious variants particularly in genes involved in chromatin modification and epigenetic processes (Sullivan et al., 2019). These genes include chromatin modifiers and genes involved in methylation or demethylation processes and genes involved in transcription regulation.

Sullivan et al. drew particular attention to altered transcription of long genes in cases of autism. In addition, Zhao et al. (2019) reported that genes implicated in autism risk harbored broad enhancer-like domains. These will be discussed further in the following.

Sullivan et al. noted that in more recent autism studies, attention was being directed away from individual genes and toward impaired gene regulatory networks that through genetic and environmental factors were impacted. Impacted regulation could then lead to altered regulation and impacted neuronal function. On the basis of studies in the mouse models of autism, they noted that there were possibilities to alter gene expression and to impact autism manifestations. Examples included studies on mice with autism manifestations due to Shank3 deficiency. Studies revealed reversibility of manifestation with enhancement of Shank3 expression. In mouse models with autistic features due to Mecp2 deficiency, specific therapeutic applications that enhanced Mecp2 function led to improvements in manifestations. Sullivan et al. also drew attention to reports that in a certain percentage of human autism cases in children, disorder manifestations improved during febrile episodes.

Quesnel Vallieres (2019) reviewed how transcriptome analyses were shedding light on potential pathways and processes involved in autism. Evidence for the role of genetic factors in autism causation emerged from twin studies and family studies. However, variable phenotypic presentation occurs even in monozygotic twin with autism and in different affected members within one family.

Different methods of transcriptome analyses have been applied in autism, and transcriptome studies have also been undertaken on postmortem brain samples in autism, and these included particularly RNA expression and sequencing studies. Quesnel Vallieres (2019) noted that transcription abnormalities were reported to occur in three different brain regions including frontal cortex, superior temporal cortex, and cerebellum. Specific studies reported alterations in expression of glutamate transporter genes and glutamate receptor genes. They also reported altered expression frequently included increased immune function genes including genes associated with microglial and astroglial functions. There was also evidence of altered regulation and expression of microRNAs in autism and of long noncoding RNAs. Quesnel Vallieres noted that a number of different studies have provided evidence for altered patterns of splicing of mRNA transcripts in autism and in certain other diseases that impact neurodevelopment. Neuronal gene transcripts were reported to be rich in small microexons 3–27 base pair in length and to have altered splicing that includes inclusion of microexons, Microexon inclusion was shown to be in part dependent on adequate functions of SRRM proteins including SRRM1 and SRRM4, proteins known to play roles in splicing, and levels of these proteins were reported to be altered in some autism brains. Additional factors that may be implicated in the altered splicing patterns in brain include RNA-binding proteins NOVA1 and polypyrimidine-binding protein PTBP1. Altered expression of SNO RNAs was also reported in some cases. SNO RNAs are small nucleolar RNAs reported to play roles in modification of other RNAs.

Tran (2019) noted that RNA processing and modification of RNA sequences are also important to consider in the context of altered levels of gene products. ADAR enzymes that change adenosine to inosine constitute the major RNA alteration mechanism. These modifications in RNA lead not only to alteration in the proteins generated but also impact RNA stability and splicing. Tran et al. identified a core set of downregulated genes in autism. These genes were found to be involved in glutamatergic signaling and synaptic pathways. Hypoediting leading to reduction in RNA editing in frontal cortex in autism was documented.

Effects of specific chromosome changes on autism manifestations

Bertero (2018) reported data on brain activity in autism cases with microdeletion of chromosome 16p11.2 and studies on mice with deletion of the corresponding genomic region. They emphasized the importance of determining how the different genetic changes identified in autism impacted brain connectivity and behavior.

Microdeletion of chromosome 16p11.2 was reported to be one of the most common chromosomal CNVs associated with autism. This chromosomal region was reported to contain 29 different genes. Hanson (2015) reported that individuals with 16p11.2 deletions had a high frequency of speech and language defects, and they had a range of intellectual abilities; however, IQ scores were generally 26 points lower than those of their family members who did not have the 16p11.2 deletion. All patients with 16p11.2 deletion met criteria for autism diagnosis.

Bertero et al. carried out resting-state function MRI (RSHMRI) studies on 19 cases with 16p11.2 deletion who ranged between 8 and 16 years of age, in which there are 10 males and 9 females. They also carried out RSFMRI on 28 control subjects who ranged between 7 and 16 years of age, in which there are 16 males and 12 females. Imaging data were processed to control motion contamination. In 16p11.2 deletion individuals, prefrontal connectivity was reduced, and multiple connectivity reductions were also observed in the lateral-temporal cortex and in the inferior parietal lobe.

Studies on mice with deletion of the genomic regions that corresponded to human 16p11.2 revealed reduced prefrontal connectivity, thalamoprefrontal connectivity, and reduced neural synchronization. Subsequent studies on mouse brain slices revealed alterations in dendritic spines.

Additional evidence for functional connectivity changes in autism

Holiga (2019) reported finding reproducible functional connectivity alterations in autism. They carried out resting-state MRI studies in individuals in four large autism study cohorts. Their studies demonstrated patterns of hypoconnectivity in some areas and hyperconnectivity in other regions. Each of the four groups studies included autistic individuals (841) and typically developing individuals (984). Clinical parameters studied included ADOS and ADI VABS (Vineland Adaptive Behavior Scale). Functional activity was expressed as degree centrality. Significant differences in degree centrality scores in four different anatomical regions were determined between ASD individuals and typically developing individual. Studies were carried out to determine if there were correlations between imaging data and clinical features.

Hyperconnectivity scores were shown to be correlated with ADI measures and VABS daily living skills but not with VABS communication subscales. Greater connectivity was associated with degree of symptom severity. Consistent effects of psychotropic medication on degree centrality scores were not demonstrated.

11. Attention deficit hyperactivity disorder

Posner (2020) reviewed ADHD and noted that this diagnosis represents an evolving construct. Recent modifications in the description of ADHD diagnostic criteria have appeared in the DSM5 version of the *Diagnostic and Statistical Manual* of mental disease and in ICD-11, the Internal Classification of Diseases.

In *DSM5*, ADHD is defined as a lifespan neurodevelopmental condition with specific diagnostic criteria for children and for adults. ADHD is a clinical diagnosis. Major diagnostic criteria in *DSM5* include inattention, hyperactivity, and impulsivity. In ICD11, major diagnostic criteria for ADHD include persistent pattern (at least for 6 months) of inattention, hyperactivity, impulsivity, onset early- to midchildhood, and symptoms that interfere with academic, occupational, or social functioning.

Posner et al. noted that when the same diagnostic criteria are applied and there are not methodological differences between studies, the incidence of ADHD across different countries is highly similar 5.01%–5.56%. Posner et al. noted that ADHD is often comorbid with other psychiatric conditions.

With respect to therapy, they noted that specific medication–based treatments with various psychostimulants have proven useful. However, they are not useful in all cases. In addition, the educational and social outcomes of specific treatment usage have proven to be areas of concern. Posner et al. noted that nonpharmacological treatment approaches have been found to be of limited use.

Etiology of attention deficit hyperactivity disorder

The high heritability of ADHD indicates roles for genetic factors. Earlier studies indicated possible relevance of variants in dopamine receptors and dopamine transporters (Faraone 2005). However, variations in these genes have not proven to be of importance in ADHD etiology across different populations. Faraone and Larson (2019) reported the polygenic heritability of ADHD was obtained in at least one-third of cases.

Premature birth and low birth weight have been shown to have some relevance to ADHD symptom development. Posner et al. noted the plausibility of the hypothesis that gene–environment interactions play roles in causation of ADHD.

Future approaches in understanding attention deficit hyperactivity disorder etiology

Posner et al. noted that ADHD diagnosis should be considered as the extreme end along a continuum of behavior variation. They also noted that ADHD is of heterogeneous etiology and that heterogeneity almost certainly exists in genetic risk factors, in brain structure, and in predisposing environmental factors.

Genome-wide association studies in attention deficit hyperactivity disorder

Demontis et al. (2020) carried out GWAS studies on 29,183 individuals with ADHD and 35,191 controls. They identified 12 genome-wide significant loci. One locus in particular had highly significant association with the disorder. This locus was detected with a single nucleotide variant rs11420276 and is located on the short arm of chromosome 1 at base pair 44,184,192. It is located in the intron of a gene ST3GAL3 that encodes ST3 beta-galactoside-alpha-2,3-sialyltransferase 3. It is not known if the variant impacts expression of this gene or of any other genes. Eleven other loci with GWAS significant scores were mapped across 12 different study cohorts.

References

Alcantara, D., O'Driscoll, M., 2014. Congenital microcephaly. Am. J. Med. Genet. C Semin. Med. Genet. 166C (2), 124–139. https://doi.org/10.1002/ajmg.c.31397.

Almannai, M., 2019. Carnitine inborn errors of metabolism. Molecules 24 (18), 3251. https://doi.org/10.3390/molecules24183251.

Almeida, M., 2016. Portuguese family with the co-occurrence of frontotemporal lobar degeneration and neuronal ceroid lipofuscinosis phenotypes due to progranulin gene mutation. Neurobiol. Aging 41 (200), e1–e5. https://doi.org/10.1016/j.neurobiolaging.2016.02.019.

Amir, R., 1999. Rett syndrome is caused by mutations in X-linked MECP2, encoding methyl-CpG-binding protein 2. Nat. Genet. 23 (2), 185–188. https://doi.org/10.1038/13810.

Bagh, M.B., Peng, S., Chandra, G., Zhang, Z., Singh, S.P., et al., 2017. Misrouting of v-ATPase subunit V0a1 dysregulates lysosomal acidification in a neurodegenerative lysosomal storage disease model. Nat. Commun. 8, 14612. https://doi.org/10.1038/ncomms14612.

Barca, E., 2018. USMG5 Ashkenazi Jewish founder mutation impairs mitochondrial complex V dimerization and ATP synthesis. Hum. Mol. Genet. 27 (19), 3305–3312. https://doi.org/10.1093/hmg/ddy231.

Beales, P., 1999. New criteria for improved diagnosis of Bardet-Biedl syndrome: results of a population survey. J. Med. Genet. 36 (6), 437–446. http://PMCID: PMC1734378.

Bertero, A., 2018. Autism-associated 16p11.2 microdeletion impairs prefrontal functional connectivity in mouse and human. Brain 141 (7), 2055–2065. https://doi.org/10.1093/brain/awy111.

Bjornsson, H.T., 2015. The Mendelian disorders of the epigenetic machinery. Genome Res. 25 (10), 1473–1481. https://doi.org/10.1101/gr.190629.115.

Blackburn, P., 2017. Maple syrup urine disease: mechanisms and management. Appl. Clin. Genet. 10, 57–66. https://doi.org/10.2147/TACG.S125962.

Bocznadi, V., 2018. Mitochondrial DNA transcription and translation: clinical syndromes. Essay. Biochem. 62 (3), 321–340. https://doi.org/10.1042/EBC20170103.

Boivin, M., 2015. Reducing neurodevelopmental disorders and disability through research and interventions. Nature 527 (7578), S155–S160. https://doi.org/10.1038/nature16029.

Braissant, O., 2013. Ammonia toxicity to the brain. J. Inherit. Metab. Dis. 36 (4), 595–612. https://doi.org/10.1007/s10545-012-9546-2.

Burkardt, D., 2019. Approach to overgrowth syndromes in the genome era. Am. J. Med. Genet. C Semin. Med. Genet. 181 (4), 483–490. https://doi.org/10.1002/ajmg.c.31757.

Carroll, B., Dunlop, E., 2017. The lysosome: a crucial hub for AMPK and mTORC1 signalling. Biochem. J. 474 (9), 1453–1466. https://doi.org/10.1042/BCJ20160780.

Chan, D., 2020. Mitochondrial dynamics and its involvement in disease. Annu. Rev. Pathol. 15, 235–259. https://doi.org/10.1146/annurev-pathmechdis-012419-032711.

Chaste, P., 2017. The yin and yang of autism genetics: how rare de novo and common variations affect liability. Annu. Rev. Genom. Hum. Genet. 18, 167–187. https://doi.org/10.1146/annurev-genom-083115-022647.

Colacurcio, D., Nixon, R., 2016. Disorders of lysosomal acidification-The emerging role of v-ATPase in aging and neurodegenerative disease. Ageing Res. Rev. 32, 75–88. https://doi.org/10.1016/j.arr.2016.05.004.

Cotter, K., 2015. Recent insights into the structure, regulation, and function of the V-ATPases. Trends Biochem. Sci. 40 (10), 611–622. https://doi.org/10.1016/j.tibs.2015.08.005.

Cyrus, S., 2019. PRC2-complex related dysfunction in overgrowth syndromes: a review of EZH2, EED, and SUZ12 and their syndromic phenotypes. Am. J. Med. Genet. C Semin. Med. Genet. 181 (4), 519–531. https://doi.org/10.1002/ajmg.c.31754.

Davenport, E., 2019. Autism and schizophrenia-associated CYFIP1 regulates the balance of synaptic excitation and inhibition. Cell Rep. 26 (8), 2037–2051. https://doi.org/10.1016/j.celrep.2019.01.092.

De Rubeis, S., 2013. CYFIP1 coordinates mRNA translation and cytoskeleton remodeling to ensure proper dendritic spine formation. Neuron 79 (6), 1169–1182. https://doi.org/10.1016/j.neuron.2013.06.039.

De Rubeis, S., 2014. Synaptic, transcriptional and chromatin genes disrupted in autism. Nature 15 (7526), 209–215. https://doi.org/10.1038/nature13772.

de Vries, P., 2018. TSC-associated neuropsychiatric disorders (TAND): findings from the TOSCA natural history study. Orphanet J. Rare Dis. 13 (1), 157. http://10.1186/s13023-018-0901-8.

Demontis, D., et al., 2020. Discovery of the first genome-wide significant risk loci for attention deficit/hyperactivity disorder. Neuropharmacology 171 (107851). https://doi.org/10.1016/j.neuropharm.2019.107851.

Deshpande, A., Weiss, L., 2018. Recurrent reciprocal copy number variants: roles and rules in neurodevelopmental disorders. Dev. Neurobiol. 78 (5), 519–530. https://doi.org/10.1002/dneu.22587.

Devlin, B., Scherer, S., 2012. Genetic architecture in autism spectrum disorder. Curr. Opin. Genet. Dev. 22 (3), 229–237. https://doi.org/10.1016/j.gde.2012.03.002.

Di Donato, N., Timms, A.E., Aldinger, K.A., Mirzaa, G.M., Bennett, J.T., Collins, S., Olds, C., Mei, D., Chiari, S., Carvill, G., Myers, C.T., Rivière, J.B., Zaki, M.S., 2018. Analysis of 17 genes detects mutations in 81% of 811 patients with lissencephaly. Genet. Med. 20 (11), 1354–1364. https://doi.org/10.1038/gim.2018.8.

Dobyns, W., 2018. MACF1 mutations encoding highly conserved zinc-binding residues of the GAR domain cause defects in neuronal migration and axon guidance. Am. J. Hum. Genet. 103 (6), 1009–1021. https://doi.org/10.1016/j.ajhg.2018.10.019.

Dogterom, M., Koenderink, G., 2019. Actin-microtubule crosstalk in cell biology. Nat. Rev. Mol. Cell Biol. 20 (1), 38–54. https://doi.org/10.1038/s41580-018-0067-1.

Domínguez-Iturza, N., 2019. The autism- and schizophrenia-associated protein CYFIP1 regulates bilateral brain connectivity and behaviour. Nat. Commun. 10 (1), 3454. https://doi.org/10.1038/s41467-019-11203-y.

D'Abate, L., 2019. Predictive impact of rare genomic copy number variations in siblings of individuals with autism spectrum disorders. Nat. Commun. 10 (1) https://doi.org/10.1038/s41467-019-13380-2.

Donovan, A., Basson, M., 2017. The neuroanatomy of autism - a developmental perspective. J. Anat. 230 (1), 4–15. https://doi.org/10.1111/joa.12542. PMID:27620360.

D'Souza, A.R., Minczuk, M., 2018. Mitochondrial transcription and translation: overview. Essays Biochem. 62 (3), 309–320. https://doi.org/10.1042/EBC20170102.

Ebrahimi-Fakhari, D., Saffari, A., Wahlster, L., Di Nardo, A., Turner, D., et al., 2016. Impaired mitochondrial dynamics and mitophagy in neuronal models of tuberous sclerosis complex. Cell. Rep. 17 (4), 1053–1070. https://doi.org/10.1016/j.celrep.2016.09.054. Erratum in: Cell Rep. 2016 Nov 15;17 (8):2162. PMID:27760312.

Ervin, D., 2014. Healthcare for persons with intellectual and developmental disability in the community. Front Public Health 2 (83). https://doi.org/10.3389/fpubh.2014.00083.

Fagan, N., 2017. Magnetic resonance imaging findings of central nervous system in lysosomal storage diseases: a pictorial review. J. Med. Imag. Radiat Oncol. 61 (3), 344–352. https://doi.org/10.1111/1754-9485.12569.

Fahrner, J.A., Bjornsson, H.T., 2019. Mendelian disorders of the epigenetic machinery: postnatal malleability and therapeutic prospects. Hum. Mol. Genet. 28 (R2), R254–R264. https://doi.org/10.1093/hmg/ddz174. PMID: 31595951.

Faraone, S., 2005. The scientific foundation for understanding attention-deficit/hyperactivity disorder as a valid psychiatric disorder. Eur. Child Adolesc. Psychiatr. 14 (1), 1–10. https://doi.org/10.1007/s00787-005-0429-z.

Faraone, S.V., Larsson, H., 2019. Genetics of attention deficit hyperactivity disorder. Mol. Psychiatry 24 (4), 562–575. https://doi.org/10.1038/s41380-018-0070-0. PMID:2989 2054.

Fassio, A., Esposito, A., Kato, M., Saitsu, H., Mei, D., Carla Marini, C., et al., 2018. De novo mutations of the ATP6V1A gene cause developmental encephalopathy with epilepsy. Brain 141 (6), 1703–1718. https://doi.org/10.1093/brain/awy092. PMID:29668857.

Fettelschoss, V., 2017. Clinical or ATPase domain mutations in ABCD4 disrupt the interaction between the vitamin B 12-trafficking proteins ABCD4 and LMBD1. J. Biol. Chem. 292 (28), 11980–11991. https://doi.org/10.1074/jbc.M117.784819.

Fine, S.A., 2019. Mitochondrial aminoacyl-tRNA synthetase disorders: an emerging group of developmental disorders of myelination. J. Neurodev. Disord. 11 (1), 29. https://doi.org/10.1186/s11689-019-9292-y.

Fons, C., Campistol, J., 2016. Creatine defects and central nervous system. Semin. Pediatr. Neurol. 23 (4), 285–289. https://doi.org/10.1016/j.spen.2016.11.003.

Freeze, H., 2015. Neurological aspects of human glycosylation disorders. Annu. Rev. Neurosci. 38, 105–125. https://doi.org/10.1146/annurev-neuro-071714-034019.

Frotscher, M., Zhao, S., Wang, S., Chai, X., 2017. Reelin signaling inactivates cofilin to stabilize the cytoskeleton of migrating cortical neurons. Front. Cell Neurosci. 11, 148. https://doi.org/10.3389/fncel.2017.00148. eCollection 2017. Review. PMID:28588454.

Fry, A., 2014. The genetics of lissencephaly. Am. J. Med. Genet. C Semin. Med. Genet. 166 (C2), 198–210. https://doi.org/10.1002/ajmg.c.31402.

Fukao, T., Sasai, H., Aoyama, Y., Otsuka, H., Ago, Y., et al., 2019. Recent advances in understanding beta-ketothiolase (mitochondrial acetoacetyl-CoA thiolase, T2) deficiency. J. Hum. Genet. 64 (2), 99–111. https://doi.org/10.1038/s10038-018-0524-x.

Gahl, W., 1987. Disorders of lysosomal membrane transport–cystinosis and Salla disease. Enzyme 38 (1–4), 154–160. https://doi.org/10.1159/000469201.

Gandal, M., 2018. Transcriptome-wide isoform-level dysregulation in ASD, schizophrenia, and bipolar disorder. Science 362 (6420), eaat8127. https://doi.org/10.1126/science.aat8127.

Garcia, A., 2018. The elegance of sonic hedgehog: emerging novel functions for a classic morphogen. J. Neurosci. 38 (44), 9338–9345. https://doi.org/10.1523/JNEUROSCI.1662-18.2018.

Ghezzi, D., Zeviani, M., 2018. Human diseases associated with defects in assembly of OXPHOS complexes. Essay. Biochem. 62 (3), 271–286. https://doi.org/10.1042/EBC20170099.

Goncalves, F., 2018. Tubulinopathies. Top. Magn. Reson. Imag. 27 (6), 395–408. https://doi.org/10.1097/RMR.0000000000000188.

Gonzalez, A., Valeiras, M., Sidransky, E., Tayebi, N., 2014. Lysosomal integral membrane protein-2: a new player in lysosome-related pathology. Mol. Genet. Metab. 111 (2), 84–91. https://doi.org/10.1016/j.ymgme.2013.12.005. PMID:24389070.

Griffiths, P.D., Bradburn, M., Campbell, M.J., Cooper, C.L., Embleton, N., et al., 2019. MRI in the diagnosis of fetal developmental brain abnormalities: the MERIDIAN diagnostic accuracy study. Health Technol. Assess. 23 (49), 1–144. https://doi.org/10.3310/hta23490.

Guarnieri, F., 2018. Disorders of neurogenesis and cortical development. Dialogues Clin. Neurosci. 20 (4), 255–266. https://doi.org/10.31887/DCNS.2018.20.4/ccardoso.

Guerrero-Castillo, S., 2017. The assembly pathway of mitochondrial respiratory chain complex I. Cell Metabol. 25 (1), 128–139. https://doi.org/10.1016/j.cmet.2016.09.002.

Hahn, J.S., Barnes, P.D., Clegg, N.G., Stashinko, E.E., 2010. Septopreoptic holoprosencephaly: a mild subtype associated with midline craniofacial anomalies. Am. J. Neuroradiol. 31 (9), 1596–1601. https://doi.org/10.3174/ajnr.A2123. PMID:20488907.

Hanson, E., 2015. The cognitive and behavioral phenotype of the 16p11.2 deletion in a clinically ascertained population. Biol. Psychiatr. 77 (9), 785–793. https://doi.org/10.1016/j.biopsych.2014.04.021.

Hart, A.R., Embleton, N.D., Bradburn, M., Connolly, D.J.A., Mandefield, L., 2020. Accuracy of in-utero MRI to detect fetal brain abnormalities and prognosticate developmental outcome: postnatal follow-up of the MERIDIAN cohort. Lancet Child Adolesc. Health. 4 (2), 131–140. https://doi.org/10.1016/S2352-4642(19)30349-9.

Hasbani, D., Crino, P., 2018. Tuberous sclerosis complex. Handb. Clin. Neurol. 148, 813–822. https://doi.org/10.1016/B978-0-444-64076-5.00052-1.

Holiga, S., 2019. Patients with autism spectrum disorders display reproducible functional connectivity alterations. Sci. Transl. Med. 11 (481) https://doi.org/10.1126/scitranslmed.aat9223 eaat9223.

Irimia, M., Weatheritt, R.J., Ellis, J.D., Parikshak, N.N., Gonatopoulos-Pournatzis, T., Babor, M., et al., 2014. A highly conserved program of neuronal microexons is misregulated in autistic brains. Cell. 159 (7), 1511–1523.

Iwase, S., 2017. Epigenetic etiology of intellectual disability. J. Neurosci. 37 (45), 10773–10782. https://doi.org/10.1523/JNEUROSCI.1840-17.2017.

Khacho, M., Slack, R., 2018. Mitochondrial dynamics in the regulation of neurogenesis: from development to the adult brain. Dev. Dynam. 247 (1), 47–53. https://doi.org/10.1002/dvdy.24538.

Kline, A., 2018. Diagnosis and management of Cornelia de Lange syndrome: first international consensus statement. Nat. Rev. Genet. 19 (10), 649–666. https://doi.org/10.1038/s41576-018-0031-0.

Kousi, M., Katsanis, N., 2016. The genetic basis of hydrocephalus. Annu. Rev. Neurosci. 39, 409–435. https://doi.org/10.1146/annurev-neuro-070815-014023.

Kruszka, P., Berger, S.I., Casa, V., Dekker, M.R., Gaesser, J., 2019. Cohesin complex associated holoprosencephaly. Brain 142 (9), 2631–2643. https://doi.org/10.1093/brain/awz210. PMID:31334757.

Kruszka, P., Muenke, M., 2018. Syndromes associated with holoprosencephaly. Am. J. Med. Genet. C Semin. Med. Genet. 178 (2), 229–237. https://doi.org/10.1002/ajmg.c.31620. PMID:29770994.

Kurolap, A., Armbruster, A., Hershkovitz, T., Hauf, K., Mory, A., et al., 2016. Loss of glycine transporter 1 causes a subtype of glycine encephalopathy with arthrogryposis and mildly elevated cerebrospinal fluid glycine. Am. J. Hum. Genet. 99 (5), 1172–1180. https://doi.org/10.1016/j.ajhg.2016.09.004.

Lewis, J.D., Meehan, R.R., Henzel, W.J., Maurer-Fogy, I., Jeppesen, P., et al., 1992. Purification, sequence, and cellular localization of a novel chromosomal protein that binds to methylated DNA. Cell 69 (6), 905–914. https://doi.org/10.1016/0092-8674(92)90610-o. PMID:1606614.

Lindstrand, A., 2019. From cytogenetics to cytogenomics: whole-genome sequencing as a first-line test comprehensively captures the diverse spectrum of disease-causing genetic variation underlying intellectual disability. Genome. Med. 11 (1), 68. https://doi.org/10.1186/s13073-019-0675-1.

Lipstein, N., 2017. Presynaptic Calmodulin targets: lessons from structural proteomics. Expert Rev. Proteomics 14 (3), 223–242. https://doi.org/10.1080/14789450.2017.1275966.

Lombardo, M., 2019. Big data approaches to decomposing heterogeneity across the autism spectrum. Mol. Psychiatr. 24 (10), 1435–1450. https://doi.org/10.1038/s41380-018-0321-0.

Lyst, M.J., Bird, A., 2015. Rett syndrome: a complex disorder with simple roots. Nat. Rev. Genet. 16 (5), 261–275. https://doi.org/10.1038/nrg3897. PMID:25732612.

Marcotte, L., 2012. Cytoarchitectural alterations are widespread in cerebral cortex in tuberous sclerosis complex. Acta Neuropathol. 123 (5), 685–693. https://doi.org/10.1007/s00401-012-0950-3.

Mercier, S., 2011. New findings for phenotype-genotype correlations in a large European series of holoprosencephaly cases. J. Med. Genet. 48 (11), 752–760. https://doi.org/10.1136/jmedgenet-2011-100339.

Mithyantha, R., 2017. Current evidence-based recommendations on investigating children with global developmental delay. Arch. Dis. Child. 102 (11), 1071–1076. https://doi.org/10.1136/archdischild-2016-311271.

Mole, S., 2019. Clinical challenges and future therapeutic approaches for neuronal ceroid lipofuscinosis. Lancet Neurol. 18 (1), 107–116. https://doi.org/10.1016/S1474-4422(18)30368-5.

Molnar, M., Kovacs, G., 2017. Mitochondrial diseases. Handb. Clin. Neurol. 145, 147–155. https://doi.org/10.1016/B978-0-12-802395-2.00010-9.

Moslehi, M., 2017. Dynamic microtubule association of Doublecortin X (DCX) is regulated by its C-terminus. Sci. Rep. 7 (1), 5245. https://doi.org/10.1038/s41598-017-05340-x.

Mukherjee, A., 2019. Emerging new roles of the lysosome and neuronal ceroid lipofuscinoses. Mol. Neurodegener. 16 (14), 4. https://doi.org/10.1186/s13024-018-0300-6 (1).

Muraresku, C., 2018. Mitochondrial Disease: advances in clinical diagnosis, management, therapeutic development, and preventative strategies. Curr. Genet. Med. Rep. 6 (2), 62–72. https://doi.org/10.1007/s40142-018-0138-9.

Nakashima, M., 2018. De novo hotspot variants in CYFIP2 cause early-onset epileptic encephalopathy. Ann. Neurol. 83 (4), 794–806. https://doi.org/10.1002/ana.25208.

Oguro-Ando, A., Rosensweig, C., Herman, E., Nishimura, Y., Werling, D., et al., 2015. Increased CYFIP1 dosage alters cellular and dendritic morphology and dysregulates mTOR. Mol. Psychiatry 20 (9), 1069–1078. https://doi.org/10.1038/mp.2014.124. PMID:25311365.

Palmer, E.E, Mowat, D., 2014. Agenesis of the corpus callosum: a clinical approach to diagnosis. Am. J. Med. Genet. C Semin. Med. Genet. 166 (C2), 184–197. https://doi.org/10.1002/ajmg.c.31405May 27. PMID:24866859.

Parrini, E., Conti, V., Dobyns, W.B., Guerrini, R., 2016. Genetic basis of brain malformations. Mol. Syndromol. 7 (4), 220–233. https://doi.org/10.1159/000448639. Review. PMID:27781032.

Pastores, G., Maegawa, G., 2013. Clinical neurogenetics: neuropathic lysosomal storage disorders. Neurol. Clin. 31 (4), 1051–1071. https://doi.org/10.1016/j.ncl.2013.04.007.

Pathania, M., 2014. The autism and schizophrenia associated gene CYFIP1 is critical for the maintenance of dendritic complexity and the stabilization of mature spines. Transl. Psychiat. 4 (3) https://doi.org/10.1038/tp.2014.16. e374.

Perenthaler, E., 2019. Beyond the exome: the non-coding genome and enhancers in neurodevelopmental disorders and malformations of cortical development. Front. Cell. Neurosci. 13 (352) https://doi.org/10.3389/fncel.2019.00352.

Pilotto, A., Blau, N., Leks, E., Schulte, C., Deuschl, C., et al., 2019. Cerebrospinal fluid biogenic amines depletion and brain atrophy in adult patients with phenylketonuria. J. Inherit. Metab. Dis. 42 (3), 398–406. https://doi.org/10.1002/jimd.12049.

Pirrozi, F., 2018. From microcephaly to megalencephaly: determinants of brain size. Dialogues Clin. Neurosci. 20 (4), 267–282. https://doi.org/10.31887/DCNS.2018.20.4/gmirzaa.

Platt, F., 2018. Lysosomal storage diseases. Nat. Rev. Dis. Prime. 4 (1), 27. https://doi.org/10.1038/s41572-018-0025-4.

Posner, J., 2020. Attention-deficit hyperactivity disorder. Lancet 395 (10222), 450–462. https://doi.org/10.1016/S0140-6736(19)33004-1.

Quesnel-Vallières, M., 2019. Autism spectrum disorder: insights into convergent mechanisms from transcriptomics. Nat. Rev. Genet. 20 (1), 51–63. https://doi.org/10.1038/s41576-018-0066-2.

Raj, B., Blencowe, B., 2015. Alternative splicing in the mammalian nervous system: recent insights into mechanisms and functional roles. Neuron 87 (1), 14–27. https://doi.org/10.1016/j.neuron.2015.05.004.

Reid, E., 2016. Advantages and pitfalls of an extended gene panel for investigating complex neurometabolic phenotypes. Brain 139 (11), 2844–2854. https://doi.org/10.1093/brain/aww221.

Reiter, J.F., Leroux, M.R., 2017. Genes and molecular pathways underpinning ciliopathies. Nat. Rev. Mol. Cell Biol. 18 (9), 533–547. https://doi.org/10.1038/nrm.2017.60. PMID:28698599.

Rodan, L., 2018. Gain-of-function variants in the ODC1 gene cause a syndromic neurodevelopmental disorder associated with macrocephaly, alopecia, dysmorphic features, and neuroimaging abnormalities. Am. J. Med. Genet. 176 (12), 2554–2560. https://doi.org/10.1002/ajmg.a.60677.

Roessler, E., Belloni, E., Gaudenz, K., Jay, P., Berta, P., Scherer, S.W., et al., 1996. Mutations in the human sonic hedgehog gene cause holoprosencephaly. Nat. Genet. 14 (3), 357–360. https://doi.org/10.1038/ng1196-357. PMID:8896572.

Roland, J., 2017. On the role of the corpus callosum in interhemispheric functional connectivity in humans. Proc. Natl. Acad. Sci. U. S. A. 114 (50), 13278–13283. https://doi.org/10.1073/pnas.1707050114.

Ruzzo, E., 2019. Inherited and de novo genetic risk for autism impacts shared networks. Cell 178 (4), 850–866. https://doi.org/10.1016/j.cell.2019.07.015.

Satterstrom, F., 2020. Large-scale exome sequencing study implicates both developmental and functional changes in the neurobiology of autism. Cell 180 (3). https://doi.org/10.1016/j.cell.2019.12.036 e23.

Schillaci, L.-A., 2018. Inborn errors of metabolism with acidosis: organic acidemias and defects of pyruvate and ketone body metabolism. Pediatr. Clin. 65 (2), 209–230. https://doi.org/10.1016/j.pcl.2017.11.003.

Schiller, S., Rosewich, H., Grünewald, S., Gärtner, J., 2020. Inborn errors of metabolism leading to neuronal migration defects. J. Inherit. Metab. Dis. 43 (1), 145–155. https://doi.org/10.1002/jimd.12194.

Schwake, M., 2013. Lysosomal membrane proteins and their central role in physiology. Traffic 14 (7), 739–748. https://doi.org/10.1111/tra.12056.

Settembre, C., 2013. Signals from the lysosome: a control centre for cellular clearance and energy metabolism. Nat. Rev. Mol. Cell Biol. 14 (5), 283–296. https://doi.org/10.1038/nrm3565.

Shapiro, E., 2017. Assessments of neurocognitive and behavioral function in the mucopolysaccharidoses. Mol. Genet. Metabol. 122 (S), 8–16. https://doi.org/10.1016/j.ymgme.2017.09.007.

Shohayeb, B., Ho, U., Yeap, Y.Y., Parton, R.G., Millard, S.S., 2019. The association of microcephaly protein WDR62 with CPAP/IFT88 is required for cilia formation and neocortical development. Hum. Mol. Genet. https://doi.org/10.1093/hmg/ddz281 pii: ddz281.

Siegmund, S., 2018. Three-dimensional analysis of mitochondrial crista ultrastructure in a patient with Leigh syndrome by in situ cryoelectron tomography. IScience 6, 83–91. https://doi.org/10.1016/j.isci.2018.07.014.

Signes, A., Fernandez-Vizarra, E., 2018. Assembly of mammalian oxidative phosphorylation complexes I-V and supercomplexes. Essay. Biochem. 62 (3), 255–270. https://doi.org/10.1042/EBC20170098.

Silva, A., 2019. Cyfip1 haploinsufficient rats show white matter changes, myelin thinning, abnormal oligodendrocytes and behavioural inflexibility. Nat. Commun. 10 (1), 3455 https://doi.org/1038/s41467-019-11119-7.

Skilrou, E., Lichter-Konecki, U., 2018. Inborn errors of metabolism with cognitive impairment: metabolism defects of phenylalanine, homocysteine and methionine, purine and pyrimidine, and creatine. Pediatr. Clin. 65 (2), 267–277. https://doi.org/10.1016/j.pcl.2017.11.009.

Sønderby, I.E., 2020. Dose response of the 16p11.2 distal copy number variant on intracranial volume and basal ganglia. Science 367 (6484), eaay6690. https://doi.org/10.1038/s41380-018-0118-1. PMID:30283035.

Srivastava, S., 2019. Meta-analysis and multidisciplinary consensus statement: exome sequencing is a first-tier clinical diagnostic test for individuals with neurodevelopmental disorders. Genet. Med. 21 (11), 2413–2421. https://doi.org/10.1038/s41436-019-0554-6.

Stence, N., 2019. Brain imaging in classic nonketotic hyperglycinemia: quantitative analysis and relation to phenotype. J. Inherit. Metab. Dis. 42 (3), 438–450. https://doi.org/10.1002/jimd.12072.

Sullivan, J.M., De Rubeis, S., Schaefer, A., 2019. Convergence of spectrums: neuronal gene network states in autism spectrum disorder. Curr. Opin. Neurobiol. 59, 102–111. https://doi.org/10.1016/j.conb.2019.04.011. Epub 2019 Jun 18. PMID:31220745.

Summar, M., Ah Mew, N., 2018. Inborn errors of metabolism with hyperammonemia: urea cycle defects and related disorders. Pediatr. Clin. 65 (2), 231–246. https://doi.org/10.1016/j.pcl.2017.11.004.

Swanson, M., 2015. Biochemical and molecular predictors for prognosis in nonketotic hyperglycinemia. Ann. Neurol. 78 (4), 606–618. https://doi.org/10.1002/ana.24485.

Thurm, A., 2016. Creatine transporter deficiency: screening of males with neurodevelopmental disorders and neurocognitive characterization of a case. J. Dev. Behav. Pediatr. 37 (4), 322–326. https://doi.org/10.1097/DBP.0000000000000299.

Toledano-Zaragoza, A., Ledesma, M.D., 2019. Addressing neurodegeneration in lysosomal storage disorders: advances in Niemann Pick diseases. Neuropharmacology 107851. https://doi.org/10.1016/j.neuropharm.2019.107851. PMID: 31734384.

Tran, S., 2019. Widespread RNA editing dysregulation in brains from autistic individuals. Nat. Neurosci. 22 (1), 25–36. https://doi.org/10.1038/s41593-018-0287-x.

Trillo-Contreras, J., 2019. AQP1 and AQP4 contribution to cerebrospinal fluid homeostasis. Cells 8 (2), 197. https://doi.org/10.3390/cells8020197.

Valente, E., 2013. Joubert syndrome and related disorders. Handb. Clin. Neurol. 113, 1879–1888. https://doi.org/10.1016/B978-0-444-59565-2.00058-7.

van Eeghen, A.M., Ortiz-Terán, L.O., Johnson, J., Pulsifer, M.B., Thiele, E.A., Caruso, P., 2013. The neuroanatomical phenotype of tuberous sclerosis complex: focus on radial migration lines. Neuroradiology 55 (8), 1007–1014. https://doi.org/10.1007/s00234-013-1184-3.

Vandervore, L., 2019. Heterogeneous clinical phenotypes and cerebral malformations reflected by rotatin cellular dynamics. Brain 142 (4), 868–884. https://doi.org/10.1093/brain/awz045.

Velinov, M., 2019. Genomic copy number variations in the autism clinic-work in progress. Front. Cell. Neurosci. 13 (57) https://doi.org/10.3389/fncel.2019.00057.

Verkman, A., 2013. Aquaporins. Curr. Biol. 23 (2), R52–R55. https://doi.org/10.1016/j.cub.2012.11.025.

Villani, G., 2017. "Classical organic acidurias": diagnosis and pathogenesis. Clin. Exp. Med. 17 (3), 305–323. https://doi.org/10.1007/s10238-016-0435-0.

Vitner, E., 2010. Common and uncommon pathogenic cascades in lysosomal storage diseases. J. Biol. Chem. 285 (27), 20423–20427. https://doi.org/10.1074/jbc.R110.134452.

Voineagu, I., 2011. Transcriptomic analysis of autistic brain reveals convergent molecular pathology. Nature 474 (7351), 380–384. https://doi.org/10.1038/nature10110.

Wang, L., 2018. Digenic variants of planar cell polarity genes in human neural tube defect patients. Mol. Genet. Metabol. 124 (1), 94–100. https://doi.org/10.1016/j.ymgme.2018.03.005.

Wasser, C., Herz, J., 2017. Reelin: neurodevelopmental architect and homeostatic regulator of excitatory synapses. J. Biol. Chem. 292 (4), 1330–1338. https://doi.org/10.1074/jbc.R116.766782.

Weiner, D., 2017. Polygenic transmission disequilibrium confirms that common and rare variation act additively to create risk for autism spectrum disorders. Nat. Genet. 49 (7), 978–985. https://doi.org/10.1038/ng.3863.

Whitley, B., 2018. Aberrant Drp1-mediated mitochondrial division presents in humans with variable outcomes. Hum. Mol. Genet. 27 (21), 3710–3719. https://doi.org/10.1093/hmg/ddy287.

Whittaker, D., 2017. The chromatin remodeling factor CHD7 controls cerebellar development by regulating reelin expression. J. Clin. Invest. 127 (3), 874–887. https://doi.org/10.1172/JCI83408.

Wiedemann, N., Pfanner, N., 2017. Mitochondrial machineries for protein import and assembly. Annu. Rev. Biochem. 86, 685–714. https://doi.org/10.1146/annurev-biochem-060815-014352.

Williams, S., 2019. An integrative analysis of non-coding regulatory DNA variations associated with autism spectrum disorder. Mol. Psychiatr. 24 (11), 1707–1719. https://doi.org/10.1038/s41380-018-0049-x.

Winkler, A.M., Greve, D.N., Bjuland, K.J., Nichols, T.E., Sabuncu, M.R., et al., 2018. Joint analysis of cortical area and thickness as a replacement for the analysis of the volume of the cerebral cortex. Cereb. Cortex. 28 (2), 738–749. https://doi.org/10.1093/cercor/bhx308. PMID: 29190325.

Woodcock, I.R., Menezes, M.P., Coleman, L., Yaplito-Lee, J., Peters, H., et al., 2018. Genetic, radiologic, and clinical variability in brown-vialetto-van laere syndrome. Semin. Pediatr. Neurol. 26, 2–9. https://doi.org/10.1016/j.spen.2017.03.001. PMID:29961509.

Yap, E.-L., Greenberg, M., 2018. Activity-regulated transcription: bridging the gap between neural activity and behavior. Neuron 100 (2), 330–348. https://doi.org/10.1016/j.neuron.2018.10.013.

Youn, Y., Han, Y., 2017. Primary cilia in brain development and diseases. Am. J. Pathol. 188 (1), 11–22. https://doi.org/10.1016/j.ajpath.2017.08.031.

Yuan, B., 2019. Clinical exome sequencing reveals locus heterogeneity and phenotypic variability of cohesinopathies. Genet. Med. 21 (3), 663–675. https://doi.org/10.1038/s41436-018-0085-6.

Zarrei, M., 2019. A large data resource of genomic copy number variation across neurodevelopmental disorders. NPJ. Genom. Med. 4 (26) https://doi.org/10.1038/s41525-019-0098-3.

Zhao, W., Gao, X., Qiu, S., Gao, B., Gao, S., et al., 2019. A subunit of V-ATPases, ATP6V1B2, underlies the pathology of intellectual disability. EBioMedicine 45, 408–421. https://doi.org/10.1016/j.ebiom.2019.06.035. PMID:31257146.

Zoghbi, H.Y., 2016. Rett Syndrome and the ongoing legacy of close clinical observation. Cell 167 (2), 293–297. https://doi.org/10.1016/j.cell.2016.09.039. PMID:2771649.

CHAPTER

8

Epilepsy and movement disorders

1. Epileptic seizures and epilepsy

Activity changes in the brain leading to epilepsy

Bromfield et al. (2006) reviewed epilepsy. They defined epileptic seizures as "abnormal excessive hypersynchronous discharge over a population of neurons." Epilepsy was defined as a condition in which recurrent epileptic seizures occurred. They noted that the hippocampus located in the temporal lobe was particularly important in epilepsy initiation. Bromfield et al. noted further that excitatory synapses occur predominantly on synapses of principal neurons, while inhibitory stimuli were often derived from interneurons. Increased excitatory neurotransmission or decreased inhibitory neurotransmission could alter action potential. Altered action potential could also result from altered ion concentrations and altered transmission through voltage-gated ion channels. Membrane potential is therefore impacted by neurotransmitters or by voltage-gated channels.

Glutamate acts as an excitatory neurotransmitter through its binding to specific glutamate receptors, AMPA, NMDA, and kainate receptors. Five different subunits form NMDA glutamate receptors: GRIN1, GRIN2A, GRIN2B, GRIN2C, and GRIN2D, and each is encoded on a different chromosome. Two different subunits form AMPA receptors: GRIA1 and GRIA2. Five different subunits form kainate receptors: GRIK1, GRIK2, GRIK3, GRIK4, and GRIK5. Neuronal potential is influenced by neurotransmitters and their ligand-gated channels and also by voltage-gated channels.

Metabotropic glutamate receptors (mGLURs) are encoded by eight different subunits: GRM1 through GRM8. Signaling through metabotropic glutamate receptors requires activity of G proteins. Bromfield et al. noted that AMPA, NMDA, and kainate agonists promoted seizure disorders.

Gamma-aminobutyric acid (GABA) constitutes the inhibitory neurotransmitter, and GABA agonists were found to reduce seizures; GABA agonists include barbiturates and benzodiazepines. They noted further that efficient clearance of neurotransmitters from the synaptic space was also important.

Bromfield et al. considered cellular environmental mechanisms of seizure induction and also neuronal intrinsic mechanisms of seizure induction. Cellular environmental mechanisms could include levels of neurotransmitters, flow of electrolytes sodium and calcium, and gene expression. With respect to inhibition, key factors included levels of GABA neurotransmitter and flow of potassium and chloride. With respect to neuronal specific factors that influence seizures, key variables included neuronal transmitter receptors and ion channel receptor types, and in case G protein depended on neurotransmitter receptors

Mechanisms and Genetics of Neurodevelopmental Cognitive Disorders
https://doi.org/10.1016/B978-0-12-821913-3.00009-3

function, alterations in the secondary messenger systems could provoke seizures. NMDA function was noted to be impacted by phosphorylation. Bromfield et al. also considered extrinsic factors that alter neuronal excitability. These included changes in extracellular ion concentration and glial cell modulation of transmitters. Altered interactions of specific neurons with interneurons were also considered to potentially play roles in seizure generation.

The potential for detecting seizures by electroencephalography (EEG) is facilitated by electropotential changes and by the similarity in orientation of pyramidal neurons in the cortex.

GABAergic interneuron hypofunction in epilepsy

Takano and Sawai (2019) noted that hypofunction of GABAergic interneurons has been considered to play roles in epilepsy Their studies on experimental animal models revealed that damaging GABAergic interneurons during development predisposed to epilepsy. Lim et al. (2018) reported that the GABAergic interneurons include a heterogeneous collection of cell types, and more than 50 different types of GABAergic interneurons occur. Diversity was found to be due to different cell intrinsic gene expression programs. Interneurons were reported to be primarily but not exclusively inhibitory.

Magliore et al. (2019) reported that the function of inhibitory neurons to restrain excitation sometimes failed. The different interneuron types are traditionally defined using specific staining techniques to detect specific proteins, parvalbumin, vasointestinal peptide, and somatostatin. They noted that optogenetic studies have been initiated to investigate interneuron functions. Different interneuron types at different levels can interact with pyramidal perisomatic regions, with cell bodies or with pyramidal processes. Magliore et al. noted that even between seizures, epileptic cortical microcircuits exhibit electrographic signals that are of fast and high amplitude. These preictal spikes were thought to be primarily glutamatergic; however, they are not necessarily suppressed by blocking ionotropic glutamate receptors.

Ion channels relevant to epilepsy

Raimondo et al. (2015) explored how transmembrane concentrations of potassium, sodium, chloride calcium, and hydrogen ions altered during seizure activity. They proposed that multiple interacting mechanisms that impacted ion concentrations could influence seizure activity. Oyrer et al. (2018) reviewed ion channels and their relation to epilepsy. They reported that 25% of genes identified as having damaging variants associated with epilepsy encoded ion channels. They noted that the lifetime prevalence of epilepsy is 3%. In addition, despite modern epilepsy treatment, seizures were not controlled in one-third of patients, and in addition, medications for epilepsy had significant side effects. Oyrer et al. noted that epilepsy types include focal, generalized, or combined forms and may be etiologically classified as genetic, structural, metabolic, infectious, or immune in origin. Specific EEG patterns are also used for classification.

Sodium channels

- SCN1A, sodium voltage-gated channel alpha subunit 1
- SCN2A, sodium voltage-gated channel alpha subunit 2
- SCN1B, sodium voltage-gated channel beta subunit 1
- SCN8A, sodium voltage-gated channel alpha subunit 8

Potassium channels

- KCNA1, potassium voltage-gated channel subfamily A member 1
- KCNA2, potassium voltage-gated channel subfamily A member 2

- KCNB1, potassium voltage-gated channel subfamily B member 1
- KCNC1, potassium voltage-gated channel subfamily c member 1
- KCNMA, potassium calcium-activated channel subfamily M alpha 1
- KDNQ2, potassium voltage-gated channel subfamily Q member 2
- KCNQ3, potassium voltage-gated channel subfamily Q member 3
- KCNT1, potassium sodium-activated channel subfamily T member 1
- KCTD7, potassium channel tetramerization domain containing 7
- HCN1, hyperpolarization-activated cyclic nucleotide–gated potassium channel

Calcium voltage-gated channels implicated in epilepsy

- CACNA1A calcium voltage-gated channel subunit alpha1 A
- CACNA1H calcium voltage-gated channel subunit alpha1 H

Oyrer et al. reported that 10% of epilepsy cases were reported to be due to variants in ligand-activated channel genes. GABA receptors that transmit the inhibitory neurotransmitter GABA have been reported to undergo mutations that lead to epilepsy.

Defects in the following GABA receptors have been reported to lead to epilepsy (Butler et al., 2018; Shen et al., 2017).

Gamma-aminobutyric acid receptors implicated in epilepsy

- GABRA2, gamma-aminobutyric acid type A receptor subunit alpha2
- GABRA5, gamma-aminobutyric acid type A receptor subunit alpha5
- GABRB3, gamma-aminobutyric acid type A receptor subunit beta3
- GABRG2, gamma-aminobutyric acid type A receptor subunit gamma2

2. Epilepsy classification

The International Alliance Against Epilepsy (ILAE) (2017) proposed a new classification of epilepsy and etiologies, and this has been reviewed by several authors. Falco-Walter et al. (2018) noted that in the new classification, seizures are defined as focal of generalized, with intact awareness or impaired awareness. In addition, focal seizures can be defined as focal remaining focal or progressing to bilateral tonic clonic seizures. Generalized seizures were defined as motor, tonic clonic, or nonmotor or absence. The ILAE approved definition for epileptic seizure is "a transient occurrence of signs or symptoms due to abnormal excessive or synchronous neuronal activity in the brain."

Etiological categories included structural, genetic, infectious metabolic, immune, and unknown. Etiology was determined through neuroimaging studies and findings that reasonably coincided with the EEG findings. Genetic etiology was determined by finding disease causing genomic variants or single-gene variants. Family history could be useful. However, some disease-causing genetic variants arise de novo.

The term "infectious etiology seizures" is applied to seizures that arise following an infection not seizures that occur during an acute infection. Infections leading to seizures could include toxoplasmosis.

Metabolic seizures referred primarily to seizures arising as a result of inborn errors of metabolism. Falco-Walter et al. included pyridoxine deficiency seizures and cerebral folate deficiency seizures in this category. Immune etiology referred to seizures that occur in individuals with antibodies to neuronal proteins or to neuronal surface molecules.

3. Genetic factors in epilepsy

Family history and twin studies provided early information on the importance of genetic factors in epilepsy. Myers et al. (2019) noted that epilepsy concordance was reported to be higher in monozygotic twins than in dizygotic twins. Increased frequency of epilepsy among family members was initially attributed to multifactorial inheritance. However, in recent years, evidence has emerged for roles of specific single-gene defects that cause epilepsy.

Early-onset infantile epilepsies include hypoxia–ischemic encephalopathies and genetic encephalopathies. Myers et al. noted that important consequences of early-onset encephalopathies included developmental delay, movement disorders, and behavioral problems. They noted that genetic forms of epilepsy could lead to generalized epilepsy and to focal epilepsy syndromes.

Sharma et al. (2019) reviewed pediatric epilepsy and noted that this disorder was primarily treated according to seizure type. However, they noted that based on newer testing methods, there are now efforts to explore treatment based on specific causative etiologies. They noted that 5% of children undergo seizures. In addition to new methods to diagnose types of seizures, improved methods are available to determine the sites of origin of focal seizures. They noted that focal seizures are now classified as aware and unaware. Distinctions are also made between focal seizures and bilateral tonic clonic seizures. Sharma et al. noted information that indicated that, of infants who presented with seizures before 18 months, 14% had likely sustained early brain injury; 54% of cases were likely of genetic origin; and 33% were of unknown etiology.

Genomic analyses in epilepsy

Thodeson and Park (2019) reviewed genomic testing in pediatric epilepsy. They noted that progress from single-gene testing to modern genomic sequencing applications led to finding that significant genetic variants were found in 30% of cases of pediatric epilepsy. They noted the importance of longitudinal reinterpretation of sequence variants. Sequence changes initially classified as variants of unknown origin (VUS) may subsequently be reclassified as additional clinical or functional information becomes available. Reclassification can also sometimes lead to downgrades of the clinical significance of variants.

Thodeson and Park considered the treatment relevance of specific genetic variants that occur in association with epilepsy. Sequence information in some cases provided information on which medications should be avoided, e.g., in SCN1A, deleterious mutation avoids carbamazepine, lamotrigine, or vigabatrin. In the case of other mutations, specific medications had been found to be beneficial.

Vitamin-responsive epilepsies

Pearl (2016) noted that vitamin-dependent epilepsies and a number of epilepsies due to inborn errors of metabolism require special targeted therapies. These disorders usually present early in life but may also be first detected in adolescents or adults with epilepsy. Newborns may present with generalized seizures, with unusual oculofacial movements or with myotonic seizures.

These were first reported in 1954; however, a specific metabolic disorder leading to these seizures was first described 50 years later and was noted to be due to mutations in ALDH7A1 (aldehyde dehydrogenase 7A1), also sometimes known as alpha amino adipic semialdehyde or antiquitin. This enzyme is involved in the detoxification of aldehydes and is also involved in lysine metabolism. Defects in the function of this enzyme were noted to present in infancy and to respond favorably to pyridoxine treatment. Pearl noted that some cases may also require folinic acid therapy. Pearl noted that abnormal brain imaging findings that may occur in this condition include hypoplastic corpus callosum, hemispheric atrophy or hypoplasia, and periventricular hyperintensities. Biomarkers include elevated alpha amino adipic acid and elevated pipecolic acid in cerebrospinal fluid (CSF). Molecular analyses are required to detect pathologic mutations in ALDH7A1. Mutations in patients are usually homozygous or compound heterozygous deleterious mutations.

Pearl noted that other mutations can lead to pyridoxine-responsive epilepsy. He noted that neonatal seizures could also occur as a result of defective function of PNPO (pyridoxamine 5′-phosphate oxidase). PNPO deficiency was reported to lead to systemic pyridoxyl-5-phosphate (PLP) deficiency and to be associated with anemia and failure to thrive.

Vitamin B6 (pyridoxine) metabolism

The molecular structure of pyridoxine is illustrated in Fig. 8.1. Wilson et al. (2019) reviewed disorders of pyridoxine metabolism. They noted that vitamin B6 absorbed by the intestine requires conversion to pyridoxal-5-phosphate, the active form of the vitamin. Pyridoxine is also converted to pyridoxal and pyridoxamine, and these three forms can be phosphorylated. Pyridoxal-5-phosphate acts as a cofactor for a number of different enzymes and participates in many different reactions. Important reactions involved neurotransmitter metabolism, folate and 1-carbon metabolism, lipid and protein metabolism, and mitochondrial function. Particularly important reactions that utilize PLP include metabolism of branched chain amino acids particularly lysine.

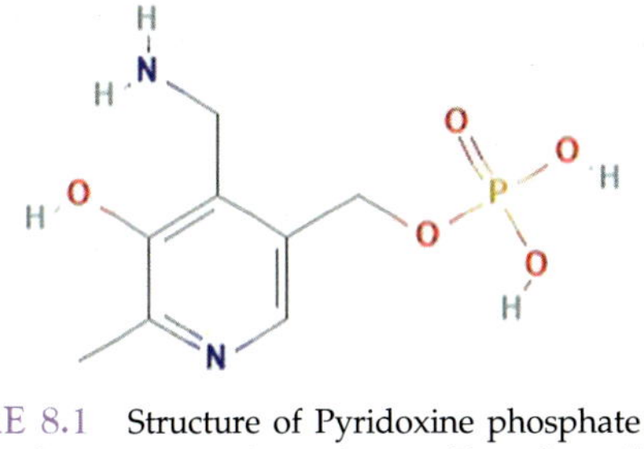

FIGURE 8.1 Structure of Pyridoxine phosphate used to treat Pyridoxine-responsive seizures. *From https://pubchem.ncbi.nlm.nih.gov/compound/Pyridoxine-phosphate.*

Deficiencies in pyridoxal-5-phosphate lead to impaired synthesis of the neurotransmitter GABA. Pyridoxal-5-phosphate deficiency was also noted to lead to impaired activity of aromatic L-amino-decarboxylase (AADC) (DCC) and lead to combined serotonin and catecholamine deficiency and to dystonia. Reduced synthesis of serine and reduced catabolism of glycine and impaired metabolism of tryptophan can also occur in pyridoxal-5-phosphate deficiency. Activity of a number of other enzymes besides ALDH7A plays important roles in PLP homeostasis and can impact brain functions and lead to epilepsy. These include PNPO and pyridoxal (PROSC) phosphate–binding protein. Pyridoxal (PDXK) kinase phosphorylates vitamin B6. Specific medications can impact generation and activity of PLP. Pyridoxal kinase can be inhibited by aminophylline and the asthma medication. Isoniazid, the medication to treat tuberculosis, can bind to and inactivate PLP. van Karnebeek et al. (2016) reviewed pyridoxine-dependent epilepsy and the important role of ALDH7A1 (antiquitin) in pyridoxine homeostasis. They emphasized that ALDH7A1 impacted lysine metabolism. They also documented a range of clinical biochemical abnormalities in this condition and also the broad phenotypic spectrum. They noted that seizure onset may be in the neonatal period but could also begin in late infancy. A variety of different seizure types occurred including generalized seizures, infantile spasm, myoclonic seizures, and partial seizures. There was some evidence that antenatal seizures occurred as evidenced by rhythmic intrauterine movements. Neurologic manifestations were reported to include dystonia, hypotonia, or hypertonia. van Karnebeek et al. noted that abnormal neuroimaging findings included cortical atrophy, dysplasia or hypoplasia of the corpus callosum, ventriculomegaly, and white matter abnormalities. van Karnebeek et al. noted that the combination of pyridoxine and folinic acid was important in treatment, as was lysine restriction.

Pyridoxine deficiency in adult life and epilepsy predisposition

Osman et al. (2020) described the case of a woman who had few seizures in childhood and developed frequent seizures in early adult life. These seizures were difficult to control. Subsequently, she was found to have pathogenic mutations in ALDH7A. Treatment with pyridoxine led her to be seizure free. Srinivasaraghavan et al. described a woman who developed seizures at age 17 years and was found to be homozygous for pathogenic ALDH7A mutations.

Cerebral folate deficiency

Pearl (2016) reported that in this condition, low levels of 5-methyl tetrahydrofolate occurred (Fig. 8.2). Pope et al. (2019) reviewed cerebral folate deficiency. Folates include a pteridine ring, a para-amino benzene ring, and glutamic acid; more than one glutamate residue may be present. Folates occur in dihydro- and tetrahydro-oxidized forms. Dihydrofolate uptake from the intestine is dependent on two different folate transporters: reduced folate transporter (RFC1), also known in humans as SLC19A1, and proton-coupled folate transporter (PCFT), known in humans as SLC46A1. Pope et al. noted that mutations in SLC46A1 have been reported to lead to systemic folate deficiency. Specific folate receptors are present on cells. Pope et al. noted that brain folate uptake is mediated by FOLR1, a high-affinity folate receptor. This gene product is a secreted protein that either anchors to membranes via a glycosyl-phosphatidylinositol linkage or exists in a soluble form. Pope noted that this receptor is expressed on choroid plexus epithelial cells. There is also evidence that PCFT (SLC46A1) plays roles in folate transport. Exosomes were reported to be present in the choroid plexus and to contain 5-methyl tetrahydrofolate that can the pass to the CSF. Pope et al. emphasized that transport of 5-methyl tetrahydrofolate into the brain is energy dependent. Folate metabolism involves several steps, and each step requires a specific enzyme. Dihydrofolate is converted to tetrahydrofolate through activity of the enzyme dihydrofolate reductase (DHFR.) Tetrahydrofolate is then converted to 5, 10-methylenetetrahydrofolate through activity of serine hydroxymethyltransferase (SHMT); 5, 10-methylenetetrahydrofolate is

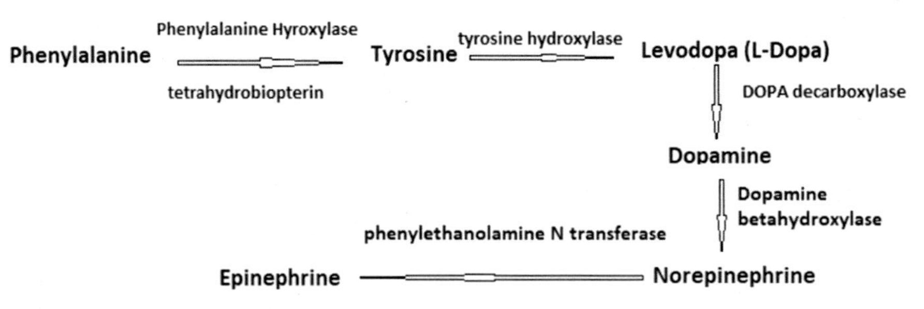

FIGURE 8.2 Folate metabolism generation of methionine. Importance in epigenetics.

converted to 5-methyl tetrahydrofolate activity of methyltetrahydrofolate reductase (MTHFR). 5-Methyltetrahydrofolate methylates homocysteine to methionine.

Pope et al. noted that several different causes of cerebral folate deficiency have been identified, which include deficiency of DHFR and MTHFR enzymes, deficiency of folate receptor FOLR1, and also deficiency of folate transporter PCFT (SLC46A1).

Specific mitochondrial enzymes involved in folate metabolism include SHMT2 serine hydroxymethyltransferase 2, MTHFD2 methylenetetrahydrofolate dehydrogenase (NADP + dependent) 2, MTHFD1 methylenetetrahydrofolate dehydrogenase, cyclohydrolase, and formyltetrahydrofolate synthetase 1. Disorders of mitochondrial folate metabolism may also be associated with seizures.

Clinical signs of cerebral folate deficiency include seizures and, in some cases, pancytopenia and anemia Other signs of cerebral folate deficiency include ataxia, white matter changes, developmental delay, and microcephaly. It is important to note that antiseizure medications can decrease blood and body folate levels Morrell (2002).

Biotin-related seizures

Biotin is an essential vitamin that must be obtained from food sources. However, the biotin form in food needs to be converted before it can be utilized in human metabolism. Biotin in food stuffs is bound to protein, and proteolysis is the first step in its processing. This leads to generation of biocytin. A further processing step involves activity of the enzyme biotinidase that can then generate free biotin. Biocytin acted upon by biotinidase generates free biotin and lysine. Free biotin then acts as a cofactor for the enzyme holocarboxylase synthase and other carboxylase enzymes. Biotinidase deficiency can lead to multiple carboxylase deficiency. Holocarboxylase synthetase (HLCS) deficiency can also be caused by mutations in the gene on chromosome 21q22 that encodes this enzyme. Deficiency of the enzyme responds to treatment with free biotin. Biotin deficiency can also arise secondarily in children due to excessive intake of avidin that is present in raw egg white (Sweetman et al., 1981). Newborn screening for biotinidase deficiency is carried out in the United States and in more than 30 countries worldwide (Strovel et al., 2017). Biotinidase defects can present with seizures in infants. Seizures can be generalized or myoclonic. Pearl (2016) noted that biotinidase deficiency can also lead to infantile spasms. Dermatologic abnormalities may also occur.

Thiamine transporter protein defects

Deficiency in a specific thiamine transporter, SLC19A3, has been shown to lead to epilepsy and sometimes to increased intracranial pressure and to neuroimaging finding of hyperintensities in the striatum and globus pallidus. Pearl noted that this condition responds to treatment with thiamine in combination with biotin.

4. Inborn errors of metabolism leading to seizures

Pearl (2016) noted that defective function of certain transporter protein could lead to seizures. Particularly important in the category is glucose transporter 1 deficiency. GLUT1 is known in humans as SLCA1. The phenotypic presentation varies widely; however, seizures were noted to be present in 90% of cases. The seizures types vary and may include absence seizures or myotonic static seizures (Larsen et al., 2015). Other manifestations in glucose transporter deficiency patients include intellectual disability and movement disorders. Significantly, Pearl (2016) noted that seizures in this disorder respond to ketogenic diet. In addition, certain medications including phenobarbital inhibit the transporter and exacerbate the seizures.

Metabolic evaluation of children with epilepsy

In 2018, van Karnebeek et al. emphasized the importance of discovery of potential metabolic causes of epilepsy in patients, since the correct diagnoses potentially lead to effective treatment. They noted that there were 74 inborn errors of metabolism that lead to epilepsy and that are potentially treatable with specific therapies. They emphasized the importance of metabolic studies on blood urine and in some cases of CSF in patients with epilepsy. In addition to epilepsy, patients with inborn errors of metabolism may have other manifestation, dysmorphic features, and organomegaly. Obtaining family history is important including information on other affected related individuals and also information on parental consanguinity. The range on metabolic disorders that can lead to epilepsy is extensive. It includes defects in amino acid metabolism, defects in carbohydrate metabolism, mitochondrial function, peroxisomal function, lysosomal storage diseases, defects in creatine metabolism, and defects in copper metabolism.

Polyamines and seizures

Moinard et al. (2005) defined ornithine decarboxylation and condensation processes that led to formation of polyamines putrescine, spermine, and spermine from ornithine. Baroli et al. (2020) reported that polyamines modulate ion channels and ionotropic glutamate receptors. They also reported that altered levels of polyamines occurred in brain disorders including disorders associated with epilepsy. They also proposed that measures that modulated polyamine levels may be useful in epilepsy therapy. Rodan et al. (2018) reported that ornithine decarboxylation is rate limiting in polyamine synthesis and that defects in ornithine decarboxylation through activity of ODC1 led to a defined neurometabolic syndrome and dysmorphology. They reported four cases with defects in ODC1.

5. Genomic studies in epilepsies

Ellis (2020) emphasized that genomic studies have come to play increasing roles in assessment of patients with neurological disorders. They particularly reviewed the role of these studies in epilepsy diagnosis and management. They noted that as of 2019, more than 140 genes had been identified as having mutations that play roles in epilepsy causation. In addition to gene mutations that exert mendelian effects that lead to epilepsy, there is now also evidence that somatic mosaicism plays roles in causation of focal epileptic seizures. In addition to mutations that lead to nucleotide changes in protein coding regions of the genome, there is also evidence that regions of repeat sequence expansions in regulatory genomic regions, including promoters and enhancers, play roles in epilepsy causation.

Ellis et al. illustrated cellular pathways implicated in epilepsy based on findings of pathogenic gene variants in epilepsy cases. They included six different categories of gene products, such as ion channels, receptors, synaptic proteins and genes in the mTOR pathway, and gene products involved in chromatin remodeling and transcription regulation.

Voltage-gated sodium channels are described as heteromeric proteins that function in the generation and propagation of action potentials in muscle and neuronal cells. They are composed of one alpha and two beta subunits. Mutations in voltage-gated sodium have been described in generalized epilepsy and in certain cardiac arrhythmias. Heyne et al. (2019) reported results of targeted genomic sequencing in 6994 cases diagnosed with epilepsy and neurodevelopmental disorder. Their studies revealed ultrarare variants in sodium ion channel genes SCN1A, SCN2A, and SCN8A and in potassium ion channel gene KCNQ2. Ultrarare variants were also detected in CDKL5 (cyclin-dependent kinase–like 5), previously reported to be mutated in X-linked West syndrome, and in STXBP1 (syntaxin-binding protein 1), previously reported to be associated with infantile epileptic encephalopathy.

Polygenic epilepsy

Studies reported by the International League Against Epilepsy Consortium on Complex Epilepsies in 2018 identified 16 significant loci in genome-wide association studies in a study of 1522 epilepsy cases and 20,677 controls. Loci with the highest level of significance in cases overall occurred at 2q24.3, and other significantly associated loci occurred at 2p16.1 and 16q12.1. When data were separated into two categories, generalized epilepsy and focal epilepsy, the 2q24.3 locus remained significant for focal epilepsy. However, in generalized epilepsy, additional loci emerged with significant risk association.

Overall, the polygenic risk burden was significantly higher in generalized epilepsy. In total, loci located within 250 kb of the GWAS-linked locus in generalized epilepsy encompassed 145 genes. Heyne et al. (2019) reported that prioritized genes included seven ion channel genes (sodium and potassium), also GABRA2 and, in addition, three transcription factor encoding genes such as ZB2, STAT4, and BCL11A; a chromatin modifier gene BRD7 (bromodomain containing 7) and STX1B (syntaxin-binding protein that plays a role in exostosis of synaptic vesicles); and PNPO that catalyzes the terminal, rate-limiting step in the synthesis of pyridoxal 5′-phosphate.

Leu et al. (2019) reported risks of a study on polygenic risk in epilepsy in cohorts derived from the United States, Europe, and Japan. Their analyses include 8396 epilepsy patients and 622,313 controls. Their study also revealed that polygenic risk was more strongly associated with generalized epilepsy than with focal epilepsy.

Inborn errors of metabolism leading to seizures

Almannai and El Hattab (2018) reviewed inborn errors of metabolism associated with seizures. Importantly, they noted that several of these disorders can be treated with disease-specific medications. Important metabolic disorders to consider in the context of epilepsy are listed in the following.

Disorders of energy metabolism include mitochondrial functional defects, disorders of pyruvate metabolism, fatty acid oxidation disorders, disorders of gluconeogenesis, glycogen storage disease and defects of gluconeogenesis, and glucose transporter defects.

- Urea cycle disorders and creatine deficiency disorders
- Lysosomal disorders, peroxisomal disorders, Zellweger syndrome, and X-linked adrenoleukodystrophy
- Purine and pyrimidine metabolic disorders
- Congenital disorders of glycosylation

6. Epilepsy types associated with specific molecular defects

Ion channel defects and epilepsy types

- SCN1A, sodium voltage-gated channel alpha subunit 1 defects in generalized epilepsy febrile seizures
- SCN1B, sodium voltage-gated channel beta subunit 1 defects in generalized epilepsy, febrile seizures
- SCN2A, sodium voltage-gated channel alpha subunit 2, defects in seizure disorders and autism spectrum
- SCN8A, sodium voltage-gated channel alpha subunit 8, defects in cognitive disability cerebellar ataxia; multiple different types of seizures have been reported to be associated with SCN8A defects
- KCNQ2, potassium voltage-gated channel subfamily Q member 2, defective in benign neonatal seizures
- KCNT1, potassium sodium-activated channel subfamily T member 1, nocturnal frontal lobe epilepsy

- KCNA2, potassium voltage-gated channel subfamily A member 2, defects in epileptic encephalopathy
- HCN, hyperpolarization-activated cyclic nucleotide-gated channels, defects in generalized epilepsy

Neurotransmitter receptor defects and epilepsy types

- GABRA1, gamma-aminobutyric acid type A alpha1 receptor subunit; mutations cause juvenile myoclonic epilepsy and childhood absence epilepsy
- GABRA3, gamma-aminobutyric acid type A receptor subunit alpha3, X linked gene; mutations can lead to seizures particularly in males
- GABRG2, gamma-aminobutyric acid type A receptor subunit gamma2; mutations can lead to epileptic encephalopathy
- GRIN1, subunit of NMDA glutamate ionotropic receptor; defects can lead to epilepsy and hyperkinetic movements
- GRIN2A, subunit of NMDA glutamate ionotropic receptor; defects can lead to focal epilepsy with paresthesias and absence of epilepsy
- GRIN2B, subunit of glutamate ionotropic receptor; rarely involved in epilepsy

Synaptic proteins that can undergo mutations leading to epilepsy and specific types

- DNM1, dynamin 1 GTP-binding protein; mutations may lead to infantile-onset epilepsy
- STXBP1, syntaxin-binding protein 1; mutations may lead to infantile epileptic encephalopathy
- STX1B, syntaxin involved in exocytosis of synaptic vesicles; fever-associated epilepsy syndromes
- PRRT2, proline-rich transmembrane protein 2; mutations in infantile epilepsy and in dyskinesia
- SNAP25, synaptosome-associated protein 25; mutations may lead to epilepsy
- SYNGAP1, synaptic Ras GTPase-activating protein 1; defects lead also to motor impairments

mTOR and regulatory pathway components that can undergo mutations leading to epilepsy

- TSC1 and TSC2 complex subunit 1, subunit 2; mutations may lead to infantile spasms, combinations of seizures including focal and multifocal seizures, and drop attacks
- mTOR, mechanistic target of rapamycin kinase (Fig. 8.3); mTOR inhibitors have antiepileptogenic properties
- DEPDC5, DEP domain–containing 5; defects may lead to excitation/inhibition imbalance that triggers epileptogenesis
- GATOR1 subcomplex subunit, familial focal epilepsy
- NPRL2, NPR2-like seizure phenotype, mostly in focal seizures
- NPTL3, NPR3-like defects mostly occur in focal seizures

Dawson et al. (2020) reported that GATOR complex components can also lead to focal cortical dysplasias that may lead to epilepsy. Baldassari et al. (2019) analyzed data on cases of epilepsy with GATOR abnormalities. They noted that various epilepsy phenotypes occurred, particularly sleep-related focal epilepsy. Intellectual disability occurred in 21% of cases. Neuroimaging revealed cortical defects in 63% of cases. In an analysis of 183 GATOR-related epilepsy cases, 83% of cases had DEPDC5 defects, 11% had defects in NPRL3, and 6% of cases had NPRL2 defects. The defective function in GATOR components limited the normal function of the GATOR complex which is repression of mTORC1.

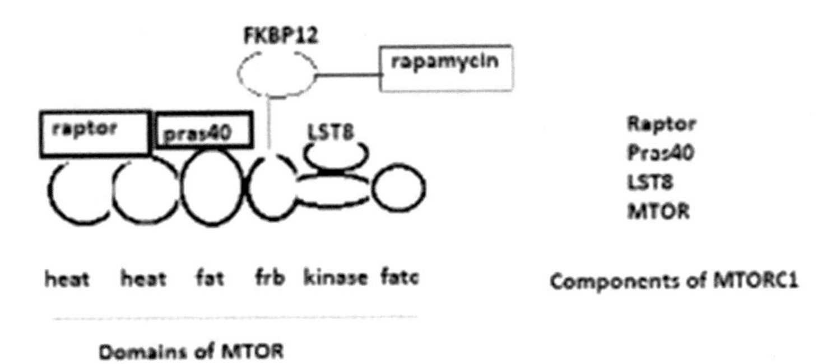

FIGURE 8.3 mTOR structure and interacting components. mTOR activity is downregulated by active TSC complex through Rheb GDP.

7. Chromatin remodeling and transcriptional regulation factors and defects leading to epilepsy

- MECP2, methyl-CpG-binding protein 2; defects can lead to epilepsy severe and include intractable myoclonic seizures and infantile spasms
- ARX, aristaless-related homeobox; defects can lead to infantile epileptic–dyskinetic encephalopathy
- CHD2, chromodomain helicase DNA-binding protein 2; defects can lead to epilepsy with myoclonic–atonic seizures
- MEF2C, DNA-binding transcription factor; defects can lead to febrile seizures or to myoclonic, focal-onset, and generalized seizures

8. Cognitive impairment and association with epilepsy

Braun (2017) noted that cognitive impairment and developmental delay in children with epilepsy are dependent on many variables; however, some of these variables may be modifiable. Braun emphasized that early diagnosis and effective treatment are essential since ongoing seizure can cause progressive structural and functional changes in brain networks. He also emphasized the importance of detecting and controlling intraictal discharges.

Braun noted that epileptogenic pathology can be initiated by structural, genetic, metabolic, and inflammatory changes. Delineation of these underlying causes may, in some cases, indicate inclusion of specific targeted therapies in treatment. Examples of targeted therapies can include appropriate dietary modifications or supplementation in cases of epilepsy of metabolic origin. Specific therapies are also utilized in treatment of seizures in tuberous sclerosis. Specific targeted therapies are also used in certain forms of epilepsy due to defects in ion channel function.

Braun noted evidence that uncontrolled epileptic seizures were correlated with specific brain area volume changes and with microstructural white matter changes. These changes correlated with measurements of cognitive decline. There are also concerns as to whether specific doses and specific types of antiepileptic drugs can impair cognition. Braun emphasized the importance of finding the optimal minimal dose of antiepileptic medications in a specific patient and the importance of avoiding polypharmacy. Braun noted evidence of the efficacy of surgical treatment in the treatment of specific cases of focal seizures.

Jung-Klawitter and Kuseyri Hübschmann (2019) described monogenic variants in specific enzymes involved in monoamine neurotransmitter biosynthesis and degradation. Defects leading to impaired neurotransmitter biosynthesis occurred in tyrosine hydroxylase (TH) and in AADC.

Defects in tetrahydrobiopterin biosynthesis resulted from mutations in GTP cyclohydrolase (GTPH), 6-pyruvoyltetrahydropterin synthase (PTPS), sepiapterin reductase (SPR), quinoid dihydropteridine reductase (QDPR, also known as DHPR). They also noted that impaired function of DnaJ heat shock protein.

Protein family (Hsp40) member C12, a chaperone involved in protein folding, also impacted tetrahydrobiopterin biosynthesis.

Monoamine neurotransmitters impacted by these defects included dopamine, epinephrine, and norepinephrine. Key clinical manifestations of these defects included movement disorders, developmental delay, hypotonia, and autonomic system dysregulation.

Defects in serotonin synthesis have also been described; these can occur due to mutations in an ATP-dependent transporter of monoamines (VMAT2) (SLC18A2), GTP cyclohydrolase (GTPH), 6-pyruvoyltetrahydropterin synthase (PTPS), QDPR, TH, dopa decarboxylase (DDC), AADC, or PNPO.

9. Neurodevelopmental disorders associated with movement abnormalities and/or cerebral palsy

The International Executive Committee defined cerebral palsy as follows (Rosenbaum et al., 2006):

"Cerebral palsy describes a group of permanent disorders of the development of movement and posture causing activity limitations that are attributed to non-progressive disturbance that occurred in the fetal or infant brain. The motor disorders of cerebral palsy are often accompanied by disturbances of sensation, perception, cognition, communication and behavior, by epilepsy and secondary musculoskeletal problems."

In a 2014 review, Colver et al. reported that the population prevalence of cerebral palsy was 2.0–3.5 per 1000 births. They noted that the prevalence has remained stable for several decades, despite apparent improvements in antenatal and perinatal care. There is a higher incidence of cerebral palsy when birth weight is below optimum, and asphyxia was likely to play roles in those cases. They noted that genetic factors are increasingly being reported as playing roles.

Colver et al. reported that in 85% of cerebral palsy cases studied using magnetic resonance imaging (MRI), abnormalities are found and abnormal MRI findings could be important indicators of predisposing genetic and metabolic conditions.

In considering management of cases with cerebral palsy, Colver et al. noted that there will continue to be increased focus on participation of these individuals in activities and increasing emphasis on quality of life as new technologies and appliances become available. Importantly, they noted that individual and societal attitudes continue to promote inclusion of individuals with cerebral palsy in societal activities.

Evidence for genetic factors in cerebral palsy

Tollånes et al. (2014) carried out a study of cerebral palsy in Norway. They noted that of 2,036,741 Norwegians born between 1967 and 2002, 3649 had a diagnosis of cerebral palsy. They noted that premature birth was a risk factor for cerebral palsy, and in their study of familial risk, they excluded cases of premature birth.

Their study revealed that in families with an affected singleton child, the risk for a subsequent child with cerebral palsy was increased between 6.4- and 13-fold. An affected parent also had increased risk for an affected child between 1.1- and 26-fold. For individuals with an affected cousin, risks were weakly increased (0.9–2.7-fold). The authors concluded that these risk patterns indicated multifactorial inheritance where multiple genes interact with each other and possibly with environmental factors.

The emerging genetic landscape in cerebral palsy

Van Eyk et al. (2018) reviewed the emerging genetic landscape in cerebral palsy. They noted that specific information supporting the possible role of genetics in cerebral palsy included evidence for higher concordance rates for cerebral palsy in monozygotic than in dizygotic twins and increased cerebral palsy risk in consanguineous families. In addition, cerebral palsy often occurs in association with inborn errors of metabolism and in individuals with congenital abnormalities.

Van Eyk et al. reported that some studies had revealed an increased frequency of cerebral palsy in association with inherited thrombophilic abnormalities, e.g., in association with factor V Leiden and plasminogen activator inhibitor 1 defects.

The adaptor-related complex 4 (AP4) is composed of different subunits, and pathogenic mutations in subunits of this complex have been reported to predispose to forms of cerebral palsy, sometimes referred to as spastic paraplegia. The implicated subunits include AP4B1, AP4E1, AP4M1, and AP4S1. The adaptor-related protein complex is reported to be involved in exosome function.

Other proteins implicated in cerebral palsy are ADD3 (adducin 3) (involved in spectrin actin in erythrocytes) and ANKRD15 (KANK1) that is reported to impact actin polymerization. Other mutant proteins associated with cerebral palsy in several reports include GAT1 glutamate decarboxylase.

However, there is likely considerable genetic heterogeneity in genetic causation of cerebral palsy as evidenced by exome sequencing studies. MacLennan et al. (2015) reported that 14% of cases of cerebral palsy had heterogeneous gene defects and a higher proportion had copy number defects.

10. Cerebral palsy spectrum disorder

Shevell (2018) proposed that in light of new information on the occurrence of cerebral palsy manifestations in a number of different disorders with different molecular defects, the name of cerebral palsy disorder should be changed to cerebral palsy spectrum disorder.

Inborn errors of metabolism that present with features of cerebral palsy

Leach et al. (2014) identified 54 inborn errors of metabolism with features of cerebral palsy including movement abnormalities. They noted that these disorders encompassed 13 different biochemical categories. Importantly, for 256 of these inborn errors of metabolism, treatments are available.

The 13 different categories of treatable included disorders of amino acid metabolism, cerebral glucose transport, creatine, fatty acid oxidation, homocysteine metabolism, lipid metabolism, lysosomal function, metal homeostasis, mitochondrial function, neurotransmitter synthesis defect, organic acid metabolism, urea cycle, and vitamin and cofactor metabolism. Leach et al. also documented specific treatments for each of these disorders.

It is also interesting to consider the range of cerebral palsy–like manifestations that occurred in these inborn errors of disorders. They included dystonia, spasticity, spastic diplegia, tetraplegia, ataxia, extrapyramidal signs, hypotonia, extrapyramidal movements, upper motor neuron signs, loss of gross motor function, hypertonia, athetosis (defined as abnormal muscle contractions with writhing), and clumsiness.

Pearson et al. (2019) presented a diagnostic approach to rule out genetic defects as causes of movement disorders. These included biochemical studies; analyses of amino acids lactate, pyruvate, and acylcarnitine and organic acid biotinidase; and thyroid hormone status assessment.

Pearson et al. noted that in some cases, specific tests would need to be carried out on cerebrospinal fluid e.g., determination of levels of neurotransmitter and their metabolites, analysis of 5-methyltetrahydrofolate, lactate, pyruvate, and amino acids. Some studies needed to be carried out on plasma, e.g., ammonia, acyl carnitine, biotinidase, urea, and amino acids.

Monoamine neurotransmitter disorders in cerebral palsy mimics and movement disorders

On the basis of studies in specific patients, Kurian et al. (2011) determined that key enzymes and cofactors involved in monoamine neurotransmitter synthesis play roles in movement disorders and cerebral palsy mimics. They determined that key cofactors in monoamine neurotransmitter biosynthesis included tetrahydrobiopterin (BH4) and pyridoxal phosphate (vitamin B6). Amino acid precursors for monoamine neurotransmitter biosynthesis included tryptophan and tyrosine.

In a 2014 review, Ng et al. noted that monoamine neurotransmitters synthesized in presynapses are transferred into synaptic vesicles by vesicular monoamine transporter (VMAT2). Vesicles are then released into the synaptic cleft. Subsequently the monoamine neurotransmitters are released from vesicles and bind to receptors on the postsynaptic membrane.

Key manifestations of impaired monoamine synthesis and availability included cognitive delay, motor impairments, epilepsy, autonomic dysfunction, and neurologic manifestations including oculogyric crisis that involved prolonged involuntary upward eye movements. Additional manifestations included abnormalities of autonomic function with altered temperature regulation. Ng noted that in some patients, behavioral abnormalities were documented.

Diagnoses were made on the basis of urine analyses of homovanillic acid and 2-hydroxyindole acetic acid and biopterin. Blood levels of phenylalanine and tyrosine were also important to measure.

Key enzymes involved in monoamine synthesis

- GCH1 GTP cyclohydrolase 1, rate-limiting enzyme in tetrahydrobiopterin (BH4) biosynthesis
- PTS (PTPS) 6-pyruvoyltetrahydropterin irreversible step in the biosynthesis of tetrahydrobiopterin
- DHPR (QDPR) catalyzes reduction of quinonoid dihydrobiopterin. PNPO, rate-limiting step in the synthesis of pyridoxal 5′-phosphate
- AADC, defects in DDC gene cause aromatic L-amino-acid decarboxylase deficiency
- MAOA, monoamine oxidase A, oxidative deamination of amines, dopamine, norepinephrine, and serotonin
- MAOB, monoamine oxidase B, oxidative deamination of biogenic and xenobiotic amines
- TH involved in the conversion of tyrosine to dopamine

Leukodystrophies

van der Knaap and Bugiani (2017) reviewed leukodystrophies and noted they can present at any point from early life on. Clinical manifestations can include psychomotor regression, irritability, ataxia, peripheral neuropathy, dysphagia, and seizures. Leukodystrophies can be due to mutation in oligodendrocytes and in myelin components or sometimes due to defects in astrocytes and microglia, axons, or due to defects in blood vessels that supply nutrients. They noted four categories of leukodystrophy: hypomyelinating, demyelinating, dysmyelinating, or disorders in which myelin lysis occurs.

In the category of demyelinating disorders, they included metachromatic leukodystrophy, globoid cell dystrophy, and Krabbe disease. They noted that myelin vacuolization can occur in mitochondrial diseases.

Hypomyelinating leukodystrophies leading to movement disorders

Inoue et al. (2019) noted that leukodystrophies are disorders that impact the white matter of the brain and can have infectious or immunological causes. Demyelinating leukodystrophies can have metabolic causes. Hypomyelinating leukodystrophies (HLDs) represent a separate category of disorders that can arise due to defects in function of one of a number of different gene products. Inoue et al. reviewed Pelizaeus–Merzbacher disease (PMD) due to disruption of functions of the proteolipid protein (PLP1). PMD was reported to be among the most common forms of leukodystrophy. This disorder was first documented by Pelizaeus in 1885 and was further analyzed by Merzbacher who noted the X-linked inheritance. Phospholipase 1 (PLP1) was first implicated in this disease by Willard and Riordan in (1985). Other interesting findings emerged including discovery of PMD cases with increased copy numbers of PLP due to chromosome duplications. PLP1 is encoded by gene that is 17 kb in length and is located on chromosome Xq22.1.

However, cases with phenotypes highly similar to PMD were found not to have PLP mutations. These became known as PMD-like disorders, first reported by Pouwels et al. in 2014. PMD is now referred to as PLP1-related disorder (Inoue, 2019). There is some degree of phenotypic heterogeneity in patients with PLP1-related disorders, and X-linked spastic paraplegia is a phenotype also due to PLP1 mutations. A spectrum of PLP1-related phenotypes has been described related to different mutations, deletions, or duplications in PLP1.

Inoue noted that the most severe forms arise due to point mutations that lead to aberrant folding of the mutant PLP1 protein that in turn lead to endoplasmic reticulum stress and the unfolded protein response. Surprisingly, null mutations leading to reduced levels of PLP1 proteins were reported to be associated with milder forms of the disease. Inoue noted that there were rare cases of PLP1 deletions that led to a milder form of PLP1 related disease, patients continued to be able to walk, and cognitive impairment was mild.

Females heterozygous for PLP1 mutations or disruptions may have some manifestations of the disorder.

Genes impacted in Pelizaeus–Merzbacher-like disorders with hypomyelinating leukodystrophy

These are listed in the following as type of hypomyelinating disorder, defective gene product, and function of that gene product.

- HLD2 (GJC2), gap junction protein2 connexin involved in myelination
- HDL3 (AIMP1), aminoacyl tRNA synthetase complex interacting multifunctional protein 1
- HDL4 (HSPD1), heat shock protein family D (Hsp60) member 1
- HLD5 (FAM126A) may play a part in the beta-catenin/Lef signaling pathway
- HLD6 (TUBB4A), tubulin beta4 A forms microtubules
- HLD7 (POLR3A), RNA polymerase III subunit
- HLD8 (POLR3B), RNA polymerase III subunit
- HLD9 (RARS1), arginyl-tRNA synthetase 1

- HLD10 (PYCR2), mitochondrial enzyme catalyzes final step in proline biosynthesis
- HDL11 (POLR1C), RNA polymerase 1, subunit C
- HDL12 (VPS11), core subunit of complexes involved in vesicular protein sorting
- HDL13 (HIKESHI), heat shock protein nuclear import factor hikeshi
- HDL14 (UFM1), ubiquitin fold modifier 1
- HDL15 (EPRS1), aminoacyl-tRNA synthetase of glutamic acid and proline
- HDL16 (TMEM106B), transmembrane protein 106B
- HDL17 (AIMP2), part of the aminoacyl-tRNA synthetase complex
- HDL18 (DEGS1), a member of the membrane fatty acid desaturase family
- HDL19 (TMEM63A), transmembrane protein 63A

Pouwels et al. (2014) reviewed HLDs and noted that these disorders are characterized by abnormalities of myelin. They reviewed aspects of myelin formation and function; in addition to supplying support and protection for axons, myelin was noted to impact nerve conduction.

Pouwels et al. documented the lineage of oligodendrocytes that generate myelin. Specific gene products involved in the generation of oligodendrocyte precursors from multipotent neural progenitor cells include

- OLIGO2, oligodendrocyte transcription factor 2,
- SOX10, SRY-box transcription factor 10,
- NKX2.2, NK2 homeobox 2, a nuclear transcription factor,
- PDGRA, platelet-derived growth factor receptor alpha, a cell surface tyrosine kinase receptor,
- NG2 (CSPG2), chondroitin sulfate proteoglycan 2

Oligodendrocytes precursors give rise to premyelinating oligodendrocytes under influence of MRF (myelin-regulating factor). The premyelinating oligodendrocytes then move to interact with axons, and myelination occurs along the axons except on the nodes of Ranvier. On the nodes of Ranvier, myelination is interrupted. Sodium channels are clustered at the nodes of Ranvier. Myelination promotes conduction of signals along the nerve fibers and acts as an electrical insulator. The nodes of Ranvier, rich in ion channels, enable rapid transmission of signal.

Myelin is composed of phospholipid, cholesterol, and proteins. Pauwels et al. noted that the HLD disorders are characterized by reduced myelin.

Charzewska et al. (2016) reviewed HLDs to be taken into account in the differential diagnosis of PMD. They include the HLD types 2 to 11 and also included the Allan Herndon Dudley syndrome due to pathogenic changes in SLC16A2, a solute carrier that functions as a transporter of thyroid hormone and facilitates the cellular uptake of thyroxine and triiodothyronine. This syndrome is associated with developmental delay, hypotonia progressing to spasticity, hyperreflexia, dystonic movements, and nystagmus.

SNAP 29 mutations have also been reported to lead to a PMB-like disorder (Llaci et al., 2019). SNAP 29 synaptosome-associated protein 29 and two other members of the SNAP family bind syntaxin and promote docking of and fusion of synaptic vesicles to the synaptic membranes.

Abnormal movements in rett syndrome and rett-like syndrome, atypical rett syndrome

Rett syndrome features include normal early growth with subsequent growth slowing. In stage I of Rett syndrome, a particularly characteristic manifestation occurs that involves the hands with initial abnormal hand movements and subsequent loss of purposeful hand movements. In this stage, growth slowing may also be observed. Stage II was reported to occur between 1 and 4 years.

In subsequent stages, motor problems increase, leading to walking difficulties and scoliosis. seizures may occur, breathing abnormalities may arise and bruxism is present, speech is lost, and cognition is increasingly impaired.

Pathological changes in the MECP2 gene are found in most cases. It is, however, important to note that MECP2 defects are not found in all patients with clinical features of Rett syndrome. Disorders with clinical manifestations of Rett syndrome but without MECP2 abnormalities are referred to as having atypical Rett syndrome or Rett-like syndrome.

Vidal et al. (2019) reported results of studies on 437 patients with a clinical diagnosis of Rett syndrome 40 patients were found to have mutation in genes that encoded proteins other than MECP2. Gene products implicated in Rett-like syndrome by Vidal et al. include the following:

- STXBP1, syntaxin-binding protein, appears to play a role in release of neurotransmitters via syntaxin
- TCF4, transcription factor 4
- SCN2A, sodium voltage-gated channel alpha subunit 2; KCNQ2, potassium voltage-gated channel subfamily Q member 2, regulation of neuronal excitability
- MEF2C, myocyte enhancer factor 2C; defect associated with cognitive disability, stereotypic movements
- SYNGAP1, synaptic Ras GTPase-activating protein1, regulates synaptic plasticity
- CDKL5, cyclin-dependent kinase like 5
- FOXG1, forkhead box G1; transcription factor plays a role in brain development

Ehrhart et al. (2018) reported that defects in genes involved in epigenetic gene regulation, chromatin shaping, neurotransmitter action, or RNA transcription/translation could give rise to Rett syndrome–like manifestations.

Bardetbiedl syndrome

Clinical characteristics of this syndrome include intellectual disability, truncal obesity, type 2 diabetes, dyslipidemia, ataxia, abnormal gait, pigmentary changes in the macular, rod–cone discovery in the retina, and neuromotor and behavioral problems. Autosomal recessive inheritance of specific gene defects leads to this condition; however, oligogenic inheritance has also been proposed for this condition (Katsanis, 2004).

Nineteen different genes have been documented as having defects in this condition. This disorder is defined as a ciliopathy, and most of the genes involved in this condition impact functions within the ciliary pathway.

11. Ataxias

In a review of ataxias in childhood, Pavone et al. (2017) noted that these can result from impairments at different levels of the nervous system. However, they emphasized that ataxias were primarily caused by defects in the circuitry that connects the basal ganglia, cerebellum, and cerebral cortex. They noted that ataxia in children can present with staggering gait, tremor impaired motor coordination, head nodding, nystagmus, and difficulty with word formation.

Acute ataxia was noted to occur in some cases in consequence of infections or as a result of ingestion of certain medications or ingestion of ethanol or toxic substances. Low intake of specific vitamins including thiamine, cobalamin, vitamin E, or folate was noted to lead to ataxia. Immune-mediated forms of ataxia have also been reported. Intermittent ataxia was reported to occur in association with specific inborn errors of metabolism including Hartnup disease and maple syrup urine disease.

Ataxia was also noted to occur in association with certain structural brain malformations such as Dandy Walker syndrome; in that syndrome, the cerebellar vermis is absent. Ataxia was also reported to occur in some patients with Arnold Chiari malformation associated with downward displacement of the cerebellar vermis.

Pavone et al. also noted that ataxia occurred in ciliary disorder Joubert syndrome. They noted that ataxias can also occur as manifestations of mitochondrial disorder.

Autosomal recessive forms of ataxia include Friedreich ataxia and ataxia telangiectasia. More than 200 different mutations in the ataxia telangiectasia gene led to ataxia, oculomotor defects, polyneuropathy, hypotonia, and oculocutaneous telangiectasia.

Friedreich Ataxia

Pandolfo (2008) reviewed the autosomal recessive disorder Friedreich ataxia and noted that it primarily impacts the nervous system and the heart. He noted that manifestations often present at puberty but may occur earlier at 2 years or later at 25 years. The presenting symptom is often gait instability; occasionally, cardiac symptoms manifest first.

The ataxia in this disorder is defined as mixed cerebellar and sensory ataxia associated with broad-based gait and progressive instability while walking. Deterioration of fine motor skills and tremor also occur. Kyphoscoliosis often develops. Later in the disorder, swallowing difficulties may develop.

Pathological changes include loss of neurons in the dorsal root ganglia and atrophy of posterior columns of the spinal cord. Changes in the brain include atrophy of the dentate nucleus. Pandolf noted that Friedreich ataxia occurs in individuals of West European, North Africa, and Middle Eastern origin.

The gene implicated in Friedreich ataxia maps to chromosome 9q21.1 and encodes a protein referred to as frataxin (FRDA) that is located in the mitochondrial matrix. Frataxin is reported to be involved in iron metabolism and to promote synthesis of mitochondrial iron–sulfur clusters that are essential for mitochondrial electron transport complex function. Iron–sulfur clusters also occur in the cytosol and in the nucleus and play roles in iron metabolism. The key disease-causing mutation in *FXN* gene is hyperexpansion of a GAA repeat. Pandolfo reported that most patients with Friedreich ataxia are homozygous for this mutation. In the normal chromosome, there are 38 repeats in the disease cases, and expansions from 70 to 1000 repeats occur. A small percentage of patients were reported to be compound heterozygotes for the repeat expansion in one frataxin gene and to have pathogenic nucleotide changes in the frataxin encoding gene on the other chromosome gene on the other chromosome. Repeat expansion reduced the amount of functional frataxin protein produced. Repeat instability occurs when the gene is transmitted from parent to offspring. In addition, there is evidence for somatic instability of the repeat.

Hartnup disease and cerebellar ataxia

This disease is characterized by cerebellar ataxia, light sensitivity of skin, skin rashes and pealing, and aminoaciduria, and it may be associated with behavioral problems. The disorder was shown by Scriver et al. (1987) to be due to defect in the transport of neutral amino acids. Subsequent studies revealed that the disease is due to mutations that impact the activity of the solute carrier SLC6A19 that is an amino acid transporter (Seow et al., 2004). Defective transport of tryptophan and niacin particularly account for the manifestations of the disease. Defective functions of SLC6A19 impact transport of neutral amino acids in the intestine and in the kidney. Urinary excretion of glutamine valine, phenylalanine, leucine asparagine isoleucine, alanine, serine, histidine, tyrosine, and tryptophan increases.

12. Neurodegeneration with brain iron accumulation

These disorders may present in childhood or in adult life. Wiethoff and Houden (2017) noted that these disorders present with pyramidal and extrapyramidal signs, ataxia, dystonia, and cognitive and behavior problems.

Hayflick et al. (2018) reported that most of these disorders result from pathological mutation in any one of three different genes *PANK2, PLA2G6,* and *C19ORF12.*

- *PANK2* encodes pantothenate kinase 2, which is expressed in mitochondria and is key to biosynthesis of coenzyme A
- *PLA2G6* encodes phospholipase A2 group VI involved in phospholipid remodeling and transmembrane ion flux
- *C19ORF12* is an open reading frame that encodes a small transmembrane protein

Klopstock et al. (2019) reported promising results in treatments of these brain iron accumulation diseases with an iron chelating agent.

Pathways involved in ataxia

Manto et al. (2020) in a review of cerebellar ataxias emphasized that basal ganglia and cerebellum are linked by pathways that ensure bidirectional communication. They noted that inherited ataxias include autosomal dominant, autosomal recessive, and X-linked ataxias. Other forms of ataxia include episodic ataxia and immune-mediated ataxia.

Autosomal dominant cerebellar ataxias

Three types of autosomal dominant cerebellar ataxias (ADCAs) have been defined. Gene products impacted in cerebellar ataxias SCAs (spinocerebellar ataxias) include proteins with function listed in the following; however, in some cases, the precise function of an impaired locus product has not yet been defined:

- ATXN1 (SCA1) function not defined, maps to 6p22.3
- ATXN2 (SCA2) membrane located protein likely involved in endocytosis maps to 12q24.12
- ATXN3 (SCA3) disorder also known as Machado–Joseph disease, maps to 14q32.12
- CACNA1A (SCA6) calcium voltage-gated channel subunit alpha1 A, maps to 19p13.13
- ATXN7 (SCA7) component of chromatin remodeling complexes maps to 3p14.1
- TBP (SCA17), Tata box–binding protein, scaffold for transcription factors, maps to 6q27

Autosomal recessive cerebellar ataxias

Manto et al. noted that there are at least 92 different genes involved in causation of autosomal recessive ataxia. These are sometimes separated into categories with or without sensory abnormalities and those with or without cognitive impairments. Well-documented forms autosomal recessive cerebellar ataxias include Friedreich ataxia and ataxia telangiectasia. Manto et al. listed most prevalent forms of autosomal recessive ataxia and noted that several forms of recessive ataxia were due to defects in genes that led to metabolic impairment.

- APTX1 (AOA1), aprataxin may play a role in single-strand DNA repair ataxia ocular apraxia
- SETX (AOA2), senataxin contains a DNA/RNA helicase domain involved in both DNA and RNA processing
- POLG (SANDO), catalytic subunit of mitochondrial DNA polymerase, ophthalmoplegia, and ataxia
- SYNE1 (ARCA1), spectrin repeat containing protein, localizes to nuclear membrane
- SPG7 is a mitochondrial metalloprotease protein, involved in membrane trafficking
- COQ8A (ARCA2) mitochondrial protein functions in an electron-transferring membrane protein complex
- APTX1 (AOA1), aprataxin may play a role in single-strand DNA repair ataxia ocular apraxia; SETX (AOA2) senataxin contains a

DNA/RNA helicase domain involved in both DNA and RNA processing
- POLG (SANDO) catalytic subunit of mitochondrial DNA polymerase, ophthalmoplegia, and ataxia
- SYNE1 (ARCA1), spectrin repeat containing protein, localizes to nuclear membrane
- SPG7 encodes a mitochondrial metalloprotease protein, involved in membrane trafficking
- COQ8A (ARCA2) mitochondrial protein functions in an electron-transferring membrane protein complex

Manto et al. also documented metabolic conditions and syndromes associated with ataxia. In Joubert syndrome, multiple gene defects each encoding a gene product involved in cilia functions CDG syndromes congenital disorders of glycosylation.

Spastic paraplegia syndromes can also occur in Biotinidase deficiency Ceruloplasmin deficiency Tay Sachs disease Alpha Mannosidosis and in Niemann Pick disease.

Genes and their products involved in other forms of autosomal recessive ataxia reported by Synofzik and Puccio (2019).

- SACS Charlevoix-Saguenay (ARSACS) multidomain sacsin protein recruits Hsp70 chaperone action
- SIL1 nucleotide exchange factor, endoplasmic reticulum (ER)
- ANO10, anoctamin transmembrane endoplasmic reticulum (ER), N-linked glycoprotein
- POLR3A catalytic component of RNA polymerase III, which synthesizes small RNAs
- POLR3B second largest subunit of RNA polymerase III synthesizes small RNAs
- STUB1 tetratricopeptide repeat and a U-box that functions as a ubiquitin ligase/cochaperone
- PMCA (four different genes now referred to as ATP2B1–ATP2B4), ATPase plasma membrane Ca^{2+} transporting
- PTPLA, PTPLB now referred to as HACD1 HACD2 conversion of long-chain fatty acids to very-long-chain fatty acids

Autosomal recessive forms of ataxia that involve proteins involved in mitochondrial function include the following:

- FXN, frataxin
- SACS, sacsin, a molecular chaperone
- SPGJ (L1CAM), a cell adhesion molecule
- AFG3L2, a mitochondrial peptidase
- COQ8A, coenzyme Q
- POLG DNA polymerase
- TWNK, twinkle mitochondrial DNA helicase thought to be involved in mitochondrial DNA replication
- ATAD3 subunits that encode membrane proteins involved in mitochondrial dynamics
- PMPC A and B that encode subunits of a mitochondrial peptidase
- PITPM1, an ATP-dependent metalloprotease that degrades postcleavage mitochondrial transit peptides

Synofzik and Puccio (2019) noted that another important mechanism disrupted in specific forms of autosomal recessive cerebellar ataxia was DNA repair. Mutations in genes involved in DNA repair that can lead to autosomal recessive cerebellar ataxia include the following:

- ATM, ATM serine/threonine kinase
- MRE11, MRE11 homolog, double-strand break (DSB) repair nuclease
- SETX, senataxin contains a DNA/RNA helicase, likely functions at DSBs
- RNF168, ring finger protein 168, involved in DNA DSB repair.
- APTX, aprataxin may play a role in single-stranded DNA repair through its nucleotide-binding activity

- PNKP polynucleotide kinase 3′-phosphatase, involved in repair of DNA damage by radiation or oxidation
- TDP1 tyrosyl-DNA phosphodiesterase 1, repair of free radical–mediated DNA DSBs.
- XRCC1, X-ray repair cross complementing 1, repair of DNA single-strand breaks
- XPA DNA damage recognition and repair factor, central role in nucleotide excision repair
- ERCC4, ERCC excision repair 4, endonuclease catalytic subunit, 5′ incision in nucleotide excision repair
- ERCC8, ERCC excision repair 8, repair of damage by ultraviolet radiation
- ERCC6, ERCC excision repair 6, chromatin remodeling factor, transcription-coupled excision repair

Ataxias due to disruption in lipid metabolic processes can occur due to the following defects and in the following disorders:

- NPC1, NPC intracellular cholesterol transporter 1 (Niemann–Pick disease)
- NPC2, NPC intracellular cholesterol transporter 2 (Niemann–Pick disease)
- CYP27A1 cytochrome P450 family 27 subfamily A member 1, synthesis of cholesterol, steroids, and lipids
- PNPLA6, patatin-like phospholipase domain containing 6, deacetylates intracellular phosphatidylcholine
- PLA2G6 phospholipase A2 group VI, phospholipid remodeling
- ARSA, arylsulfatase A, hydrolyzes cerebroside sulfate to cerebroside and sulfate
- GALC, galactosylceramidase, hydrolyzes the galactose ester bonds of ceramides and sphingosines
- GBA, glucosylceramidase beta, cleaves the beta-glucosidic linkage of glycosylceramide GBA2 microsomal beta-glucosidase
- HEXA, hexosaminidase subunit alpha, catalyzes the degradation of the ganglioside GM2
- GLB1, galactosidase beta 1, catalyzes the hydrolysis of a galactose residue from ganglioside substrates
- PHYH, phytanoyl-CoA 2-hydroxylase, defective in certain peroxisomal diseases
- PEX7, peroxisomal biogenesis factor 7
- HSD17B4, hydroxysteroid 17-beta dehydrogenase 4, peroxisomal beta-oxidation pathway for fatty acids

Synofzik et al. (2019) noted that delineation of specific mechanism disrupted in ataxias was potentially important in designing therapies.

Stroke-like episodes in children

MELAS is a specific syndrome associated with mitochondrial encephalopathy, lactic acidosis, and stroke-like episodes that arise as a result of impaired mitochondrial function. This syndrome was reviewed by El-Hattab et al. (2018). A specific mutation in mitochondrial DNA m.3243 $A > G$ that alters mitochondrial tRNA was reported by Goto et al. (1992) to give rise to this syndrome.

The key features of the disorder include stroke-like episodes that occur in children and adults before the age of 40 years, seizures, myopathy, and peripheral neuropathy. Other features of the disorder include recurrent headaches and recurrent vomiting episodes. El-Hattab et al. noted that clinical manifestations during the stroke-like episodes include headache, visual impairment, loss of speech, and increased motor weakness.

Histological studies on muscle reveal ragged red fibers, indicative of impaired mitochondrial function. Lactic acidemia is a key feature of the disorder. Other body systems may also be affected in the disorder. Defective function can occur in the gastrointestinal tract, heart, and endocrine systems. El-Hattab et al. noted that there is considerable phenotypic variability in this disorder. This variability may be in part due to mitochondrial heteroplasmy, i.e., all the mitochondria in cells in a particular tissue and system may not manifest the characteristic DNA mutation.

13. Other abnormal movements that occur in specific disorders

Stereotypies

These include repetitive or ritualistic movements. They frequently occur in individuals with autism. Mackenzie (2018) noted that the pathophysiology of stereotypies involved at least three different pathways: overactive dopaminergic pathway, underactive cholinergic pathway, and underactive GABAergic inhibitory pathway.

Abnormal movements may occur in the Brown–Vialetto–Van Laere sensory–motor disorder that is associated with bulbar dysfunction and also with deafness. This disorder arises due to defective function of riboflavin transporters SLC52A2 and SLC52A3 (RFVT2 and RFVT3).

Chorea

It is characterized by abnormal movements that flow across different parts of the body. It is defined as hyperkinetic movement disorders. Hermann and Walker (2015) reviewed this disorder. This disorder may arise in children during streptococcal infection and is referred to as Sydenham's chorea. It arises as a result of cross-reactivity of antibodies to streptococcal components that also cross-react with basal ganglia neurons.

Specific genetic forms of chorea occur. One of the best studied is Huntington's chorea although this seldom presents in children. Specific gene defects that can give rise to chorea in children include deleterious mutations in GLUT1 (SLC2A1), a glucose transporter and mutations in PRRT2 (proline-rich transmembrane protein 2), a transmembrane protein expressed in brain. Deleterious mutations in this gene product can give rise to episodic dyskinesia or chorea.

Chorea may also occur in cases of Lesch Nyhan syndrome due to defective function of hypoxanthine phosphoribosyltransferase 1 (HGPRT). Chorea may also occur in individuals with inborn errors of metabolism associated with ataxia. Specific mitochondrial disorders in children may also be associated with chorea.

Dystonia

This is described as a disorder in which involuntary abnormal postures tend to be sustained (Tisch, 2018). At least 26 different forms of dystonia are recognized, and genomic defects predisposing to these are being recognized.

- DYT1, one form of torsion dystonia maps to chromosome 9q34.11
- DYT2 is an autosomal recessive form of dystonia due to defects in the HPCA gene product hippocalcin, a neuron-specific calcium-binding protein
- DYT3 is an X-linked dystonia sometimes referred to as torsion dystonia-Parkinsonism like
- DYT4 is defined as whispering torsion dystonia due to autosomal dominant defects in TUBB4A that leads to HLD that impacts basal ganglia
- DYT5 is defined as a dopamine-responsive dystonia due to mutations in GTP cyclohydrolase
- DYT6 is a torsion dystonia due to defect in THAP1 domain protein possibly involved in cell cycle regulation
- DYT7 cervical dystonia map to chromosome 18p
- DYT9 also defined as choreoathetosis with episodic ataxia is due to mutations in SLC2A1, a major glucose transporter also known as GLUT1
- DYT10 is paroxysmal kinesigenic choreoathetosis due to defects in PRRT2 (proline-rich transmembrane protein)

- DYT11 myotonic dystonia due to defect in SCGE (sarcoglycan epsilon), a transmembrane protein in muscle
- DYT12 is dystonia–parkinsonism rapid onset due to defects in ATP1A3, sodium potassium ATPase
- DYT13 autosomal dominant torsion dystonia, maps to chromosome 1p36
- DYT14 is a dopamine-responsive dystonia due to mutations in GTP cyclohydrolase GCH1
- DYT15 is a myoclonic dystonia, autosomal dominant maps to 18p11
- DYT16 autosomal recessive torsion dystonia due to defects in PRKRA protein activator of initiation factor
- DYT17 autosomal recessive torsion dystonia maps to chromosome 20
- DYT18 paroxysmal exercise-related dystonia related to GLUT deficiency (SLC2A1)
- DYT19 episodic kinesigenic dystonia also associated with chorea, maps to 16q
- DYT20 paroxysmal dyskinesia autosomal dominant maps to 2q31
- DYT21 pure torsion dystonia, autosomal dominant maps to chromosome 2q
- DYT23 autosomal dominant impacts neck especially, torticollis, 9q34
- DYT24 dystonia impacts neck, larynx, upper arms defects in ANO3 (anoctamin), a chloride channel
- DYT25 adult dystonia impacts neck defects in GNAL, a guanine nucleotide–binding protein
- DYT26 myoclonic dystonia, autosomal dominant due to defects in KCTD17 potassium channel tetramerization domain

Tourette syndrome

This syndrome is defined by repetitive uncontrollable movements (tics) and uncontrollable utterances. Genomic and gene sequencing studies in Tourette syndrome were reported (Wang et al., 2018). They identified genes that were recurrently found to have pathogenic variants; these included WWC1 WW and C2 domain containing 1

- CELSR3 cadherin EGF LAG seven-pass G-type receptor 3
- OPA1, OPA1 mitochondrial dynamin-like GTPase
- NIPBL cohesin loading factor
- FN1, fibronectin1
- FN2, fibronectin 2

CELSR3 variants were found in singleton Tourette syndrome cases and also in multiplex cases (child and one parent). Wang et al. noted that WWC1 and CELSR3 proteins play roles in the determination of cellular polarity. In addition, 15 of 292 de novo deleterious variants found in multiplex Tourette syndrome families were characterized as having roles in the determination of cellular polarity.

Genome-wide association studies by Yu et al. (2019) indicated that Tourette syndrome was a polygenic disorder; they also determined that variants in the non–protein-coding regions of the gene likely modulated gene expression that impacted corticostriatal pathways.

Aspects of treatment of movement disorders

Jinnah et al. (2018) reviewed treatment of inherited rare movement disorders. They noted that individual rare disorders occur at low frequencies, e.g., in the European Union at fewer than 50 individuals per 100,000 population. However, more than 7000 different rare diseases occur, and collectively, they are reported to impact 6%–8% of the population. Jinnah et al. noted that approximately 80% of rare diseases are genetically determined.

They focused their report on rare movement disorders in which there was strong clinical evidence for efficacy with specific treatments and tabulated evidence of different approaches to therapy that included one or more of the following:

1. Reduction of toxic or damaging products that resulted from specific enzyme deficiencies
2. Dietary intervention to bypass specific enzyme deficiencies
3. Use of specific vitamin supplements to improve functions of specific impaired proteins or enzymes
4. Avoidance of certain triggers or targets that accentuated defects; these could include some nonspecific factors, e.g., avoidance of stress low food intake
5. They documented specific small molecules including pharmacological products useful in the treatment of movement disorders

Jinnah et al. emphasized that treatments designed to treat specific underlying disease mechanisms can be very beneficial in some situations. They noted that in addition, mainstay treatments of movement disorders include physiotherapy, ambulatory aids, and balance training.

It is important to note that in a number of different types of disorders due to deficiency of particular enzymes or proteins or transporters, movements disorders may occur along with other manifestations including epilepsy and or developmental delay. A number of such genetic disorders are in part responsive to vitamin supplementation. Examples include the following:

Ataxia dystonia due to defective activity of alpha tocopherol transfer protein (TTPA), which is responsive to vitamin E supplementation

Biotin thiamine–responsive basal ganglia disease due to deficiency of thiamine transporter deficiency (SLC19A3), which is responsive to biotin and thiamine supplementation

Biotinidase deficiency, associated with ataxia, dystonia and spastic paresis, which is responsive to biotin

Cerebral folate deficiency due to defects in folate receptors or defects in folate transporter SLC19A3, which is responsive to folinic acid

Cerebral cobalamin deficiency can result from defects in a number of different genes and is associated with atonia, dystonia, or spasticity and is responsive to cobalamin (vitamin B12) therapy.

Pyruvate dehydrogenase complex deficiency can result from defects in several genes and can lead to dystonia ataxia and seizure, which is responsive to thiamine and to ketogenic diet

A movement disorder that is particularly important to correctly diagnose is Wilson's disease. This disorder is associated with dystonia, tremor, chorea, myoclonus, and potential cognitive impairment. It is due to defective function or deficiency of the copper transporter ATP7B. This disorder is frequently responsive to penicillamine or trientine (triethylenetetramine), and increased zinc intake may also be helpful.

It is also important to correctly diagnose disorders due to defects in gene products in the dopamine catecholamine synthesis pathway. These gene defects can lead to impaired developments and movement disorders sometimes diagnosed as cerebral palsy. In addition, dystonia and oculogyric crises may occur in these patients. Disease manifestations in these disorders may respond to therapy with levodopa.

Certain calcium ion channel disorders, e.g., CACNA1A defects lead to episodic ataxia and dystonia that can be treated with a specific small molecule 4-aminopyridine acetazolamide.

The mTOR pathway and its critical functions in the brain

In a 2016 review of the mTOR signaling cascade, Crino noted the key roles played by mTOR in the integration of oxygen, nutrients, and growth factor levels to regulate cell growth and proteins synthesis. He noted too evidence that altered mTOR signaling plays in specific neurological disorders. These include developmental brain malformations and neurodevelopmental disorders.

Specific questions still arise concerning the impact of mTOR on specific cellular function including functions in neuroprogenitor cells, neurons, and astrocytes. Crino noted that in addition to impacts on neuronal growth, the mTOR pathways also regulate excitatory and inhibitory neurotransmission.

In addition to the mTORC1 complex, an mTORC2 complex exists, and this complex has some different functions than mTORC1. Genes that encode specific proteins in the mTORC2 have been reported to undergo mutations that lead to structural brain abnormalities.

Crino reported that the two different mTOR complexes regulate different cellular processes and operate in different cellular processes and operate to different degrees in different cell types. mTORC1 occurs in the cytoplasm and endoplasmic reticulum and was reported to be bound to lysosomal membranes; in addition, mTORC1 proteins were reported to occur in the nucleus. The mTORC2 complex was reported to be involved in cytoskeletal maintenance and in cell migration.

The mTORC1 and mTORC2 complexes have three main proteins in common: mTOR, DEPTOR, and MLST8. In addition, there are proteins that are unique to a specific mTOR complex.

Proteins common to mTORC1 and mTORC2

- mTOR, mechanistic target of rapamycin kinase member of the family phosphatidylinositol kinase-related kinases, encoded by a gene on 1p36.2
- DEPTOR DEP domain (disheveled and plekstrin)—containing mTOR interacting protein, the encoding gene maps to 8q24.12
- MLST8, mTOR-associated protein, LST8 homolog the encoding gene maps to 16p13.3, LST8 is reported to be a phosphorus/phosphate sensing molecule

Proteins unique to the mTORC1 complex include the following:

- RAPTOR (RPTOR) regulatory-associated protein of mTOR complex 1, encoded on 17q25.3, it also associates with eukaryotic initiation factor 4E-binding protein 1 and ribosomal protein S6 kinase.
- PRAS40 (AKT1S1) proline-rich substrate of AKT (MIM 164730) that binds 14-3-3 protein when phosphorylated encoded on 19q13.33
- FKBP1A prolyl isomerase 1A FKBPs are involved in numerous biologic processes, including protein folding, receptor signaling, interacts with cellular signaling molecules and intracellular ion channels encoded by a gene on 20p13

Proteins unique to the mTORC2 complex include the following:

- PROTOR1/2 (PRR5) proline-rich complex regulates PDGF growth factor signaling encoded by a gene on chromosome 22q13.31
- RICTOR RPTOR independent companion of mTOR complex 2 encoded on 5p13.1
- SIN1 (mSIN1) (MAPKAP1) Map kinase-associated protein, stress-activated protein kinase encoded by a gene on9q33.3

Liu and Sabatini (2020) noted that RAPTOR constitutes the regulatory scaffold protein in mTORC1 that is primarily involved in energy metabolism, nutrient usage, growth factor responses, and protein synthesis. They reported that mTORC2 was primarily involved in cytoskeletal arrangements and cell proliferation.

Important impacts of mTORC1 on protein synthesis include activation of S6 kinase and SKAR (POLDIP) DNA polymerase delta interacting protein 3, to facilitate translation of spliced transcripts. mTORC1 was also reported to facilitate ribosome biogenesis.

mTORC1 was also shown to facilitate metabolism through impact on activity of ATF4-activating transcription factor 4, MTHFD2 (methylenetetrahydrofolate dehydrogenase

[NADP + dependent] 2), methenyltetrahydrofolate cyclohydrolase, and through activity of S6K1 (S6 kinase) that stimulates nucleotide synthesis.

mTORC1 stimulates activity of PPARG (peroxisome proliferator–activated receptor gamma), thereby promoting lipid synthesis. mTORC1 stimulation of HIFA (hypoxia-inducible factor 1 subunit alpha) stimulates aerobic glycolysis. mTORC1 was reported to suppress autophagy.

mTORC2 was reported to stimulate protein kinase activity and cytoskeletal rearrangement and mobility.

The tuberous sclerosis complex is composed of TSC1, TSC2, and TBC1D7 (TBC1 domain family member 7). The kinases AKT1 and AMPK differentially phosphorylate specific phosphorylation sites on the TSC complex and therefore differentially influence TSC complex activity toward RHEB. AKT phosphorylation on the TSC complex inhibits its ability to block RHEB activity. The activated TSC complex phosphorylated by AMPK catalyzes the conversion of RHEB GTP to its inactive form RHEB GDP that thereby acts as a brake on the ability of RHEB GTP to activate mTORC1 activity.

Phosphorylation of TSC subunits at specific sites promoted dissociation of TSC from the lysosome.

mTOR complexes and neuronal function

Liu and Sabatini reported that mTOR complexes play key roles in neuronal functional including establishment of cortical architecture and also in neuronal circuitry remodeling in response to experience. Loss of RAPTOR, the regulatory subunit of mTORC1, and loss of RICTOR, the regulatory component of mTORC2, were reported to be associated with microcephaly.

Hyperactive mTOR signaling can occur due to loss of function of mTOR regulators, defects in TSC1 and TSC2 function, defects in function of AMPK activator STRADA, impaired functions of GATOR and KICSTOR, and regulators of mTORC1.

Hyperactive mTOR function can also result from activating mutations of RHEB. RHEB mutations were reported to lead to altered mTORC activity with megalencephaly and intellectual disability.

Increased activity of the kinase PIK3 (phosphatidyl inositol kinase 3) and increased activity of BDNF (brain-derived neurotrophic factor) can also lead to mTORC1 hyperactivity.

Liu and Sabatini noted that loss-of-function mutations and/or deletion of TSC1 can lead to constitutive mTORC1 hyperactivity and result in disruption of the cortex lamina organization and to generation of epileptic foci. In addition, large neurons and giant astrocytes can form.

Inactivating mutations in AMPK or in the regulators GATOR and KICTOR or in RHEB can also lead to hyperactive mTORC1 and to epilepsy.

Liu and Sabatini reported that impaired TSC gene expression or lack of expression of 4EBP2 (EIF4EBP2), the transcription initiation factor, leads to dysregulation of synaptic function. They noted that mTORC1 also impacts brain function via its impact on protein translation and autophagy regulation.

Active mTOR suppresses autophagy through down regulation of ULK1 (unc-51 like autophagy-activating kinase 1 encoded on 12 q24.33). Hyperactive mTOR can then over suppress autophagy, and some level of autophagy is reported to be necessary for synaptic pruning that occurs in brain development. Constitutive mTORC1 hyperactivity was shown to compromise synaptic pruning in model organisms.

Liu and Sabatini noted that optimal levels of mTOR and BDNF promote optimal mRNA translation and function at synapses and thereby impact learning and memory.

A number of agents have been identified that impact mTOR activity. Some agents, notably Rapamycin and Rapalogs, decrease mTORC1 activity. Agents that activate mTORC1 activity have also been identified, which include NV5138.

Proteins that function in the mTOR pathways

- mTOR mechanistic target of rapamycin kinase 1p36.22
- AKT1 AKT serine/threonine kinase, mediator of growth factor−induced neuronal survival14q32.33
- RHEB Ras homolog, mTORC1 binding 7q36.1, vital in growth regulation and cell cycle progression
- PIK3CA, phosphatidylinositol-4,5-bisphosphate 3-kinase catalytic subunit alpha, 3q26.32
- PIK3R2, phosphoinositide-3-kinase regulatory subunit 2, 9p13.11
- PIK3R1, phosphoinositide-3-kinase regulatory subunit 1, 5q13.1
- DEPDC5, DEP domain−containing 5, GATOR1 subcomplex subunit, 22q12.2-q12.3
- NPR3 like, GATOR1 complex subunit, 16p13.3
- NPR2 like, GATOR1 complex subunit 3p21.31
- SZT2 subunit of KICSTOR complex 1p34.2
- STK11 also known as LKB1 subunit of KICSTOR complex 19p13.3
- AMPK also known as PRKAA1
- SZT2, protein kinase AMP-activated catalytic subunit alpha 1 5p13.1
- PRKAA2, protein kinase AMP-activated catalytic subunit alpha 2, 1p32.2
- PRKAB1, a regulatory subunit of the AMP-activated protein kinase (AMPK) 12q24.23
- PRKAB2, regulatory subunit of the AMPK 1q21.1
- PRKAG2, protein kinase AMP-activated noncatalytic subunit gamma 2, 7q36.1
- PRKAG1, protein kinase AMP-activated noncatalytic subunit gamma 1, 12q13.12
- STRADA STE20 forms a heterotrimeric complex with serine/threonine kinase 11 (LKB1) 17q23.3
- TSC1, TSC complex subunit 1, hamartin, stabilizes the GTPase activating protein tuberin, 9q34.13
- TSC2, its gene product tuberin, is able to stimulate specific GTPases, 16p13.3
- TBC1D7 functions as a subunit of the tuberous sclerosis TSC1−TSC2 complex 6p24.1

References

Almannai, M., El-Hattab, A.W., 2018. Inborn errors of metabolism with seizures: defects of Glycine and serine metabolism and cofactor-related disorders. Pediatr. Clin. 65 (2), 279−299. https://doi.org/10.1016/j.pcl.2017.11.007.

Baldassari, S., Picard, F., Verbeek, n, Kempen, Brilstra, E., et al., 2019. The landscape of epilepsy-related GATOR1 variants. Genet. Med. 21 (2). https://doi.org/10.1038/s41436-018-0060-2 (PMID).

Baroli, G., Sanchez, J.R., Agostinelli, E., Mariottini, P., Cervelli, M., 2020. Polyamines: the possible missing link between mental disorders and epilepsy (Review). Int. J. Mol. Med. 45 (1), 3−9. https://doi.org/10.3892/ijmm.2019.4401.

Braun, K.P.J., 2017. Preventing cognitive impairment in children with epilepsy. Curr. Opin. Neurol. 30 (2), 140−147. https://doi.org/10.1097/WCO.0000000000000424.

Bromfield, E., Cavazos, J., Sirven, J., 2006. An Introduction to Epilepsy. https://pubmed.ncbi.nlm.nih.gov/20821849/.

Butler, K.M., Moody, O.A., Schuler, E., Coryell, J., Alexander, J.J., Jenkins, A., Escayg, A., 2018. De novo variants in GABRA2 and GABRA5 alter receptor function and contribute to early-onset epilepsy. Brain 141 (8), 2392−2405. https://doi.org/10.1093/brain/awy171.

Charzewska, A., Wierzba, J., Iżycka-Świeszewska, E., Bekiesińska-Figatowska, M., Jurek, M., Gintowt, A., Kłosowska, A., Bal, J., Hoffman-Zacharska, D., 2016. Hypomyelinating leukodystrophies — a molecular insight into the white matter pathology. Clin. Genet. 90 (4), 293−304. https://doi.org/10.1111/cge.12811.

Colver, A., Fairhurst, C., Pharoah, P.O.D., 2014. Cerebral palsy. In: The Lancet, vol. 383. Lancet Publishing Group, pp. 1240−1249. https://doi.org/10.1016/S0140-6736(13)61835-8. Issue 9924.

Dawson, R.E., Nieto Guil, A.F., Robertson, L.J., Piltz, S.G., Hughes, J.N., Thomas, P.Q., 2020. Functional screening of GATOR1 complex variants reveals a role for mTORC1 deregulation in FCD and focal epilepsy. Neurobiol. Dis. 134. https://doi.org/10.1016/j.nbd.2019.104640.

Ehrhart, F., Sangani, N.B., Curfs, L.M.G., 2018. Current developments in the genetics of Rett and Rett-like syndrome. Curr. Opin. Psychiatr. 31 (2), 103–108. https://doi.org/10.1097/YCO.0000000000000389.

El-Hattab, A.W., Suleiman, J., Almannai, M., Scaglia, F., 2018. Mitochondrial dynamics: biological roles, molecular machinery, and related diseases. Mol. Genet. Metabol. 125 (4), 315–321. https://doi.org/10.1016/j.ymgme.2018.10.003.

Ellis, C.A., Petrovski, S., Berkovic, S.F., 2020. Epilepsy genetics: clinical impacts and biological insights. Lancet Neurol. 19 (1), 93–100. https://doi.org/10.1016/S1474-4422(19)30269-8.

Falco-Walter, J., Scheffer, I., Fisher, R.S., 2018. The new definition and classification of seizures and epilepsy. Epilepsy Res. 139. https://doi.org/10.1016/j.eplepsyres.2017.11.015.

Goto, 1992. Mitochondrial myopathy, encephalopathy, lactic acidosis, and stroke-like episodes (MELAS). Neurology. https://doi.org/10.1212/wnl.42.3.545.

Hayflick, S.J., Kurian, M.A., Hogarth, P., 2018. Neurodegeneration with brain iron accumulation. In: Handbook of Clinical Neurology, vol. 147. Elsevier B.V, pp. 293–305. https://doi.org/10.1016/B978-0-444-63233-3.00019-1.

Hermann, A., Walker, R.H., 2015. Diagnosis and treatment of chorea syndromes. Curr. Neurol. Neurosci. Rep. 15 (2) https://doi.org/10.1007/s11910-014-0514-0.

Heyne, H.O., Artomov, M., Battke, F., Bianchini, C., Smith, D.R., Liebmann, N., Tadigotla, V., Stanley, C.M., Lal, D., Rehm, H., Lerche, H., Daly, M.J., Helbig, I., Biskup, S., Weber, Y.G., Lemke, J.R., 2019. Targeted gene sequencing in 6994 individuals with neurodevelopmental disorder with epilepsy. Genet. Med. 21 (11), 2496–2503. https://doi.org/10.1038/s41436-019-0531-0.

Inoue, K., Disease, P.-M., 2019. Molecular and cellular pathologies and associated phenotypes. Adv. Exp. Med. Biol. 1190. https://doi.org/10.1007/978-981-32-9636-7_13 (PMID).

Jung-Klawitter, S., & Hübschmann, K. (n.d.). Analysis of Catecholamines and Pterins in Inborn Errors of Monoamine Neurotransmitter Metabolism-From Past to Future. Cells (Vol. 8). https://doi.org/10.3390/cells8080867.Review. PMID:31405045.

International League Against Epilepsy Consortium on Complex Epilepsies. Genome-wide mega-analysis identifies 16 loci and highlights diverse biological mechanisms in the common epilepsies., 2018. Genome-wide mega-analysis identifies 16 loci and highlights diverse biological mechanisms in the common epilepsies. Nat. Commun. 9 (1), 5269. https://doi.org/10.1038/s41467-018-07524-z.

Jinnah, H.A., Albanese, A., Bhatia, K.P., Cardoso, F., Da Prat, G., de Koning, T.J., Espay, A.J., Fung, V., Garcia-Ruiz, P.J., Gershanik, O., Jankovic, J., Kaji, R., Kotschet, K., Marras, C., Miyasaki, J.M., Morgante, F., Munchau, A., Pal, P.K., Rodriguez Oroz, M.C., 2018. International parkinson's disease movement disorders society task force on rare movement disorders. Mov. Disord. 33 (1), 21–35. https://doi.org/10.1002/mds.27140. Epub 2017 Sep 1.

Katsanis, N., 2004. The oligogenic properties of Bardet-Biedl syndrome. Hum. Mol. Genet. 13 (1). https://doi.org/10.1093/hmg/ddh092.

Klopstock, T., Tricta, F., Neumayr, L., Karin, I., Zorzi, G., Fradette, C., Kmieć, T., Büchner, B., Steele, H.E., Horvath, R., Chinnery, P.F., Basu, A., Küpper, C., Neuhofer, C., Kálmán, B., Dušek, P., Yapici, Z., Wilson, I., Zhao, F., Vichinsky, E., 2019. Safety and efficacy of deferiprone for pantothenate kinase-associated neurodegeneration: a randomised, double-blind, controlled trial and an open-label extension study. Lancet Neurol. 18 (7), 631–642. https://doi.org/10.1016/S1474-4422(19)30142-5.

Kurian, M., Gissen, P., Smith Jr., M., Clayton, P.T., 2011. The monoamine neurotransmitter disorders: an expanding range of neurological syndromes. Lancet Neurol. 10 (8) https://doi.org/10.1016/S1474-4422(11)70141-7. (Review. PMID).

Larsen, J., Johannesen, k J., Tang, S., Marini, C., et al., 2015. The role of SLC2A1 mutations in myoclonic astatic epilepsy and absence epilepsy, and the estimated frequency of GLUT1 deficiency syndrome. Epilepsia 56 (12), 203–208. https://doi.org/10.1111/epi.13222. PMID:26537434.

Leach, E.L., Shevell, M., Bowden, K., Stockler-Ipsiroglu, S., van Karnebeek, C.D.M., 2014. Treatable inborn errors of metabolism presenting as cerebral palsy mimics: systematic literature review. Orphanet J. Rare Dis. 9, 197. https://doi.org/10.1186/s13023-014-0197-2.

Leu, C., Stevelink, R., Smith, A.W., Goleva, S.B., Kanai, M., Ferguson, L., Campbell, C., Kamatani, Y., Okada, Y., Sisodiya, S.M., Cavalleri, G.L., Koeleman, B.P.C., Lerche, H., Jehi, L., Davis, L.K., Najm, I.M., Palotie, A., Daly, M.J., Busch, R.M., 2019. Polygenic burden in focal and generalized epilepsies. Brain 142 (11), 3473–3481. https://doi.org/10.1093/brain/awz292.

Lim, L., Mi, D., Llorca, A., Marín, O., 2018. Development and functional diversification of cortical interneurons. Neuron 100 (2), 294–313. https://doi.org/10.1016/j.neuron.2018.10.009.

Liu, G.Y., Sabatini, D.M., 2020. mTOR at the nexus of nutrition, growth, ageing and disease. Nat. Rev. Mol. Cell Biol. 21 (4), 183–203. https://doi.org/10.1038/s41580-019-0199-y.

Llaci, L., Ramsey, K., Belnap, N., Claasen, A.M., Balak, C.D., Szelinger, S., Jepsen, W.M., Siniard, A.L., Richholt, R., Izat, T., Naymik, M., De Both, M., Piras, I.S., Craig, D.W., Huentelman, M.J., Narayanan, V., Schrauwen, I., Rangasamy, S., 2019. Compound heterozygous mutations in SNAP29 is associated with Pelizaeus-Merzbacher-like disorder (PMLD). Hum. Genet. 138 (11–12), 1409–1417. https://doi.org/10.1007/s00439-019-02077-7.

Mackenzie, K., 2018. Stereotypic movement disorders. Semin. Pediatr. Neurol. 25, 19–24. https://doi.org/10.1016/j.spen.2017.12.004.

MacLennan, A.H., Thompson, S.C., Gecz, J., 2015. Cerebral palsy: causes, pathways, and the role of genetic variants. Am. J. Obstet. Gynecol. 213 (6), 779–788. https://doi.org/10.1016/j.ajog.2015.05.034.

Magloire, V., Mercier, M.S., Kullmann, D.M., Pavlov, I., 2019. GABAergic interneurons in seizures: investigating causality with optogenetics. Neuroscientist 25 (4), 344–358. https://doi.org/10.1177/1073858418805002.

Manto, M., Gandini, J., Feil, K., Strupp, M., 2020. Cerebellar ataxias: an update. Curr. Opin. Neurol. 33 (1), 150–160. https://doi.org/10.1097/WCO.0000000000000774.

Moinard, C., Cynober, L., de Bandt, J.P., 2005. Polyamines: metabolism and implications in human diseases. Clin. Nutr. 24 (2), 184–197. https://doi.org/10.1016/j.clnu.2004.11.001.

Morrell, M.J., 2002. Folic acid and epilepsy. Epilepsy Curr. 2 (2), 31–34. https://doi.org/10.1046/j.1535-7597.2002.00017.x.

Myers, K.A., Johnstone, D.L., Dyment, D.A., 2019. Epilepsy genetics: current knowledge, applications, and future directions. Clin. Genet. 95 (1), 95–111. https://doi.org/10.1111/cge.13414.

Ng, J., Heales, S.J., Kurian, M.A., 2014. Clinical features and pharmacotherapy of childhood monoamine neurotransmitter disorders. Paediatr. Drugs 16 (4), 275–291. https://doi.org/10.1007/s40272-014-0079-z. Review.

Osman, C., Foulds, N., Hunt, D., Jade Edwards, C., Prevett, M., 2020. Diagnosis of pyridoxine-dependent epilepsy in an adult presenting with recurrent status epilepticus. Epilepsia 61 (1), e1–e6. https://doi.org/10.1111/epi.16408.

Oyrer, J., Maljevic, S., Scheffer, I.E., Berkovic, S.F., Petrou, S., Reid, C.A., 2018. Ion channels in genetic epilepsy: from genes and mechanisms to disease-targeted therapies. Pharmacol. Rev. 70 (1), 142–173. https://doi.org/10.1124/pr.117.014456.

Pavone, P., Praticò, A.D., Pavone, V., Lubrano, R., Falsaperla, R., Rizzo, R., Ruggieri, M., 2017. Ataxia in children: early recognition and clinical evaluation. Ital. J. Pediatr. 43. https://doi.org/10.1186/s13052-016-0325-9.

Pearl, P.L., 2016. Amenable treatable severe pediatric epilepsies. Semin. Pediatr. Neurol. 23 (2), 158–166. https://doi.org/10.1016/j.spen.2016.06.004.

Pearson, T.S., Pons, R., Ghaoui, R., Sue, C.M., 2019. Genetic mimics of cerebral palsy. Mov. Disord. 34 (5), 625–636. https://doi.org/10.1002/mds.27655.

Pope, S., Artuch, R., Heales, S., Rahman, S., 2019. Cerebral folate deficiency: analytical tests and differential diagnosis. J. Inherit. Metab. Dis. 42 (4), 655–672. https://doi.org/10.1002/jimd.12092.

Pouwels, P.J., Vanderver, A., Bernard, G., Wolf, N.I., Dreha-Kulczewksi, S.F., et al., 2014. Hypomyelinating leukodystrophies: translational research progress and prospects. Ann. Neurol. 76 (1), 5–19. https://doi.org/10.1002/ana.24194.

Raimondo, J.V., Burman, R.J., Katz, A.A., Akerman, C.J., 2015. Ion dynamics during seizures. Front. Cell. Neurosci. 9. https://doi.org/10.3389/fncel.2015.00419. OCTOBER.

Rodan, L.H., Anyane-Yeboa, K., Chong, K., Klein Wassink-Ruiter, J.S., Wilson, A., Smith, L., Kothare, S.V., Rajabi, F., Blaser, S., Ni, M., DeBerardinis, R.J., Poduri, A., Berry, G.T., 2018. Gain-of-function variants in the ODC1 gene cause a syndromic neurodevelopmental disorder associated with macrocephaly, alopecia, dysmorphic features, and neuroimaging abnormalities. Am. J. Med. Genet. 176 (12), 2554–2560. https://doi.org/10.1002/ajmg.a.60677.

Rosenbaum, Paneth, Leviton, Goldstein, B., et al., 2006. A report: the definition and classification of cerebral palsy. Dev. Med. Child Neurol. 109 (6).

Scriver, C., Mahon, B., Levy, H., Clow, C., Reade, T., et al., 1987. The Hartnup phenotype: mendelian transport disorder, multifactorial disease. Am. J. Hum. Genet. 40 (5).

Seow, H.F., Bröer, S., Bröer, A., Bailey, C.G., Potter, S.J., Cavanaugh, J.A., Rasko, J.E.J., 2004. Hartnup disorder is caused by mutations in the gene encoding the neutral amino acid transporter SLC6A19. Nat. Genet. 36 (9), 1003–1007. https://doi.org/10.1038/ng1406.

Sharma, P., Hussain, A., Greenwood, R., 2019. Precision in pediatric epilepsy. F1000Res 8 pii: F1000 Faculty Rev-163. doi:10.12688/f1000research.16494.1. eCollection 2019. Review. PMID:30800292.

Shen, D., Hernandez, C.C., Shen, W., Hu, N., Poduri, A., Shiedley, B., Rotenberg, A., Datta, A.N., Leiz, S., Patzer, S., Boor, R., Ramsey, K., Goldberg, E., Helbig, I., Ortiz-Gonzalez, X.R., Lemke, J.R., Marsh, E.D., Macdonald, R.L., 2017. De novo GABRG2 mutations associated with epileptic encephalopathies. Brain 140 (1), 49–67. https://doi.org/10.1093/brain/aww272.

Shevell, M., 2018. Cerebral Palsy to Cerebral Palsy Spectrum Disorder: Time for a Name Change? Neurology. https://doi.org/10.1212/WNL.0000000000006747. Review PMID:30568002.

Strovel, E.T., Cowan, T.M., Scott, A.I., Wolf, B., 2017. Laboratory diagnosis of biotinidase deficiency, 2017 update: a technical standard and guideline of the American College of Medical Genetics and Genomics. Genet. Med. 19 (10) https://doi.org/10.1038/gim.2017.84.

Sweetman, L., Surh, L., Baker, H., Peterson, R.M., Nyhan, W.L., 1981. Clinical and metabolic abnormalities in a boy with dietary deficiency of biotin. Pediatrics 68 (4), 553–558.

Synofzik, M., Puccio, H., Mochel, F., Schöls, L., 2019. Autosomal recessive cerebellar ataxias: paving the way toward targeted molecular therapies. Neuron 101 (4), 560–583. https://doi.org/10.1016/j.neuron.2019.01.049.

Takano, T., Sawai, C., 2019. Interneuron dysfunction in epilepsy: an experimental approach using immature brain insults to induce neuronal migration disorders. Epilepsy Res. 156 https://doi.org/10.1016/j.eplepsyres.2019.106185.

Thodeson, D.M., Park, J.Y., 2019. Genomic testing in pediatric epilepsy. Cold Spring Harbor Molecul. Case Studies 5 (4). https://doi.org/10.1101/mcs.a004135.

Tisch, 2018. Recent Advances in Understanding and Managing Dystonia. https://doi.org/10.12688/f1000research.13823.1.

Tollånes, M.C., Wilcox, A.J., Lie, R.T., Moster, D., 2014. Familial risk of cerebral palsy: population based cohort study. BMJ 349 (g4294). https://doi.org/10.1136/bmj.g4294.

van der Knaap, M.S., Bugiani, M., 2017. Leukodystrophies: a proposed classification system based on pathological changes and pathogenetic mechanisms. Acta Neuropathol. 134 (3), 351–382. https://doi.org/10.1007/s00401-017-1739-1. Epub 2017 Jun 21. Review.

Van Eyk, C., 2018. The emerging genetic landscape of cerebral palsy. Handb. Clin. Neurol. 147, 331–342. https://doi.org/10.1016/B978-0-444-63233-3.00022-1.

van Karnebeek, C.D.M., Sayson, B., Lee, J.J.Y., Tseng, L.A., Blau, N., et al., 2018. Metabolic evaluation of epilepsy: a diagnostic algorithm with focus on treatable conditions. Front Neurol. 9, 1016. https://doi.org/10.3389/fneur.2018.01016. eCollection 2018.

van Karnebeek, C.D., Tiebout, S.A., Niermeijer, J., Poll-The, B.T., Ghani, A., 2016. Pyridoxine-dependent epilepsy: an expanding clinical spectrum. Pediatr. Neurol. 59, 6–12. https://doi.org/10.1016/j.pediatrneurol.2015.12.013. Review.

Vidal, S., Brandi, N., Pacheco, P., Maynou, J., Fernandez, G., Xiol, C., Pascual-Alonso, A., Pineda, M., Maria del Mar, O.C., Garcia-Cazorla, À., del Carmen Serrano Munuera, M., García, S.C., Troncoso, M., Fariña, G., García Peñas, J.J., Fournier, B.G., León, S.R., Guitart, M., Baena, N., 2019. The most recurrent monogenic disorders that overlap with the phenotype of Rett syndrome. Eur. J. Paediatr. Neurol. 23 (4), 609–620. https://doi.org/10.1016/j.ejpn.2019.04.006.

Wang, S., Mandell, J.D., Kumar, Y., Sun, N., Morris, M.T., Arbelaez, J., Nasello, C., Dong, S., Duhn, C., Zhao, X., Yang, Z., Padmanabhuni, S.S., Yu, D., King, R.A., Dietrich, A., Khalifa, N., Dahl, N., Huang, A.Y., Neale, B.M., 2018. De novo sequence and copy number variants are strongly associated with tourette disorder and implicate cell polarity in pathogenesis. Cell Rep. 24 (13), 3441–3454. https://doi.org/10.1016/j.celrep.2018.08.082 e12.

Wiethoff, S., Houlden, H., 2017. Neurodegeneration with brain iron accumulation. Handb. Clin. Neurol. 145, 157–166. https://doi.org/10.1016/B978-0-12-802395-2.00011-0. Review.

Willard, H.F., Riordan, J.R., 1985. Assignment of the gene for myelin proteolipid protein to the X chromosome: implications for X-Linked myelin disorders. Science 230 (4728), 940–942. https://doi.org/10.1126/science.3840606.

Wilson, M., Plecko, B., Mills, P.B., Clayton, P.T., 2019. Disorders affecting vitamin B_6 metabolism. J. Inherit. Metab. Dis. 42 (4), 629–646. https://doi.org/10.1002/jimd.12060.

Yu, D., Sul, J.H., Tsetsos, F., Nawaz, M.S., Huang, A.Y., Zelaya, I., Illmann, C., Osiecki, L., Darrow, S.M., Hirschtritt, M.E., Greenberg, E., Muller-Vahl, K.R., Stuhrmann, M., Dion, Y., Rouleau, G.A., Aschauer, H., Stamenkovic, M., Schlögelhofer, M., Sandor, P., 2019. Interrogating the genetic determinants of Tourette's syndrome and other tiC disorders through genome-wide association studies. Am. J. Psychiatr. 176 (3), 217–227. https://doi.org/10.1176/appi.ajp.2018.18070857.

CHAPTER

9

Health and well-being

1. Health and well-being in the Anthropocene

In 2015, a Rockefeller Lancet Commission report was generated by Whitmee et al. and entitled "Safeguarding Human Health in the Anthropocene." The executive summary in this report emphasized that although human health had steadily improved over the decades, earth's natural systems were increasingly threatened. Authors noted, "we have been mortgaging the health of future generations to realize the economic and development gains of the present."

This report identified three categories of challenges to be addressed: overreliance on gross domestic product as a measure of human progress, failure to take into account the impact of present gains on future health, and thirdly not taking into account the inequality and evidence for increased burdens placed on the poor and on developing nations.

The report emphasized the need for transdisciplinary research and that this would increasingly be essential to address planetary and human health–related problems, particularly as the size of urban populations increased.

Metrics indicative of improved population health, when considered on a worldwide scale, included increased life expectancy, decreased percentages of deaths of children younger than 5 years of age, and decreased poverty rates.

Nevertheless, considering these factors on a worldwide basis was noted to fail to take into account great differences between economically advantaged and economically disadvantaged populations and countries.

The metrics that revealed increases in improvements in human health were in stark contrast to metrics related to environmental health. This report used graphics to illustrate different metrics during periods between 1800 and 2000. The graphics illustrated dramatic increases in energy use, increases in water use (Landrigan, 2017), increases in water shortages, increases in carbon dioxide emissions, and increases in ocean acidity.

The Rockefeller Lancet Commission report generated a definition of planetary health: "Planetary health is the health of human civilization and the state of the natural systems on which it depends."

The report also noted the importance of addressing unacceptable inequities and the need for integrated political approaches.

Pollution and effects on human health

Landrigan in a 2017 report, drew attention to the damaging effects on human health. Attention was focused particularly on industrial, vehicular, and chemical pollution. The deleterious pollution effects particularly impacted low-

https://doi.org/10.1016/B978-0-12-821913-3.00002-0

income and middle-income countries. Pollution was also noted to have effects on natural systems.

Mental well-being in the Anthropocene

White (2020) reported on approaches to mental well-being in the Anthropocene and focused in part on the capabilities approach and on concepts of justice with respect to resource use. White noted that questions have been raised with respect to possible preoccupation with treatment of mental disorders in contrast to promoting psychological well-being (Ryff et al., 2016; Ryff and Psychological, 2014), also referred to as eudaimonic concepts.

Six aspects of psychological well-being were defined, which included autonomy (self-determination), capacity to manage one's life, positive relations with other people, and self-acceptance. Collectively, these concepts are included in the capabilities approach. The capabilities approach in economics was promoted by Amartya Sen, and it included access to economic transactions and political participation. The capabilities approach also emphasizes the importance of people being able to have a concept that they are living lives that are valuable and worthwhile (Nobel Prize, 1998). The key concepts that feed into human capabilities assessments included bodily health, senses including emotion and thought, affiliations, play, and some sense of political and material control.

The World Health Organization definition of a human health is "A state of complete, physical, mental and social well-being and not the mere absence of infirmity." The WHO comprehensive mental action plan defined additional concepts relevant to health and well-being, social, cultural economic, political, and environmental factors. Additional factors noted to influence health and well-being included national policies, environmental factors, working conditions, and community and social support.

White (2020) noted that key factors of importance included housing, justice, and employment. Disturbing environmental factors noted to impact well-being include violence, marginalization, and discrimination. White also noted that a series of different instruments have been developed to measure capabilities. These include the Oxford Capability Questionnaire of mental health OCAP MH.

Attention has been placed on specific strain induced by living in overcrowded urban settings. Research studies have also revealed the importance of access and exposure to natural settings. In 1984, Wilson presented the Biophilia hypothesis and the natural affinity of humans to the natural world and to living systems.

2. The Lancet one health commission report (2019)

This report stressed that a symbiotic relationship between environment, human, and animal health would be necessary for continued sustenance and evolution on our planet.

The report drew attention to the impact of industrialization, globalization, urbanization, and human dominance on planetary degradation. Planetary degradation was predicted to predispose to the transfer of pathogens between animals and humans, and these would likely frequently be viral pathogens. Bacterial pathogens were predicted to be and have been shown to have become increasingly resistant to antimicrobial agents. Methods of prevention of zoonoses will likely meet with opposition from different parties involved in facilitating marketing of live animals, often wild live animals, in proximity to marketing of plant products. Such markets also serve as places for community interactions and are considered important places for cultural exchange.

Planetary degradation is further complicated by armed conflicts that increase poverty and also precipitate extensive migration.

Significant evidence has been gathered on the dimensions of the problem of degradation and its consequences that include decreased availability of clean water and decreased food production. Humans have clearly learned over many centuries of evolution that certain calamities require adaptations. In considering evolution, much attention has been focused on genetic changes in promoting adaptation. However, there is also evidence for social and cultural adaptations that promote survival. Questions arise as to what extent political and economic forces need to change to facilitate necessary social and cultural adaptations. In addition to the striking need for changes in behaviors to promote human physical health, there is abundant evidence that the environmental degradation is harmful to mental health.

Adverse conditions, environmental stress, poverty, and parental stress have deleterious effects on both physical and mental child development.

Humans have a long history of struggling and adapting to survive. Nevertheless, despite the urgency for change, there is resistance to change. A report by Rull in 2011 noted "Our social and economic systems are too reluctant to even acknowledge, let alone abandon destructive practices." Rull questioned further, is there could be benefits of changes in economic approaches, for example, could capitalism incorporate natural capital into its cost–benefit analyses?

Questions inevitably arise concerning global governance and global cooperation to address problems, and some of these were addressed in the Lancet University of Oslo Commission on Global Governance in 2014 https://www.thelancet.com/commissions/global-governance-for-health. The report of this commission noted that domains that need to be addressed include health inequities that exist in populations and inequities in trade and investments and factors that generate armed conflicts (Lucey et al., 2017).

In some areas, global cooperation was achieved and led to eradication of a significant problem. Lucey et al. (2017) noted the great achievement in eradicating smallpox globally by 1980. Smallpox was an infection that occurred in different parts of world and was documented from the 16th century on. Smallpox epidemics sometimes reached pandemic proportions.

In addition to worldwide cooperation needed to confront infectious disease and to prevent pandemics, broad attention needs to be focused on aspects of pollution of water, air, and soil. A reported generated by Landrigan (2017) revealed that all countries and all life forms are impacted by pollution of one sort or another.

It is worthwhile considering the possibility that we can gain some measure of hope and perhaps gain insights into approaches that promoted human adaptations and survival, through studies of histories of different populations and by becoming familiar with the literature of social change as promoted by Pannikar (2012). Pannikar noted "Literature does not happen outside history but within it."

The comprehensive report "Our Common Future," a report of the Report of the World Commission on Environment and Development, was generated in 1987 https://digitallibrary.un.org/record/139811 under the leadership of Gro Harlem Brundtland who noted that the key goals of the report included proposals for long-term environmental strategies for sustainable development and proposals for cooperation among countries at different stages of economic development to achieve objectives and deal with environmental concerns.

Brundtland wrote, "Since the answers to fundamental and serious concerns are not at hand, there is no alternative to keep trying to find them. Brundtland emphasized the need for coordinated political action and responsibility and multilateralism."

3. Promoting child health, child development, and child well-being

In this chapter, I will explore environmental factors and education in childhood as mechanisms that influence neurodevelopment.

Environmental factors to be explored will include nutritional factors, living conditions, potential exposure to damaging substances, and social factors. Educational factors will include home influences, aspects of schooling not only academic teaching but also aspects of teaching that enhance personal skills, sense of achievement, and interaction skills. The impact of education and skills development on arts and physical activity will also be considered in the context of their impacts on brain plasticity.

Black et al. (2017) reviewed aspects of child development in different countries in the world. They noted that Lancet Commissions in 2007 and 2011 reported on child health and well-being in developing countries. These reports revealed that in developing countries, 39% of children under the age of 5 years were at risk to not achieve their developmental potential, due to poverty and adverse child experiences. These reports and studies in the report of Black et al. in 2017 emphasized the need for nurturing care that included adequate nutrition, care giving, safety, and early learning. Black et al. particularly focused on the needs of children under age 3 years when rapid brain development was occurring.

Black et al. defined normal child development as "an ordered progression of motor, cognitive, language, socioemotional, and self-regulation skills."

Nurturing care was noted to include attention to nutritional needs, emotional support and provision of opportunities for play, and exploration in a safe environment.

Adverse environments that occurred in the presence of extreme poverty were found to be more frequently associated with family stress, child neglect, and increased risk for child abuse.

Nutrition and brain development

Georgieff et al. (2018) noted that both preclinical and clinical human studies revealed the important roles of optimal nutrition during late fetal and early neonatal life and that specific nutrients are required for optimal brain growth. They emphasized particularly the remarkable changes in growth and differentiation that occur in the brain particularly those staring in 23–24 weeks postconceptionally and continue during the first 2 years of life. Growth during that period involved rapid neuronal and glial growth and conformational changes in the brain. Increased connectivity of circuits also occurred during this period, partly in response to increased sensory perception.

Georgieff et al. noted that adverse perinatal events that could include deprivation of essential nutrients were shown to impact hippocampal and striatal integrity and could affect connectivity of these structures with the prefrontal cortex.

Georgieff et al. emphasized the high metabolic demand of the developing brain that was reported to use 60% of blood oxygen and caloric intake.

It is, however, also important to note that there is evidence of plasticity in the brain, and if adequate nutrition is supplied, some levels of catchup growth can be achieved.

Georgieff et al. documented that optimal brain growth and development required both macronutrients and micronutrients. Adequate levels of protein are required to achieve optimal linear body growth and brain development. Long-chain polyunsaturated fatty acids are reported to be important macronutrients for brain growth in some studies.

Micronutrients for optimal brain development include particularly iodine, B vitamins, and zinc. Adequate maternal iron intake during pregnancy was reported to be critical for fetal brain growth and in the postnatal period.

Linear growth of the infant and child and adequate length measurements are important to follow throughout infancy and childhood. Stunting of child growth is considered to be related largely to nutritional inadequacy. Stunting is defined as child length two standard deviations or more below the average population child for an individual in the same population and of the same sex.

Comprehensive information of worldwide studies on child growth were reported in 2017 by Campisi et al. These studies revealed that growth stunting occurred in 62.7 million children in South Asia; in sub-Saharan Africa, growth stunting occurred in 57.2 million children. Stunting was reported in 5.4 million child in Latin American and Caribbean children and occurred in 5.1 million children in Europe and Central Asia.

There is growing evidence for importance of adequate intake of vitamins A and D and of calcium for bone growth. Necessity for adequate thyroid function and that for adequate intake of iodine to support thyroid function are well documented to be critical for growth and cognitive development (Andersson et al., 2012).

Brain plasticity

Mateos-Aparicio and Rodriguez-Moreno (2019) defined brain plasticity as changes in brain, including reorganization of structure, connections, or functions that occur in response to intrinsic of extrinsic stimuli. Synaptic plasticity can refer to alterations in signal transmission. Ideas of Ramon y Cajal documented in 1892 and 1894 contribute to concepts that increase in the numbers of connections contributed to brain function. An Italian neurophysiologist Tanzi was reported to propose in 1893 that if in a specific neuronal pathway repetitive activity occurred, this reinforced existing connections.

Mateos-Aparicio and Rodriguez-Moreno noted that Polish neurophysiologist Konorski in 1948 proposed that learning led to morphological changes in neural connections.

Mateos-Aparicio and Rodriguez-Moreno noted that concepts of plasticity were later incorporated into concepts of synaptic facilitations and synaptic depression. Another concept relevant to synaptic activity is homeoplastic plasticity, considered to balance neural activity. Plasticity was considered to be associated with modifications of axons, dendritic branches, and dendritic spines. Plasticity was also shown to be related to calcium-dependent sensors that impacted neurotransmitter receptor activity at specific sites. Mateos-Aparicio and Rodriguez-Moreno emphasized that synaptic plasticity is essential to brain development that occurs in response to stimulation and activity.

Kolb et al. (2017) noted four special factors in brain plasticity. They drew attention to stem cell in the subventricular zone of the lateral ventricles and cell in the hilus of the dentate gyrus. In humans, subventricular zone stem cells were noted to be activated in response to specific cerebral perturbations. However, cells in the dentate gyrus were reported to generate new neurons throughout life.

Kolb et al. reported on three different types of plasticity in the human brain. They noted that expectant plasticity initiated in postnatal development, for example, in response to special sounds, including parental speech. Experience-independent plasticity was defined as a function of connectivity between different brain systems. In experience-dependent plasticity new ensembles of connections were activated by a specific stimulus.

Kolb et al. noted the impact of stress on the brain that involves communications of the brain with other body systems through neural and endocrine mechanisms. They noted that in adults, stress was reported to result in changes in the size and structure of neurons in the hippocampus, amygdala, and prefrontal cortex.

They noted that early life stress predisposed individuals to maladaptive behaviors and psychopathology. One form of early stress that has been studied in detail includes impact of reduced parent–child interaction in children adopted into institutions at an early age.

Kolb et al. noted evidence that nutrient deficiencies have global impacts on the developing brain and on brain circuits.

Klingberg et al. (2014) reported that brain structural developmental changes and learning are key factors in child development. They emphasized the importance of specific brain changes, including changes in frontoparietal networks and in corticostriatal white matter tracts and changes in activity of striatal dopamine receptors.

Klingberg documented specific cognitive neuroscience theories pertinent to development. These included (1) maturation changes driven by gene expression, (2) environmental factors that included skill learning, and (3) interactive specialization stimulated by the influence of input into specific brain regions.

Development of working memory during development was reported to involve aspects of environmental experience and skill learning. Klingberg proposed that increase in working memory depended on extent of prenatal and postnatal development of cortical connections.

Neural mechanisms involved in increasing cognitive capacity during development included synaptic pruning and myelination of axons to improve connectivity. Specific studies in monkeys that included BOLD analysis of neural activity and EEG studies revealed that performance of specific tasks led to increased frontoparietal coherence. Striatal dopaminergic activity, particularly DRD2 receptor activity, was shown to play key roles in alterations of neural plasticity in response to training.

Environmental toxicants and the developing brain

Known agents damaging to the brain include heavy metals. Davis et al. (2019) reviewed other damaging environmental toxicants. They included pesticides, bisphenol A, polycyclic aromatic hydrocarbons, polychlorinated biphenyls, phthalates, flame retardants, and tobacco smoke.

Lanphear (2015) reviewed the impact of specific toxins and the critical vulnerability at different stages of development. He noted that numerous toxins that impact the developing brain have been implicated in intellectual deficits and mental disorders in children.

Lanphear noted that specific environmental disasters have served to focus attention on the impact of toxins. He drew attention to an epidemic of paint-based lead poisoning in Australia that led to paralysis and blindness. He also drew attention to mercury poisoning and fetal damage in a fishing village in Japan. Milk powder contaminated with arsenic was reported to increase rate of developmental impairment in Japan in 1953.

Herbert Needleman in the USA (1990) documented the damaging effect of lead in children and his efforts played a role in legislation leading to promotion of lead-free gasoline and lead-free paint in the United States.

Lanphear drew attention to the changing patterns of childhood morbidities in the United States and increased rates of diagnosis of autism spectrum disorders and attention hyperactivity disorder. Definitive conclusions regarding causes of the increased rates of these disorders have as yet not been derived at.

Lanphear described molecular and cellular aspects of increased vulnerability of the developing brain to damaging substances. He noted that during development, the immature blood–brain barrier is more permeable than in later life. In addition, different types of cells with different trajectories of proliferation and differentiation occur, and toxins present at different times may impair different cell types. Lead was reported to impact cell division and cell migration. Mercury was reported to impact synapses. A number of different toxins were reported to impact epigenetic modifications; these include bisphenol A.

Susceptibility to damage by organophosphates is increased in fetuses and infant due to lower concentrations of the enzyme PON1 in early life. PON1, paraoxonase 1, hydrolyzes thiolactones and xenobiotics, including paraoxon, a metabolite of the insecticide parathion.

Lanphear noted difficulties inherent in determining levels of exposure to damaging agents in pregnant woman, infants, and children. He noted, however, that some progress is being made in development of biomarkers of exposures.

Sripada (2017) noted that although factory jobs decreased poverty over several decades, factories increased industrialization and urbanization and increased accumulation of wastes and substances potentially harmful to children and to their brains. Sipada noted problems of bioaccumulation of certain toxins in food stuffs including fish and shellfish.

Sripada also drew attention to the fact that the nutritional status of the child could influence the level of sensitivity to toxins.

Although leaded gasoline use has declined substantially, there is evidence that lead poisoning is still a problem in children living in lower-socioeconomic areas in older housing. Lead may also be derived from batteries.

Sripada also drew attention to the harmful impact of air pollution and evidence that particulate matter pm2.5, and other airborne materials have been shown to trigger neuroinflammation.

Arsenic has also been shown to negatively affect the brain and behavior. Arsenic was reported to be emitted from coal burning power plants, by mining and smelting facilities. It is reported to be a particular problem in some regions of India.

Additional studies on nutrition and brain development and possible recovery from early deprivations

Prado and Dewey (2014) reviewed nutrition and brain development in early life. They considered the cellular and molecular consequences of under nutrition during pregnancy and during infancy and also possibilities for recovery from early nutritional deprivation. They also reviewed data from animal studies.

Animal studies revealed that nutrient deficiency during prenatal and early postnatal life impaired neuron and axon proliferation and also impaired dendritic growth, synapse formation, and myelination. Specific form of nutrient deficiency that led to these deleterious effect included protein energy malnutrition and iron deficiency. Other forms of deficiency that impact neuron proliferation axon and dendrite growth and myelination included deficiencies of iodine, thyroid hormone, and B vitamins. Deficiency of zinc and deficiency of choline also impacted neuronal growth and synapse formation.

Prado and Dewey also emphasized the importance for brain growth of stimulation from the environment. They noted the additive effects of nutrient deficiency and deficiency in experiential input.

A number of studies have been carried out to investigate the potential for recovery from nutrients deficiency in early life. Specific studies indicated that restoration of adequate nutrition and stimulation at least by 2 years of age was important if detrimental effects are to be avoided.

Johnson et al. (2016) reviewed poverty and the developing brain. They considered environmental mediators that impacted pregnant women and infants and young children and that impacted brain development. These included stress and micronutrient deficiencies. The list of important micronutrients that should be available in adequate quantities included vitamin B12, folate, retinoic acid, omega-3-fatty acids, zinc, and iron.

Johnson et al. also drew attention to data particularly from animal studies that maternal caregiving regulates expression of certain genes. These particularly included genes in the hippocampus and included the glucocorticoid receptor genes and nerve growth factor genes.

Johnson et al. also reviewed studies of specific brain areas that showed evidence of vulnerability in deprived children. The left occipitotemporal and perisylvian regions were shown to be impacted. These regions are later to be involved in reading and language, indicating that deprivation may subsequently lead to decreases in language and reading skills.

They noted that there is evidence that early deprivation can lead to decreased hippocampal volumes. Early-life caregiving was shown to be important in determining amygdala structure and function. The amygdala was reported to be important in emotional learning and motivation. Deprivation, stress, and negative parenting were reported to potentially impair optimal development of the prefrontal cortex.

Johnson et al. stressed that although young individuals are more vulnerable to negative effects of poverty, their systems are malleable to intervention. They noted evidence of the value of preschool programs in facilitating recovery from some aspects of deprivation.

Environmental stewardship and the rights of children

In a publication in 2019, Makuch et al. focused on children's rights and their need for protection against factors and processes that had adverse environmental impacts. They proposed that treaties generated regarding sustaining development should take into account quality of life in children in this and in future generations. Therefore, environmental rights should include not only aspects of clean air, water, and sanitation but also developmental, recreational, and health benefits that ensue from protection of environments.

Makuch et al. emphasize that factors leading to environmental degradation breach both environmental and human rights. Both human rights laws and environmental law needed to be taken into account in specific regulatory frameworks. They noted that concepts of sustainable development and rights of children were articulated in the 1987 Brundtland commission report "Our common future."

The Brundtland report emphasized how degradation risked undermining children's rights and the need for common understanding and sense of responsibility.

Makuch et al. (2019) drew attention to article 27 of the United Nations Convention on the Rights of the Child, which promoted the environmental rights of the child. They noted support for the premise that promotion of the child's environmental rights was necessary for physical, mental, spiritual, moral, and social development.

An important aspect of physical and mental health in children was access to open space and natural areas. Makuch et al. emphasized that, however, the open spaces were sometimes not healthful areas since the land was contaminated.

Makuch et al. drew attention to the transnational effects of certain practices such as the "slash and burn" fires used to clear forests for agricultural purposes. Slash and burn fires in Indonesia greatly increased air pollution in Malaysia, Singapore. Thailand, and the Philippines.

Several authors including Carson and Pratt (1965) and Barlow (2004) have written of the sense of wonder that the natural world can inspire in children. Barlow wrote "children are born with a sense of wonder and affinity with Nature. Properly cultivated, these values can mature into ecological literacy and eventually into sustainable patterns of living."

Neighborhood and social determinants of health

Crime and fear of crime can have significant negative impacts on health and well-being. Lorenc et al. (2012) reviewed aspects of crime and health and well-being in specific neighborhoods. They documented environmental stresses present in specific neighborhoods including poor housing, overcrowding, noise, high traffic flow, crime, and threat of crime, and they also investigated protective factors including degree of social integration.

The impact on individuals was evaluated. This included physical injury and psychological trauma. Personal victimization was shown to be particularly damaging to mental health. The fear of crime also impacted health by limiting physical activity and limiting social interactions.

Bonner et al. (2018) published a chapter entitled, "The individual growing into society" in a book entitled *Social Determinants of Health*. Bonner emphasized the complex interactions of genetic factors, and prenatal and postnatal factors in influencing development.

Important factors determining development and maturation of the nervous system, muscular system, and metabolic processes that ultimately contribute to independent living were considered to be adequate nutrition, secure environments, and parental nurturing. Adverse early life experiences were also noted to negatively impact development of the immune system.

Bonner stressed the importance of sensory stimulation in promoting increased efficiency of neuronal pathways through appropriate synaptic pruning. Also noted was evidence that specific input played roles in appropriate development of the limbic system that includes the amygdala that impacts emotional behaviors. Fear was noted to be particularly harmful to appropriate development of the limbic system.

Resilience

Taub and Boynton Jarrett (2017) reviewed aspects of resilience and factors that could improve outcomes. These included positive appraisal of child function, anticipatory guidance fostering of aspects of self-care, and establishments of household routines.

4. Proposing solutions

Many scientists, politicians, and authors have generated important documents regarding environmental degradation and factors hazardous to life on earth. It is, however, important to draw attention to those who document possible solutions.

An especially important publication documenting possible pathways to solution was edited and published in 2017 by Hawken and Drawdown in Hawken, 2017. The publication was entitled Drawdown; eight specific sections were presented, and in each, problems were presented along with possible solutions. Sections included Energy, Food, Women and Girls, Buildings and Cities, Land Use, Transport, Materials, and the final section was entitled Coming Attractions.

The section on energy documented technologies and strategies to supplement energy production to move away from major reliance on fossil fuels.

In the section on food, Hawken and Drawdown documented strategies to reduce food waste, development of multistrata agroforestry, improved methods of cultivation of key food staples, regenerative organic agriculture, composting, improved methods of irrigation, and alterations in management of grazing patterns.

In a section on women and girls, Hawken and Drawdown explored important roles of women in agriculture in low-income countries. Concepts of access to family planning were discussed as was education of girls.

The section on buildings and cities described new approaches to use renewable energy sources, new concepts of architecture and insulation of buildings, concepts of walkable cities, and resources for safe cycling and use of led lighting.

The section on land use included consideration of application of methods to promote carbon capture from air and methods to maintain soil moisture, decreased use of synthetic pesticides, strategies for grazing cattle within forests, and methods for improved maintenance of coastal wetlands. Strategies to counteract desert encroachment and destruction of peat lands included learning about land use from indigenous peoples and honoring their land rights.

The section on transport promoted use of electric vehicles, alteration in concepts of town planning, and increased use of Internet resources to cut down on business travel include air travel.

The section on materials emphasized household recycling, paper recycling, water saving, and development of alternative forms of cement that increased life span of buildings.

The role of the arts in empowering health and well-being

Narrative

In 2017, Puchner wrote, "From ancient epics to modern novels, some narratives changes history and influenced mindsets of generations." In his book *The Written World*, Puchner discussed influences on the life and actions of Alexander the Great through being tutored on Homer's Iliad. Puchner noted that this poem had influenced entire societies.

He noted that it was not the only epic poem that influences societies; "The Book of Songs," a collection of poems, was reported to have had great influence in China and East Asia. Much has been written about the power of storytelling, and Scheherazade is said to have saved her life by telling exciting stories to her would be assassin. Puchner wrote, "We learnt how repressive regimes sort to control through destruction of books and suppression of the press."

In a chapter in a book entitled *The goals and achievements of narrative*, Ritivoi (2018) contributed a chapter entitled, "Narrative in Dark Times" wrote of Hannah Arendt and her celebration of the those who brought light in dark times (Bonner, 2018; Ritivoi, 2018).

Editors of the book *The goals and achievements of narrative* (Bonner 2018) presented aspects of the value of narrative in offering insights into unfamiliar worlds and the power of stories to promote deeper understanding and compassion. They noted too that the stories can deepen insights into questions that have no easy answers.

In considering positive impacts of creativity on experiences and responses, it is perhaps also appropriate to think of architecture and also of light. In his book *The Creators* (1992), Boorstin (2017) wrote of the great spiritual impact of Gothic Architecture and its use of light with extension "from material to immaterial."

Evidence on the role of the arts in improving health and well-being

Fancourt and Finn (2019) authored a World Health Organization report entitled, "What is the evidence on the role of arts in improving health and well-being." This report gathered evidence generated between 2000 and 2019. https://www.euro.who.int/__data/assets/pdf_file/0020/412535/WHO_2pp_Arts_Factsheet_-v6a.pdf?ua=1. Data were gathered under two broad themes related to health, the first was Prevention and Promotion and the second theme was Management and Treatment.

Data gathered under the Prevention and Promotion theme demonstrated how arts could impact within social determinants of health and through support of child development through encouraging health-promoting behaviors and through enhancing caregiving.

Data collected under the second category Management and Treatment determined how arts could be beneficial to people with developmental defects, neurological problems, or mental illness. Information was also obtained on the value of arts in supporting individuals with acute illness and on the value of arts in end-of-life care.

In their report, Fancourt and Finn sort to define arts. They included art objects physical or experiential and included preforming arts, visual arts and crafts, literature, writing, literature festivals, visits to museums, concerts, exhibitions and fairs, and digital and electronic arts.

Fancourt and Finn noted extension in the WHO definition of health from the original definition, "health as a state of complete physical mental and social well-being and not merely the absence of disease and infirmity." They noted that later definitions included an individual and social perspective and the definition included integration within society, contribution to society acceptance and trust within society, individual understanding of society, and belief in the potential of society.

They noted further that the WHO concept of health had been further expanded to include managements of chronic mental and physical illness and continued opportunities for chronically ill persons to participate in society.

In stressing the health-promoting components of arts activities, Fancourt and Finn noted that they included esthetic engagement, emotion, imagination, cognitive, and sensory activation and one might include activities that promoted understanding of society.

In summary, arts activities potentially have physiological, psychological, and behavioral benefits.

Social determinants of health and well-being

Parnham in Chapter 3 in a book entitled *Social Determinants of Health* (Bonner 2018) noted that positive psychology expanded concepts related to human thriving including expression of gratitude, undertaking acts of kindness, and nurturing relationships. Some writers proposed that meaning and purpose are key aspects of well-being, and others emphasize physical, mental, and emotional relationships and societal and spiritual aspects of well-being.

5. Aspects of positive psychology relevant to well-being

Positive psychology approaches and interventions aim to build well-being and not simply to reduce ill being (Seligman, 2019). Positive psychology concepts, as developed by Seligman, proposed to evaluate and promote individual strengths and to assess qualities that promote individual strengths based on the premise that character strengths promote happiness and life satisfaction. Peterson et al. initially defined broad categories of strengths to promote well-being in adults. They included six broad categories of strengths, and each of these included subcategories.

Broad categories and sub-categories included:

1. Wisdom and knowledge with subcategories love of learning and creativity
2. Courage with subcategories perseverance and bravery
3. Humanity skills with subcategories kindness and love
4. Justice with subcategories fairness and teamwork
5. Temperance with subcategories forgiveness and mercy
6. Transcendence with subcategories gratitude, hope, humor, and appreciation of beauty

Seligman also stressed the value of positive psychology principals in child education in a 2019 review. Programs designed by Seligman et al. have been applied in almost 200 schools in different parts of the world. Analyses of data revealed that well-being is higher, and students also do better in academic courses. The positive psychology approaches also led to increased self-discipline in students. He stressed that pursuit of engagement and meaning are predictive of life satisfaction.

Seligman concluded his 2019 review with the statement: "Arriving at the good is a lot more than just eliminating the bad".

In 2007, Peterson et al. investigators reviewed aspects of strengths of character, orientations to happiness, and life satisfaction, and they also traced connections of these concepts to ideas generated by earlier philosophers. In considering ancient Greek philosophers, they noted that Epicurus promoted Hedonism and linked human obligation to experiences of pleasure. The concept of Eudaimonia (Eudomonia) was promoted by Aristotle and considered identifying virtues and living in accordance with these virtues. John Stuart Mill (1806–73), Bertrand Russell (1872–1970), and Abraham Maslow (1908–70) were reported to promote development of a fully functioning individual. Theories of Ryff and Singer (1996) and Ryan and Deci (2000) on self-determination were reported to promote identification of one's strengths or virtues and living in accordance with these to promote welfare of others. Csikszentmihalyi (1990) promoted the concept of Flow, when an individual became deeply emerged in a specific activity and immersion produced a great sense of satisfaction.

In the 2007 report, Peterson et al. documented results of surveys conducted in the United States and in Switzerland that gathered information on character strengths and life satisfaction. Data from the United States revealed that life satisfaction was most highly correlated with character strengths that included zest, hope, love, gratitude, and curiosity. Information from the Swiss survey indicated that strengths considered to be most highly correlated with life satisfaction included zest, hope, love, curiosity, and perseverance.

The authors concluded that the results of these surveys revealed that life satisfaction qualities were to some degree linked to strengths that are particularly valued in a particular nation.

Eudaimonia and physiological well-being

Ryff and Psychological (2014) explored Eudaimonia in the context of physiological well-being and positive psychology. Eudaimonia was discussed by Aristotle and other Greek philosophers and was defined as bringing one's best to activities in accordance with virtue and striving for the best that is within us.

Ryff and Psychological reported that studies in positive psychology involved investigated the personality correlates of well-being, investigating the connections between well-being and physical health and how well-being related to work, family, and community life.

Ryff and Psychological identified six specific areas and concepts included in assessment of well-being:

1. Environmental mastery, a sense of competence in managing one's environment
2. Personal growth, sense of continued development, growing, expanding, and open to new experiences
3. Positive relations with others, has trusting and empathetic relations
4. Purpose in life, feeling of meaning, and goals in life
5. Self-acceptance, acceptance of aspects and qualities of self, both good and bad, satisfied with past life
6. Self-determination, regulating behavior from within and resisting social pressure

Ryff and Psychological noted that there are strong connections between sense of well-being and positive health behaviors. She also noted the benefits of having individuals compose autobiographical narrative and document goal orientations and developmental changes.

References

Andersson, M., Karumbunathan, V., Zimmermann, M.B., 2012. Global iodine status in 2011 and trends over the past decade. J. Nutr. 142 (4), 744–750. https://doi.org/10.3945/jn.111.149393. PMID:22378324.

Barlow, D., 2004. Confluence of streams. An introduction to the ground-breaking work of the Center for Ecoliteracy. Dancing Earth Vol. 226 (Sept/Oct). https://centerforhealthjournalism.org/resources/sources/zenobia-barlow.

Black, M.M., Walker, S.P., Fernald, L.C.H., Andersen, C.T., DiGirolamo, A.M., Lu, C., et al., 2017. Early childhood development coming of age: science through the life course. Lancet 389 (10064), 77–90. https://doi.org/10.1016/S0140-6736(16)31389-7. PMID:27717614.

Bonner, A., 2018. Social Determinants of Health: An Interdisciplinary Approach to Social Inequality and Wellbeing. Bristol University Press, Policy Press. https://doi.org/10.2307/j.ctt22p7kj8.

Boorstin, D., 2017. Castles of Eternity. The Creators, 1st1. Random House, New York.

Bruntland, G., 1987. Our Common Future (Brundtland Report) (PDF, 1 MB, 20.03.1987) Report of the World Commission on Environment and Development, 1st. WHO, Zurich.

Campisi, S.C., Cherian, A.M., Bhutta, Z.A., 2017. World perspective on the epidemiology of stunting between 1990 and 2015. Horm. Res. Paediatric. 88 (1), 70–78. https://doi.org/10.1159/000462972.

Carson, R., Pratt, C., 1965. A Sense of Wonder 1965. Harper Collins Publisher, New York.

Csikszentmihalyi, M., 1990. Flow: The Psychology of Optimal Experience. New York: Harper and Row. ISBN 0-06-092043-2.

Davis, A.N., Carlo, G., Gulseven, Z., Francisco Palermo, F., Lin, C.H., Nagel, S.C., 2019. Exposure to environmental toxicants and young children's cognitive and social development. Rev. Environ. Health 34 (1), 35–56. https://doi.org/10.1515/reveh-2018-0045. PMID: 30844763.

Fancourt, D., Finn, S., 2019. What is the evidence on the role of the arts in improving health and well-being? Health Evidence Network Synthesis Report 67. WHO Regional office for Europe, Copenhagen.

Georgieff, M.K., Ramel, S.E., Cusick, S.E., 2018. Nutritional influences on brain development. Acta Paediatric., Int. J. Paediatric. 107 (8), 1310–1321. https://doi.org/10.1111/apa.14287.

Hawken, P., 2017. Drawdown: The Most Comprehensive Plan Ever Proposed to Reverse Global Warming. Penguin Books.

https://www.thelancet.com/commissions/global-governance-for-health.

https://www.euro.who.int/__data/assets/pdf_file/0020/412535/WHO_2pp_Arts_Factsheet_v6a.pdf?ua=1. (Accessed 22 January 2021).

Johnson, S.B., Riis, J.L., Noble, K.G., 2016. State of the art review: poverty and the developing brain. Pediatrics 137 (4). https://doi.org/10.1542/peds.2015-3075.

Klingberg, T., 2014. Childhood cognitive development as a skill. Trends Cognit. Sci. 18 (11), 573–579. https://doi.org/10.1016/j.tics.2014.06.007.

Kolb, B., Harker, A., Gibb, R., 2017. Principles of plasticity in the developing. Brain Dev. Med. Child Neurol. 59 (12), 1218–1223. https://doi.org/10.1111/dmcn.13546. PMID:28901550.

Landrigan, P., 2017. Air pollution and health. Lancet Public Health 2 (1), e4–e5. https://doi.org/10.1016/S2468-2667(16)30023-8. Epub 2016 Nov 26.

Lanphear, B.P., 2015. The impact of toxins on the developing brain. Annu. Rev. Publ. Health 36. https://doi.org/10.1146/annurev-publhealth-031912-114413 (PMID).

Lorenc, T., Clayton, S., Neary, D., Whitehead, M., Petticrew, M., Thomson, H., Cummins, S., Sowden, A., Renton, A., 2012. Crime, fear of crime, environment, and mental health and wellbeing: mapping review of theories and causal pathways. Health Place 18 (4), 757–765. https://doi.org/10.1016/j.health place.2012.04.001.

Lucey, D.R., Sholts, S., Donaldson, H., White, J., Mitchell, S.R., 2017. One health education for future physicians in the pan-epidemic "Age of Humans". Int. J. Infect. Dis. 64, 1–3. https://doi.org/10.1016/j.ijid.2017.08.007.

Makuch, K., Zaman, S., Aczel, 2019. Tomorrow's Stewards: The Case for a Unified International Framework on the Environmental Rights of Children Health Hum Rights, vol. 21.

Mateos-Aparicio, P., Rodriguez-Moreno, 2019. A the impact of studying brain plasticity. Front. Cell. Neurosci. 13 https://doi.org/10.3389/fncel.2019.00066.PMID: 30873009.

Needleman, H.L., Schell, A., Bellinger, D., Leviton, A., Allred, E.N., 1990. The long-term effects of exposure to low doses of lead in childhood: an 11-year follow-up report. N. Engl. J. Med. 322 (2), 83–88. https://doi.org/10.1056/NEJM199001113220203.

Nobel Prize, 1998. The Possibility of Social Choice. https://www.nobelprize.org/prizes/economic-sciences/1998/sen/lecture/. (Accessed 22 January 2021).

Pannikar, K.N., 2012. Literature as a history of social change. Soc. Sci. 40 (3/4), 3–15.

Peterson, C., Ruch, W., Beermann, U., Park, N., Seligman, M.E.P., 2007. Strengths of character, orientations to happiness, and life satisfaction. J. Posit. Psychol. 2 (3), 149–156. https://doi.org/10.1080/17439760701228938.

Prado, E.L., Dewey, K.G., 2014. Nutrition and brain development in early life. Nutr. Rev. 72 (4), 267–284. https://doi.org/10.1111/nure.12102. 28.

Puchner, M., 2017. The Written World: The Power of Stories to Shape People, History, Civilization. Random House http://ISBN 9780812998931.

Ritivoi, A.D., 2018. Narrative in dark times. The Goals and Achievements of Narrative. Bristol University Press.

Rull, V., 2011. Sustainability, capitalism and evolution. EMBO Rep. 12 (2), 103–106. https://doi.org/10.1038/embor.2010.211. PMID: 21233854.

Ryan, R.M., Deci, E.L., 2000. Self-determination theory and the facilitation of intrinsic motivation, social development, and well-being. Am. Psychol. 55 (1), 68–78. https://doi.org/10.1037//0003-066x.55.1.68. PMID: 11392867.

Ryff, C.D., Heller, A.S., Schaefer, S.M., van Reekum, C., Davidson, R.J., 2016. Purposeful engagement, healthy aging, and the brain. Curr. Behav. Neurosci. Rep. 3 (4), 318–327. https://doi.org/10.1007/s40473-016-0096-z.

Ryff, C., Psychological, 2014. Well-being revisited: advances in the science and practice of. Eudaim. Psych. Psychosom 83 (1). https://doi.org/10.1159/000353263 (PMID).

Ryff, C.D., Singer, B., 1996. Psychological well-being: meaning, measurement, and implications for psychotherapy research. Psychother Psychosom. 65 (1), 14–23. https://doi.org/10.1159/000289026. PMID: 8838692.

Seligman, M., 2019. Positive Psychology: A Personal History Annu Rev Clin Psychol, vol. 15. https://doi.org/10.1146/annurev-clinpsy-050718-095653 (PMID).

Sripada, K., 2017. Beginning with the Smallest Intake\: Children's Brain Development and the Role of Neuroscience in Global Environmental Health Neuron, vol. 95. https://doi.org/10.1016/j.neuron.2017.08.009 (PMID).

Traub, F., Boynton-Jarrett, R., 2017. Modifiable resilience factors to childhood adversity for clinical pediatric practice. Pediatrics 139 (5). https://doi.org/10.1542/peds.2016-2569.

White, R.G., 2020. Mental wellbeing in the Anthropocene: socio-ecological approaches to capability enhancement. Transcult. Psychiatr. 57 (1), 44–56. https://doi.org/10.1177/1363461518786559.

Whitmee, S., Haines, A., Beyrer, C., Boltz, F., Capon, A.G., de Souza Dias, B.F., 2015. Safeguarding human health in the Anthropocene epoch: report of the rockefeller foundation-lancet commission on planetary health. Lancet 386 (1007), 1973–2028. https://doi.org/10.1016/S0140-6736(15)60901-1. PMID:26188744.

Wilson, E.O., 1984. Biophilia. Harvard University Press, Cambridge, MA. ISBN 0-674-07442-4.

CHAPTER

10

Brain and mind

1. Cognitive neuroscience

In introducing cognitive neuroscience, Postle (2015) noted that it involved synthesis of information from neurobiology, psychology, electrophysiology, neuroimaging, functional neurology, genetics, and genomics. A goal of cognitive neuroscience is the understanding of the biological basis of mental phenomena and phenomena of consciousness.

Historical aspects of cognitive neuroscience involved studies to localize certain functions to specific topographic regions of the brain. Experimental studies, particularly studies in animals, were carried out to define the neural basis of specific behaviors. Studies were also carried out to delineate the cellular basis of activities in the brain. Studies in humans included analyses of functional deficits that resulted from injuries that led to damage of specific brain structures.

Postle noted that from 1988, brain imaging tomography and functional scanning during the performance of specific tasks were carried out, and there was progress in determining how neurons in specific regions were activated during performance of specific tasks.

Neurophysiological experiments were designed and carried out to determine correlations in brain activity and performance of certain functions. First functions to be analyzed included motor control, vision, hearing, and speech. Subsequent studies explored specific changes in the brain or in specific brain regions and their impact on behavior.

Increasingly, cellular and molecular studies yielded information on neurons, synapses, neurotransmitters, ion passage and neuronal polarization, electrical activity signal transduction, and flow of information. Studies on electrical gradients in neurons revealed that at rest the electrical potential in a neuron is negative relative to potential in extracellular fluid, and neurons have low levels of sodium ions (Na) relative to extracellular fluid and higher levels of potassium ions (K) relative to extracellular fluid. On neuronal activation, there is an efflux of K ions from the neuron and an influx of Na sodium ions and calcium ions. These changes in ion concentration generate action potential.

Consciousness

Postle reviewed aspects of neurological processes involved in consciousness. The brain stem and the reticular system in the brain stem have been shown to play important roles in consciousness. This system is sometimes referred to as the reticular activating system (RAS) and is composed of neurons that project particularly to the thalamus and the hypothalamus. Postle noted that specific nuclei that have connections with the thalamus deliver acetylcholine and glutamate to the thalamus. These nuclei also deliver signal to the hypothalamus and the

Mechanisms and Genetics of Neurodevelopmental Cognitive Disorders
https://doi.org/10.1016/B978-0-12-821913-3.00006-8

forebrain. He noted that there is redundancy in the nuclei and fibers of the reticular system, and this redundancy plays roles in recovery from coma.

There is evidence that activity of the reticular system nuclei can be inhibited by GABA signaling from the hypothalamus and forebrain. Waxing and waning of activity of the reticular system was reported to play roles in the sleep cycle.

There is also evidence that the pons is important in consciousness. Damage to the pons has been shown to lead, in some cases, to the so-called locked-in syndrome when the only sign of consciousness is that the patient can blink eyes.

Activation, attention, awareness

Kok (2019) noted that work in the 1970s led to identification of the suprachiasmatic nucleus in the hypothalamus that was determined to play a key role in determining the 24-h circadian rhythm. Patton and Hastings (2018) reported on the subregions of suprachiasmatic nucleus that produced different neuropeptides; they also reported on how the subregions communicate and on the roles of neurotransmitter γ-aminobutyric acid (GABA) in this nucleus.

Kok noted that states of consciousness were not uniform, since variations in alertness were found to be in part dependent on activities in the anterior and posterior hypothalamus, activity in the pons, and activity in the visual cortex. EEG studies were reported to indicate that specific neural dynamics, leading to altered patterns occurred during sleep.

Kok traced the history of concepts of attention and noted that attention is modulated by perception and that memory also plays roles. Attention also requires alertness, and alertness involves the brain stem and the reticular activating system. Specific network activity involved in attention was proposed to include the higher cortex centers, posterior parietal cortex, dorsolateral prefrontal cortex, ventrolateral prefrontal cortex, the pulvinar nucleus of the thalamus, and the inferior temporal region.

Arend et al. (2008) reported that the pulvinar is particularly important for visual attention.

The cortical visual system includes the V1 region in the occipital lobe that send signal to the V2, V3, and V4 regions in the inferotemporal lobe. Visual attention also includes the region that controls eye movements, the colliculus, located in the midbrain.

Kok noted that aspects of visual attention and the regions involved were obtained using functional magnetic resonance imaging (fMRI) technologies.

Attention also requires alertness, and alertness involves the brain stem and the reticular activating system. Focused attention was also shown to involve the anterior cortex and the anterior cingulate region.

Free-running mind, the mind in an unfocused state

The default-mode network is considered to reflect the "free-running mind." Kok noted that studies reveal that in the default-mode network, there are activity connections between posterior and anterior midline cortical structures. The core structures involved were reported to include the medial prefrontal cortex, the precuneus gyrus, and the angular gyrus of the parietal lobe.

A different network is also proposed to be involved in active cognitive processes. This network is referred to as the salience network. Seeley (2019) reported that this network includes anterior cingulate and ventral anterior insular (i.e., frontoinsular) cortices. This network also includes nodes in the amygdala, hypothalamus, ventral striatum, thalamus, and specific brain stem nuclei.

2. Perception

Kok (2019) emphasized that perception involves complex processes and that each involves primary and higher-order cortical functions. In considering vision for example, he noted involuntary processes, e.g., involvement of physical energy in light transmitted to the retina, followed by electrical impulses to the visual cortex, connections to the visual cortex, the posterior parietal cortex, and then to the inferior temporal cortex. He noted that visual perception is sometimes then linked to memory, e.g., remembering faces. Stimulation of visual cortex may also trigger connections to areas involved in detection of other senses.

He noted that complex interactions between different cortical regions are required in the planning and executing of specific movements, e.g., regions involved in memory, cerebellar regions, and motor cortex.

3. Memory systems

In consideration of aspects of cortical involvement in memory, Kok noted both excitatory and inhibitory patterns. He noted that memories are stored in populations of neurons in distributed networks and involve establishment of connections between neurons and differences in the strength of establishment of connections between neurons and differences in the strength of synaptic connections.

Memory processes were noted to involve activation, consolidation, storage, and retrieval. Kok emphasized that neural systems involved in memory are distributed throughout the brain and particularly involve the cortex. Different forms of memory are defined, and basic distinctions include long-term and short-term memories.

The cortical regions involved in memory storage are related to the type of memory stored. For example, memories related to specific movements are stored in the motor cortex. He noted that to some extent, the anterior and posterior axes of the brain differ in function with action memories located in more anterior regions, while perceptive memories are often located in posterior regions.

Kok noted evidence that new information initially recruits the limbic system, hippocampus, amygdala, cerebellum and basal ganglia, and circuits in the limbic system and subcortical pathways Permanent memories tend to be stored in the neocortex.

He noted that processes in memory include acquisition and consolidation, and the steps that follow include storage and retrieval. Acquisition requires attention, and if this is insufficient, consolidation does not occur. The consolidation steps require structural changes and generation of connections.

Kok noted that memory defects have been documented in patients with temporal lobe damage and also in patients with damage to the medial dorsal nucleus of the thalamus that projects to the cortex. Various forms of disorders of information retrieval occur and are thought to involve corticocortical pathways.

Memory collections and consolidation

Josselyn and Frankland (2018) reviewed neuronal assemblies, also defined as engrams and aspects of allocation and consolidation of memories. They concentrated on aspects of establishment of fear memories through studies primarily in rodent brain and noted evidence that the lateral nucleus of the amygdala was particularly important in fear conditioning. The dorsal hippocampal region was found to be particularly important in encoding spatial memories.

They noted that in a particular region, the most excitable neurons were most often selected for memory allocations. They noted the critical role of cAMP response element–binding protein (Crebbp) played a critical role in memory establishment, the transcription factor Arc was also noted to be important, and the ion channel KCNQ2 was also important.

Particular cells within the specific brain region were selected for memory storage. In the hippocampus, place cells were fired in response to stimuli related to environmental space. Olfactory memories were noted to be stored in cells in the piriform cortex, located between the insula and the temporal lobe. Memories related to taste situated in the insular cortex that lies deep within lateral sulcus of each hemisphere.

There was also evidence that stimuli that simultaneously evoked stimuli in different brain regions could generate memories that were linked together.

Chaaya et al. (2018) reviewed roles of the amygdala and hippocampus in consolidation of contextual fear memories. In tracing the processes of establishment of memories, they traced features from stimulation of excitatory glutamate receptors, activation of glutamate-activated ionotropic channels, and voltage-gated calcium channels, followed by signaling that included protein kinase activity, phosphorylation in the phosphatidyl inositol kinase, and MAP-activated kinase protein signaling MAPK. This signaling led to phosphorylation of CREB that promoted transcription and translation of intermediate-early genes including ARC activity CFOS, EGR1 transcription factor, and other early response genes.

Chaaya et al. noted that in many cases, the stimulation and activation of specific neurons led to structural alteration in neurons, sometimes referred to as neuronal plasticity. These alterations included increases in dendritic spine density. Consolidation of memories was reported to be related to structural alterations in neurons.

They noted that although the dorsal hippocampus has been designated as the region particularly involved in spatial and contextual memories, it is also involved in consolidation of fear memories.

4. Emotions

Kok (2019) noted that emotion and cognition originate from the neural systems that are separate but interacting. Emotion may also be accompanied by somatic changes. He noted further that there has been increasing emphases on the neuroaffective network, and neurotransmitters have been shown to also modulate affect. There is evidence that the orbitofrontal cortex is involved in regulating reactions to emotions.

Kok emphasized that different forms of emotions exist and that different structures and circuits are involved in processing different emotion. Emotions can be positive or negative and differ in type or intensity. Primary emotions are thought to have bioevolutionary roots, e.g., fight or flight. Secondary emotions are thought to be the product of later development and learning. Moods and mood changes are also known to influence emotions.

Kok explored aspects of empathy and noted that it is expressed early in humans even in infancy and is manifested by facial expression and language.

He noted that primary emotions are associated with the limbic systems, hypothalamus, and brain stem, while secondary emotions are attributed to the neocortex. Conscious or explicit emotions are considered to represent affective and cognitive processes.

Kok defined the major components of emotions as evaluation, explicit feeling, or processing, followed by action.

Relationships between affect and cognition have been analyzed, and there is evidence for involvement of different interacting systems, e.g., emotions that trigger autonomic responses and neurotransmitter products and emotions that trigger motor systems.

Kok addressed interactions between memory and emotions. He noted the role of the amygdala in memory and evidence that strong emotions generate strong memories. Investigations have provided evidence that links between emotion and memory that involves the neurotransmitter noradrenaline.

Kok defined primary emotions as the product of implicit learning and noted that they are regulated by a circuit that includes the hypothalamus and the amygdala. He defined secondary emotions as the product of explicit learning and noted that they are regulated by circuits that are more extensive and includes higher cortical areas and that hypothalamus and amygdala are included.

Emotion regulation

Etkin et al. (2015) defined emotions as "cognitive, subjective physiologic and motor changes that arise in response to specific stimuli." They noted that valuation plays a key role in emotion. Neuroimaging studies have revealed particular regions involved in emotion regulation; these include the ventral anterior cingulate, the ventromedial prefrontal cortex, and the lateral prefrontal and parietal cortices.

5. Imagination

Horvath (2018) reviewed relevance of visual imagination and generations of art and narrative. Horvath drew attention to the work of Lessing (1984) who emphasized the relationship of seeing and generations of the mental process of imagination. Horvath also drew attention to visual imaging in story telling in art.

Beaty et al. (2018) carried out studies on a trait referred to as openness to experience. Several studies reported that this personality trait is associated with imaginative, creative, and abstract cognitive processes (De Young, 2014 Kaufman et al., 2013). They also noted that several studies have reported the association of imagination with activity of the default-mode network.

Beaty et al. also noted that several studies have revealed association of the default-mode networks with other brain networks. Divergent thinking processes were also shown to involve connections with the default-mode network, executive control network, and the salience network.

Beaty et al. carried out studies that included fMRI in individuals defined as having this trait. In their studies, openness to experience, epitomized by imaginative and creative thought, was found to be associated with default-mode networks correlated with activity of large-scale connective brain regions. Their studies revealed that increased function of the default-mode networks during imaginative and creative cognition likely supported dynamic and efficient cooperation of other brain systems.

Beaty et al. (2018) concluded that the creative and imaginative thinking capacity of individuals with the openness to experience trait was associated with increased activity and engagement of several brain networks and areas.

Pearson (2019) reviewed human imagination in the context of the cognitive neuroscience of visual mental imagery. He noted that mental imagery potentially involves all five senses, but that visual imagery has tended to dominate research.

Pearson noted that mental imagery includes large neural networks that span the primary sensory areas and the frontal areas. He explored the hypothesis that mental imagery is "vision in reverse" as images that the individual previously stored in memory are reactivated. However, in addition, there are visual images generated of scenes that the person has not previously imaged.

Pearson noted that there is evidence that the strength and vividness of imagery differs in different individuals; in some individuals, it is absent, while other individuals have "photographic memories." He noted that mental imagery can also play roles in psychopathology.

Literature and storytelling; mind and synthesis

It seems we are barely at the start of explorations of activities of the mind in synthesis of literature and storytelling. Kazuo Ishiguro (2017) wrote about stories, "but for me the essential thing is they communicate feelings, that they appeal to what we share as human beings across our borders and divides."

Neuroaesthetics

In a 2016 review, Pearce et al. (2016) defined neuroaesthetics as the cognitive neuroscience of esthetic experience. They noted increases in the popularity of neuroaesthetic studies but also criticism of the concept from humanities and sciences. They noted further that a number of researchers have demonstrated that esthetic features of environments and objects influence peoples' attitudes and behaviors.

Pearce et al. also defined the goals of this form of cognitive neuroscience as particularly the analysis of the evolutionary underpinnings of esthetic responses and analyses of emergent states following esthetic exposure, including sensory, motor, and emotional responses.

Specific studies have been designed to investigate brain activity changes in response to neuroaesthetic experiences. Such changes involve attention, memory, emotion, and social cognition and potentially involve different brain region and networks.

Reybrouck et al. (2018) reviewed brain connectivity in response to music. Specific fMRI studies reveal signaling in the default-mode network when listening to favorite music. They also noted reports of connectivity between audiosignaling networks, default-mode network, and a reward circuit in the inferotemporal cortex.

Poetry, neural circuitry, and psychophysiology

Wassiliwizsky et al. (2017) reported that their studies provided evidence that when poetry reading evokes powerful emotional responses, brain areas of primary reward were stimulated. They particularly noted stimulation of the nucleus accumbens.

Literary fiction reading and theory of mind

Kidd and Castano (2013) reported that literary fiction reading was noted to stimulate the skill known as theory of mind. Theory of mind is the ability to understand the mental state in others and is thought to be important in establishments of complex social relationships.

They noted that reading fiction expands knowledge of the lives of others. Literature reading has been deployed in some programs designed to promote empathy. They noted that literary fiction reading has been employed in prisons to promote life skills.

6. Creative cognition and brain network dynamics

Creativity is defined differently by different investigators. Since the reports on networks during creativity are those reported by Beaty et al. in 2016 and 2017, I will present their definition of creativity: "Creative cognition can be understood as a set of mental processes that support generation of novel and useful ideas."

In 2016, Beaty et al. reported that fMRI and connectivity analyses had been carried out to analyze brain activity during creative thought. Studies revealed activity in the default network that involved the midline and posterior parietal regions also known to be involved in self-generated thought, social cognition, and autobiographical retrieval. During creative thought, the executive control network was also shown to be involved. This network is reported to also regulate attention, integration of working memory, and task switching and is reported to include prefrontal and anterior–inferior parietal regions. The studies reported in 2016 led Beaty et al. to conclude then that creative cognition involved dynamic interaction between default-mode and network and the executive control network. They also proposed that memory retrieval likely played important roles in idea generation.

Although both default-mode and executive control networks were shown to be involved in creative though Beaty et al. noted that the extent to which the executive control network is involved might be different, dependent on whether task-specific goals constituted part of the creative activity.

In a 2018 publication, Beaty et al. (2018) reported results of brain imaging studies on individuals involved in specific creative thinking tasks. Their study led to the identifications of three main networks involved in the activity: the default-mode network, the executive control network, and the salience network. They reported that coordination between the three networks was necessary for creative thinking. They confirmed that the default-mode network involved the cortical midline and posterior parietal regions. The executive control network involved the lateral prefrontal and anterior–inferior parietal region. The salience network was reported to include the bilateral insular and anterior cingulate cortex.

Beaty et al. proposed that variations in thinking ability in different individuals were related to individual variation in ability to simultaneously engage executive control and salience networks.

Their fMRI study aimed to define a creative connectome. They concluded that network-based approaches can investigate the interplay of multiple neurocognitive processes.

Possible role of the thalamus in cognition

Rikhye et al. (2018) noted that theories of cognition primarily had a corticocentric view. In considering evolutionary changes that occur in the transition to species with increased cognitive abilities, they noted that major evolutionary transitions occur in the cortex, the basal ganglia, and the thalamus.

Intelligences

In 1983, Gardner (1983) published *Frames of Mind*. He explored human potentials and drew attention to the different forms of intelligence in humans. Gardner documented the following intelligences:

- Human potential, multiple intelligences
  - Linguistic intelligence, musical intelligence
    - Logical mathematical intelligence
    - Bodily kinesthetic intelligence
    - Personal intelligence: projecting inward understanding and interpreting feelings, and also personal intelligence projecting outward understanding temperaments and intentions in others

Gardner defined linguistic intelligence as including sensitivity to the meaning of words, to the order of words, sensitivity to sounds, rhythms, inflections, and appreciation of function of words. He also documented skills in using words to describe inner feeling and assessment of how words convey concepts.

Bodily kinesthetic intelligence was defined as skill in positioning of body parts as in sports and dance; sense of timing, and skills in fine movements.

Gardner promoted marshalling biology to understand intelligences.

In writing of new discoveries and phenomena in science, specific linguistic skills and logical skills are advantageous. Gardner did note in *Frames of mind* that the different intelligences can interact and build upon one another.

Gardner expanded knowledge and concepts of psychology and understanding of human abilities in subsequent works describing different individuals with minds that were extraordinary, each in a different way in his book *Extra-ordinary minds* (1997). Gardner (1997) explored aspects of leadership skills "Leading Minds: An Anatomy of Leadership."

Linking different intelligences

Gardner's works lead me to consider aspects of different intelligences and examples of extraordinary minds. In addition, it seems important to consider the crossover and interactions in different forms of intelligence. Poets have used words to describe scenes, feelings, emotions, their emotions in ways that echo feeling felt in all of us, and feelings that we have not had the courage or skill to convey. Clearly these contributions required skills in language and also personal intelligences, both inward and outward.

In Shakespeare, it is the skill with language that captures us as well as the range of emotions, contemplations, and inner conflicts, which we recognize as existing in our own lives. Also, in Shakespeare, we experience deep and astonishing imagery: "The quality of mercy is not strained but droppeth as a gentle rain from heaven".

Dickens used words and stories to move us between different social strata, and he could move us from individuals' outer lives to their inner lives.

George Eliot gives us insight into individuals' motivations desires to reach beyond expectation to achieve new levels of understanding and engagement.

It is interesting to consider advantages of occurrence of different intelligences in a particular individual. Clearly, spatial intelligence and mathematical intelligence are advantageous in solving problems in understanding molecular structure of complex molecules. This was demonstrated in the 19th century by Pasteur, for example. He used polarized light to reveal different diffraction pattern of crystal of tartaric acid and to identify different forms of tartaric acid.

In the 1913, Bragg and Bragg (1913) William Bragg and his son determined that patterns revealed when X-rays were passed through a crystal were determined by the molecular properties arrangement of atoms in the crystal.

Dorothy Hodgkin et al., (1956) initially used X-ray crystallography to define the molecular structure of the protein pepsin (Fig. 10.1). Subsequently, she used X-ray crystallography to reveal the atomic structure of the complex molecule vitamin B12 C63 H88 N14 O14 P Co.

In the years after World War II, X-ray crystallography became a key tool in protein chemistry

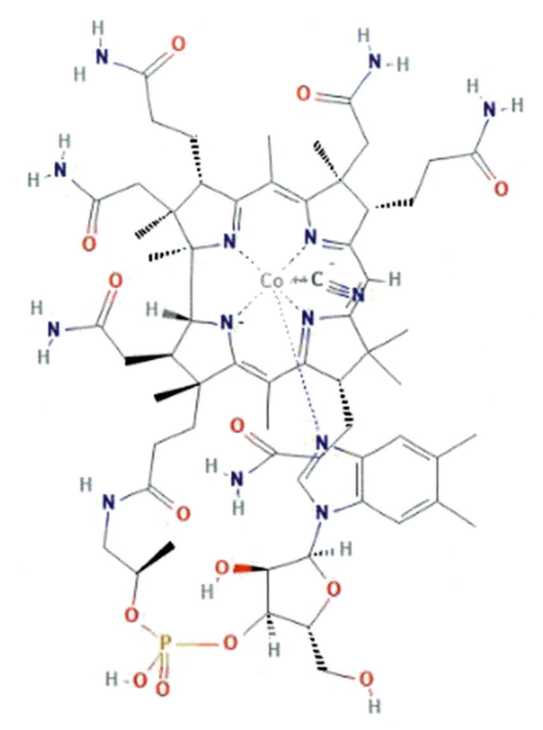

FIGURE 10.1 **Structure of vitamin B12.** Dorothy Hodgkin resolved the structure of vitamin B12. *From #https://pubchem.ncbi.nlm.nih.gov/compound/Vitamin-B12#section=Structures.*

and molecular biology. Clear data derived using this technique required both spatial and mathematical intelligence.

Following purification of insulin by Fred Sanger (1959), Dorothy Hodkin (1974) used X-ray crystallography to define the structure of insulin crystals.

7. Five different minds and the future

In 2006, Howard Gardner published an article on synthesizing minds and a book *5 Minds for the Future*. The five minds considered included the disciplined mind, the synthesizing mind, the creating mind, the respectful mind, and the ethical mind.

In the following sections, "different minds" and processes will be considered also in that light.

The synthesizing mind

In 2008, Gardner (2008) published a paper entitled "The synthesizing mind: making sense of a deluge of information." He defined the synthesizing mind as follows:

> The synthesizing mind is capable of assembling a large amount of information, data, knowledge- evaluating its accuracy and relevance for the task at hand and putting together that information in a succinct format that will be useful for a particular audience.

Gardner presented examples of the synthesis of information in different fields of endeavor and presented examples from across different ages.

One well-known example of synthesis includes what has been referred to as "The Modern Synthesis." This referred to works in the early 20th century to combine insights from Charles Darwin's theory of evolution and information on genetic heredity as presented by Gregor Mendel. Thus, the Modern Synthesis sought to bring together concepts of mutations and their inheritance and influence on phenotypic characteristics and the role of mutations and their inheritance in driving evolution of new characteristics.

A group of researchers, referred to as biometricians, in the early 1900s studied inheritance in populations through application of statistics and mathematically derived evidence of quantitative inheritance.

Thus, different schools of thought arose. In 1918, R.A. Fisher (1918) resolved the conflicts in the different schools through demonstration that continuous variation of a specific trait could result from mutations at distinct genetic loci. Fisher's paper was entitled "The Correlation between Relatives on the Supposition of Mendelian Inheritance."

In lectures in 1924, Fisher (1924) presented aspects of the biometrical study of heredity, and he also presented the advantages of incorporating concepts of Mendelism into the study of heredity. He noted particularly that in certain rare diseases in humans, concepts of Mendelism could clearly be applied.

In 1943, Julian Huxley (1943) published his book, *Evolution: The modern synthesis*.

He noted especially studies that had revealed the role of gene mutations in altering specific physical characteristics in organisms and that these characteristics were then inherited in subsequent generations leading to evolutionary differences.

Huxley noted that it speedily became apparent that Mendelism applied to the heredity in many-celled and in single-celled organisms, both plants and animals.

Gardner noted in describing synthesis processes that metaphors often serve useful purposes; one example noted was the use of the concept of "The tree of Life" as a key metaphor used in description of evolution.

Gardner drew attention to other important early examples of synthesis in science. Specific examples included the work of Linnaeus in taxonomy and classification of plants and animals. He also drew attention to the important taxonomic classification of elements in the periodic table generated by Mendeleev.

Synthesis in art

Gardner noted the use in writing of description of place or happenings as a metaphor for specific emotions without specifically stating the emotions. He noted that TS Eliot's (1963) poetry provided examples of this.

Conclusions regarding emotions can be drawn for example from description of scenes in the poetry of TS Eliot. I will present examples of this that I found in TS Eliot.

> What might have been and what has been
>
> Point to one end, which is always present
>
> Footfalls echo in the memory
>
> Down the passage which we did not take
>
> Toward the door we never opened
>
> Into the rose-garden. **Burnt Norton *I lines* 6–14**

Gardner noted that in art, synthesis can involve illustration of a particular scene to illustrate large-scale tumultuous events, e.g., Picasso's painting Guernica to indicate the widespread violent destruction of the Spanish Civil War.

It is perhaps also interesting to think of changes in art that coincided with societal changes, for example, the changes from Heroic Painting and paintings of dignitaries in the 17th century and then transition to both political and art changes in the 18th century and particularly in France in the 19th century and in the times of the Impressionists when incidents and moments in the lives of ordinary citizens were portrayed.

It is perhaps also important to note that there are examples of individuals who have drawn on information from diverse sources and arrive at synthesis and generate plans to solve specific large-scale problems in society. In his book, The Modern Mind, published in 2000, Peter Watson (2001) described process of extensive information gathering adopted by William Beveridge (1944) to generate evidence and compile documents between 1941 and 1942 to support his arguments related to "Social Insurance: General Consideration in the United Kingdom." Beveridge drew up plans for measure to particularly attack WANT that could be associated with unemployment, squalor, and disease. He gathered information not only through review of documents and multiple interviews but also through consultations with experts in other fields, e.g., the economist Maynard Keynes.

8. Mind, ideas, and synthesis

In the course of evolution, it seems likely that ideas arose as responses to interacting and coping with environmental conditions and factors. In Peter Watson's (2005) book, *Ideas: a history of thought and invention from fire to Freud* published in 2005, he traces aspects of the history of ideas such as fire making, tool making, and progression to concepts of spirituality to law. Perhaps the greatest forward moves came with development of concepts to test the validity of ideas through discussion and debate but particularly through the design of experiments.

In considering aspects of synthesizing minds, it is particularly enlightening to review the book *Convergence* by Peter Watson (2016) published in 2016. Watson particularly emphasized the coming together of the different branches of science.

Synthesis and convergence

In the introduction to his book *Convergence*, published in 2016, Peter Watson (2016) writes of the surprising achievements of Mary Somerville (1846) who in the mid-19th century published several notable books, perhaps most importantly, "On the connection of the Physical Sciences." She presented mathematics as discipline unique and relevant to all physical sciences.

Watson also emphasized other unifying ideas that came together in the 1850s, which included ideas regarding electricity and magnetism, the idea of conservation of energy, aspects of optics, and astronomy. Also remarkable in the late 19th century was the coming together of ideas on deep-time geology, paleontology, and concepts of evolution.

In tracing convergence and synthesis concepts further, Watson described the fusion of physics and chemistry, and fusion of chemistry and biology followed later by concepts of fusion of physics and biology. He noted elements of synthesis of concepts in other discipline, for example, the joint consideration of concepts in sociobiology and economics.

Synthesis in physics, chemistry, and astronomy

Watson (2000) in his book *A Terrible Beauty* (2000) explored the evolution of thoughts and ideas that had shaped the 20th century. He identified the golden age of physics as beginning in 1919 with the splitting of the atom by Rutherford. Related ideas in physics were also developed and tested by Niels Bohr in Copenhagen and Max Planck and student in Germany.

Ideas generated in physics and mathematics also intersected with ideas in aerodynamics and engineering. Concepts in physics and engineering were fused, leading to the development of telescopes by Hubble in 1929 that enabled more detailed studies of stars and galaxies.

Among the most fruitful examples of fusion of ideas in mathematics and engineering in the 20th century led to the development of computers. Watson recounted the cross-fertilization of experience and ideas between Robert Lewinski a Polish engineer who had worked in Germany prior to the start of World War II on a code signaling machine the Enigma, and ideas of Alan Turing, a brilliant mathematician assigned to the Code Breaker operation at Bletchley park in England during World War II.

Brilliant mathematicians working at the Institute for Advanced Study at Princeton in the United States including Von Neumann also explored development of electronic Equipment for computation.

Perhaps the greatest examples of development and synthesis of concepts and ideas from mathematics, physics, and engineering led to advances in astronomy, cosmology, and space exploration.

Synthesis of ideas in chemistry, biology, medicine, and genetics

In 1949, Needham (1949) edited a series of paper on contributions of Gowland Hopkins to developing the field of biochemistry. In 1898, Gowland Hopkins was appointed to a position as demonstrator at Guys Hospital in London. He initially specialized in chemical studies of urine and the relevance of urinalysis in medical diagnosis. Later, Hopkins began studies on metabolism in human and investigations of dietary components and their relevance to health.

An early example of the synthesis of concepts in genetics and chemistry came with the work of Archibald Garrod who demonstrated that abnormal levels of a specific chemical in urine occurred in humans with a particular inherited disease. Garrod (1909) then formulated the concepts of inborn errors of metabolism.

In the early years of the 20th century, William Osler promoted the synthesis of observation and experimentation and insights from literature and publications to advance clinical medicine. In 1906, Osler (1906) wrote:

> to correlate the vast stores of knowledge that they may be quickly available for the prevention and cure of disease to carefully observe the phenomena of life in all it phases normal and perverted, to call to aid the science of experimentation, to cultivate the reasoning faculty to be able to know the true from the false-These are our methods. To prevent disease, to relieve suffering and to heal the sick- this is our work.

Genetic concepts were developed particularly in the 20th century and arose through the generation and exchange of information and synthesis of ideas. Information accrued on diseases and traits that followed Mendelian inheritance patterns on other traits that followed polygenic inheritance and on the realization that some genetic variation had minimal effects on phenotypes and were defined as polymorphic variants (Harris, 1969).

In addition, in the 20th century, information accrued on the structure and variation in chromosomes. In 2015, Malcolm Ferguson-Smith (2015) described the history of cytogenetics and the analysis of chromosomes. Analysis of chromosome structure became more detailed and more clinically useful as ideas were shared regarding methods for chromosome analysis.

Increasingly techniques that facilitated isolation and purifications of DNA, RNA, and proteins and exchange of ideas led to clearer understanding of gene expression, its regulation, and nucleic acid changes that altered gene expressions and gene products (Judson, 1979).

It is interesting to consider modes of information exchange that made the synthesis of ideas possible. Exchange of ideas initially took place through joint meeting, through publication of experimental data in journals, exchange occurred through telephone and telegraph, and more recently idea exchange through the Internet became predominant.

In 1974, Lewis Thomas (1974) wrote: "An active field of science is like and immense intellectual anthill: the individual almost vanishes in the mass of minds tumbling over each other, carrying information from place to place. Passing it around at the speed of light."

References

Arend, I., Rafal, R., Ward, R., 2008. Spatial and temporal deficits are regionally dissociable in patients with pulvinar lesions. Brain 131 (8), 2140–2152. https://doi.org/10.1093/brain/awn135.

Beaty, R.E., Benedek, M., Silvia, P.J., Schacter, D.L., 2016. Creative cognition and brain network dynamics. Trends Cognit. Sci. 20 (2), 87–95. https://doi.org/10.1016/j.tics.2015.10.004.

Beaty, R.E., Chen, Q., Christensen, A.P., Qiu, J., Silvia, P.J., Schacter, D.L., 2018. Brain networks of the imaginative mind: dynamic functional connectivity of default and cognitive control networks relates to openness to experience. Hum. Brain Mapp. 39 (2), 811–821. https://doi.org/10.1002/hbm.23884.

Beveridge, W., 1942. Social Insurance and Allied Services. http://HMSO.

Beveridge, W., 1944. Full Employment in a Free Society. George Allen and Unwin LTD.

Bragg, W.H., Bragg, W.L., 1913. The reflexion of X-rays by crystals. Proc. R. Soc. Lond. A. 88 (605), 428–438. https://doi.org/10.1098/rspa.1913.0040.

Chaaya, N., Battle, A.R., Johnson, L.R., 2018. An update on contextual fear memory mechanisms: transition between Amygdala and Hippocampus. Neurosci. Biobehav. Rev. 92, 43–54. https://doi.org/10.1016/j.neubiorev.2018.05.013.

N.d. Retrieved October 24, 2020, from: https://pubchem.ncbi.nlm.nih.gov/compound/Vitamin-B12#section=Structures.

De Young, C.G., 2014. Openness/Intellect: A dimension of personality reflecting cognitive exploration. APA Handbook of Personality and Social Psychology: Personality Processes and Individual Differences 4, 369–399.

Eliot, 1963. Eliot TS Collected Poems 1909–1962. Harcourt, Brace, and Company.

Etkin, A., Büchel, C., Gross, J., 2015. The neural bases of emotion regulation. Nat. Rev. Neurosci. 16 (11) https://doi.org/10.1038/nrn4044. PMID: 26481098.

Ferguson-Smith, M., 2015. History and evolution of cytogenetics. Mol. Cytogenet. 8 (19) https://doi.org/10.1186/s13039-015-0125-8.

Fisher, 1918. The correlation between relatives on the supposition of mendelian inheritance. Trans. R. Soc. Edinb. 52, 399–433.

Fisher, 1924. The Biometrical Study of Heredity Lectures Delivered at the London School of Economics.

Gardner, H., 1983. Frames, & Mind. The theory of Multiple Intelligences First Published.

Gardner, H., 1997. Extraordinary Minds: Portraits Of 4 Exceptional Individuals And An Examination Of Our Own Extraordinariness by Howard Gardner. Basic Books, Harper Collins, New York. ISBN-13: 978-0465045150.

Gardner, H., 2006. Multiple Intelligences Revised. http:// ISBN13:9780465047680.

Gardner, H., 2008. The Synthesizing Mind: Making Sense of the Deluge of Information in Globalization and Socialization Pontifical Academy of. Social Sciences Extra Series.

Garrod, A., 1902. The incidence of alkaptonuria: a study in chemical individuality. Lancet 1616–1629. https://doi.org/10.1016/s0140-6736(01)41972-6.

Garrod, A., 1909. The Inborn Errors of Metabolism. Hodder.

Harris, H., 1969. Enzyme and protein polymorphism in human populations. Br. Med. Bull. 25 (1), 5–13. https://doi.org/10.1093/oxfordjournals.bmb.a070670.

Hodgkin, D., Kamper, J., Mackay, M., Pickworth, J., Trueblood, K.N., White, J.G., et al., 1956. Structure of vitamin B12. Nature 178 (4524), 64–66. https://doi.org/10.1038/178064a0.

Hodkin, D., et al., 1974. Insulin, its chemistry and biochemistry. Proc R Soc Lond B Biol Sci. 186 (1084), 191–215. https://doi.org/10.1098/rspb.1974.0046.

Horváth, G., 2018. Visual imagination and the narrative image. Parallelisms between art history and neuroscience. Cortex 105, 144–154. https://doi.org/10.1016/j.cortex.2018.06.007.

Huxley, J., 1942. Evolution. The Modern Synthesis. MIT press, Boston. ISBN-13: 978-0262513661.

Huxley, J., 1943. Evolution: The Modern Synthesis. Published Harper and Bothers Publishers.

Ishiguro, K., 2017. My Twentieth Century Evening and Other Small Breakthroughs. The Nobel Lecture.

Josselyn, S.A., Frankland, P.W., 2018. Memory allocation: mechanisms and function. Annu. Rev. Neurosci. 41, 389–413. https://doi.org/10.1146/annurev-neuro-080317-061956.

Judson, H.F., 1979. The Eighth Day of Creation. Simon and Shuster.

Kaufman, S.B., 2013. Opening up openness to experience: a four-factor model and relations to creative achievement in the arts and sciences. J. Creativ. Behav. 47 (4), 233–255. https://doi.org/10.1002/jocb.33.

Kidd, D.C., Castano, E., 2013. Reading literary fiction improves theory of mind. Science 342 (6156), 377–380. https://doi.org/10.1126/science.1239918.

Kok, A., 2019. Functions of the Brain A Conceptual Approach to Cognitive Neuroscience, first ed.

Lessing, G, 1984. An essay on the limits of painting and Poetry, 1st. John Hopkins University Press, Baltimore.

Needham, J., 1949. In: Needham, J. (Ed.), Hopkins and Biochemistry 1861-1947, 1st edn. W. Heffner and Sons Limited, New York. ASIN : B00A04I2F6.

Osler, 1906. Address: Chauvinism in Medicine Aequanimitas.

Patton, A., Hastings, M., 2018. The suprachiasmatic nucleus. Curr. Biol. 28 (15) https://doi.org/10.1016/j.cub.2018.06.052. PMID:30086310.

Pearce, M.T., Zaidel, D.W., Vartanian, O., Skov, M., Leder, H., Chatterjee, A., Nadal, M., 2016. Neuroaesthetics: the cognitive neuroscience of aesthetic experience. Perspect. Psychol. Sci. 11 (2), 265–279. https://doi.org/10.1177/1745691615621274.

Pearson, J., 2019. The human imagination: the cognitive neuroscience of visual mental imagery. Nat. Rev. Neurosci. 20 (10), 624–634. https://doi.org/10.1038/s41583-019-0202-9.

Peter, W., 2000. A Terrible Beauty: A History of the People and Ideas that Shaped the Modern Mind Orion. Books, London.

Postle, B., 2015. Essentials of Cognitive Neuroscience, first ed. ISBN-13: 978-1118468067.

Reybrouck, M., Vuust, P., Brattico, E., 2018. Brain connectivity networks and the aesthetic experience of music. Brain Sci. 8 (6) https://doi.org/10.3390/brainsci8060107.

Rikhye, R.V., Wimmer, R.D., Halassa, M.M., 2018. Toward an integrative theory of thalamic function. Annu. Rev. Neurosci. 41, 163–183. https://doi.org/10.1146/annurev-neuro-080317-062144.

Sanger, F., 1959. Chemistry of insulin; determination of the structure of insulin opens the way to greater understanding of life processes. Science 129 (3359), 1340–1344. https://doi.org/10.1126/science.129.3359.1340.

Seeley, W.W., 2019. The salience network: a neural system for perceiving and responding to homeostatic demands. J. Neurosci. 39 (50), 9878–9882. https://doi.org/10.1523/JNEUROSCI.1138-17.2019. The Official Journal of the Society for Neuroscience.

Somerville, M., 1846. On the Connection of the Physical Sciences 1834, 1846, 7th ed. Key and Biddle, Philadelphia.

Thomas, 1974. Chapter 20 Natural Sciences in A Long Line of Cells Collected Essays.

Wassiliwizky, E., Koelsch, S., Wagner, V., Jacobsen, T., Menninghaus, W., 2017. The emotional power of poetry: neural circuitry, psychophysiology and compositional principles. Soc. Cognit. Affect Neurosci. 12 (8), 1229–1240. https://doi.org/10.1093/scan/nsx069.

Watson, P., 2000. A Terrible Beauty: A History of the People and Ideas that Shaped the Modern Mind. Orion.

Watson, P., 2001. The Modern Mind. Harper Collins.

Watson, P., 2005. Ideas: A History of Thought and Invention, from Fire to Freud. Harper Collins, New York.

Watson, P., 2016. Convergence the Idea at the Heart of Science. Simon Schuster.

CHAPTER

11

Psychiatric disorders

1. Shared heritability of common disorders of the brain

In 2018, the Anttila et al., 2018 published a study that revealed the shared heritability of 10 common disorders of the brain. The total study includes 265,218 cases and 784,643 controls.

Bipolar disorder shared risk variants with schizophrenia, obsessive compulsive disorder, and major depressive disorder.

Attention hyperactivity disorder shared genetic variants with major depressive disorder, bipolar disorder, and schizophrenia.

Autism spectrum disorder shared risk variants with schizophrenia and major depressive disorder.

Major depressive disorder shared genetic risk variants with schizophrenia and obsessive–compulsive disorder.

It is clear that risk variants for schizophrenia shared significant correlation with most psychiatric disorders.

Posttraumatic stress disorder shared no significant correlation with the other common brain disorders.

Investigators in this report noted clinical evidence that the diagnostic boundaries between disorders are porous and that a specific patient received different diagnoses at different times and that there was evidence that patients progressed from one diagnosis to another. They concluded that interconnected genetic etiologies underlie psychiatric disorders.

In 2018, Working Group of the Psychiatric Genomics Consortium, 2018 The Bipolar and Schizophrenia working group of the Psychiatric Genomics Consortium (2018) defined schizophrenia and bipolar disorder as disorders that are distinct but share symptomatology. They reported data on 53,555 cases that included 20,129 cases of bipolar disorders and 33,426 schizophrenia cases. Both disorders were reported to have heritability estimates between 60% and 80%. They noted that diagnostic features defined in the late 19th century were still used. Bipolar disease was defined as primarily characterized by episodic mood disorders, while schizophrenia is defined as a mood disorder with delusions and sometimes with hallucinations.

This group reported that the two disorders share genetic factors but that genetic differences also exist between the two disorders. Their genetic analyses were carried out in cases with matched ancestry. One such difference is that fact that defined genomic copy number variants occur in some cases of schizophrenia but are absent in bipolar disease. In a search for divergent loci, they identified a locus on chromosome 1 within an intron of the DARS2 gene. This gene maps on 1q25.1 and encodes a mitochondrial enzyme that specifically aminoacylates aspartyl-tRNA. Another divergent locus mapped to an intron of

Mechanisms and Genetics of Neurodevelopmental Cognitive Disorders
https://doi.org/10.1016/B978-0-12-821913-3.00001-9

the gene on chromosome 20q13.13 that encodes ARFGEF2, ADP ribosylation factor guanine nucleotide exchange factor 2. For both of these divergent loci, the minor allele was more common in individuals with bipolar disease.

A specific variant in the gene on 17q21.31 that encodes DCAKD was found to be more highly associated with schizophrenia than with bipolar disorder. DCAKD defined as a dephospho-CoA kinase domain–containing protein.

Wu et al. (2020) undertook genome-wide association studies (GWASs) in five different psychiatric disorders that included 65,967 cases of schizophrenia, 41,653 cases of bipolar disease, 46,350 cases of autism spectrum disorders, 55,374 cases of attention deficit hyperactivity disorders, and 688,809 cases of depressive disorder. They noted overlap in risk variants in seven different genes across disorders.

The SORCS3 variant was associated with all five diseases. SORCS3 10q25.1 encodes a type I receptor transmembrane protein that is a member of the vacuolar protein sorting 10 receptor family.

For the other genes, risk variants were associated with four out of five diseases.

Specific genes with risk variants identified in four out of five psychiatric diseases

Wu et al. (2020) reported that specific genes with risk variants occurred in four of five different psychiatric diseases studies. The products of these genes and the chromosome location of the gene that encodes each product are listed in the following:

- GLT8D1 3p21.1, glycosyltransferase 8 domain–containing 1
  - GABR1 6p22.1, gamma-aminobutyric acid type B receptor subunit 1
    - HIST1H1B 6p22.1, H1.5 linker histone, cluster member; the linker histone, H1, interacts with linker DNA between nucleosomes and functions in the compaction of chromatin into higher order structures
    - HIST1H2B 13q12.11, histone pseudogenes
    - KCNB1 20q13.13, potassium voltage-gated channel subfamily B member 1
    - DCC 18q21.2 encodes a netrin 1 receptor mediates axon guidance of neuronal growth cones

Comorbidity of psychiatric disorders

In a study carried in Denmark that included total of 5,940,778 persons, including 2,958,293 men and 2,982,485 women, mean [SD] age at beginning of follow-up is 32.1 [25.4] years. Cumulatively the study involved 83.9 million person-years; Plana-Ripoll et al. (2018) determined that all mental disorders were associated with an increased risk of all other mental disorders. The study included individuals living in Denmark between 2000 and 2016. The authors concluded that comorbidity within mental disorders is pervasive, and the risk persists over time.

2. Architecture of psychiatric diseases

In 2019, Sullivan and Geschwind (2019) reviewed the genetic, genomic, cellular, and diagnostic architecture of psychiatric disorders. They emphasized that genetic architecture included the overall risk variants, the degree of risk they conferred, and the risk frequency in the population. They noted that over the past decades, studies have revealed polygenic risk for psychiatric disorders, and this includes multiple risk alleles of small effect and also variants with larger effects.

Sullivan and Geschwind presented data on the heritability of psychiatric disorders in monozygotic twins. In monozygotic twins, the heritability of bipolar disease was noted to be 0.85; heritability of schizophrenia was reported to be 0.81. In contrast, the heritability for major depressive disorder was reported as 0.37.

Bipolar disease was defined as a disorder in which manic depressive episodes were the key feature. Key manifestations of schizophrenia were reported to be delusions and hallucinations.

Sullivan and Geschwind emphasized the great impact that these disorders have on life and noted that they are ranked in the fifth highest position in the WHO list of disability-causing diseases.

GWASs examine nucleotide variants at a particular locus to determine whether particular nucleotide variants at that locus occur more frequently in patients with psychiatric disease than in the general population. Geschwind and Sullivan reported that through GWASs, 241 loci had been associated with 10 psychiatric disorders. Importantly, 22 loci were reported to be associated with two or more psychiatric disorders.

Rare variants established by DNA sequencing have been identified in some cases of autism. In addition, several rare genomic copy number variants have been found to increase risk for specific psychiatric disorders particularly autism spectrum disorders and schizophrenia.

In 2020, Satterstrom et al. (2020) reported results of exome sequencing in 35,584 samples that included 11,986 cases with autism. They identified 53 genes with variants associated with autism and 49 genes with disruptive de novo variants primarily associated with developmental delay. Most of the genes with damaging exome changes were noted to be involved in genes that play roles in regulation of gene expression or in neuronal communication. Implicated genes were noted to be important in neurodevelopment.

3. Schizophrenia

In 2019, Marder and Cannon (2019) reviewed schizophrenia. They noted that the heritability factor for this disorder was 80% and indicated that there were also environmental factors implicated in schizophrenia. They distinguished different phases in schizophrenia, including a prodromal phase, often in teenage year or early adult years when individuals manifested delusional beliefs and withdrawal.

They noted evidence that in the prodromal phase, there is evidence for loss of gray matter in the prefrontal and hippocampal regions. In this phase, altered levels of immunologic markers were often reported. Immunologic factors, including complement alteration, are reported to impact functional connectivity in several brain regions including prefrontal cortex, temporal cortex, thalamus hippocampus, and cerebellum.

Marder and Cannon noted that in the prodromal phase of schizophrenia and in the first episode of psychosis alterations in dopamine levels have been reported.

Immune hypothesis of schizophrenia and complement system

Woo et al., 2020 reviewed genetic factors in schizophrenia that potentially relate to the hypothesis that immune factors play roles in this disorder. She noted the value of considering GWASs that provided information on common nucleotide variants associated with the disorder and also rare structural genetic variants and noted also the importance of functional studies.

Several GWASs provided evidence of association of schizophrenia with variants in the HLA MHC region and particularly with variants in complement genes that map in that region. Structural variants in the complement C4 regions and increased expression of C4 due to these variants were subsequently identified (Sekar et al., 2016). Woo et al. (2019) noted that 23 new studies on the complement C4 schizophrenia association had been carried out since 2008 and overall indicated overexpression of C4.

In the Psychiatric Genetic Consortium 2 GWAS, by Pardinas et al. (2018) with 11,260 cases and 24,502 controls, 145 schizophrenia associated loci were identified. The most striking association was in the C4 complement locus in the major histocompatibility region on chromosome 6, $p = 5 \times 10^{-44}$.

Pouget et al. (2019) examined data outside the MHC region in schizophrenia cases and identified significant GWAS-associated loci near other genes encoding immune function related products. These included the following:

- CLU, clusterin-secreted chaperone, involved in cell death processes
  - DPP4, dipeptidyl peptidase 4, exopeptidase
    - EGR1, early growth response 1, transcriptional regulator
    - HSPD1, heat shock protein D1, mitochondrial expressed immune function regulator
    - NFATC3, nuclear factor of activated T cells

These genes are also expressed in brain. In addition, the GWASs revealed schizophrenia-associated loci that were noted to be expressed in T lymphocytes and B lymphocytes. They proposed that genes associated with schizophrenia were expressed in brain and were immune function related.

Schizophrenia risk and complement C4

Imaging studies by Cannon et al. (2015) revealed progressive reduction in brain cortical thickness associated with psychosis development in adolescents. Earlier studies by Garey et al. (2010) reported decreased dendritic spine density particularly in pyramidal neurons in the cortices of patients with schizophrenia.

In a number of different GWAS, the strongest genetic locus associated with schizophrenia was reported to be in the major histocompatibility complex (MHC) region on human chromosome 6p. Further mapping revealed the strongest genetic association to be with the C4 complement within the MHC complex region. Association of schizophrenia with the CSMD1 encoding locus was also reported. CSMD1 maps to chromosome 8p23.2 and is reported to be a regulator of C4 complement.

Sekar et al. (2016) reported that two isoforms of C4 complement are produced C4A and C4B. The number of C4 complement genes in the MHC regions varies between 1 and 3. There are long and short forms of these genes that differ with respect to inclusion of a retroviral HERV insertion in intron 9. The variable insertion generates C4AL, C4AS, C4BL, and C4BL. Sekar et al. developed an assay to investigate haplotypes of C4A and C4B in chromosome 6p. They also studied C4 expression in brain and determined that the HERV insertion serves as an enhancer. The RNA expression levels increased also as a function of the C4A and C4B gene copy number.

Sekar et al. analyzed the C4 region structure and haplotypes in schizophrenia patients. In addition, they studied genes in the region just proximal to MHC that included BAK1 (BCL2 antagonist) and SYNGAP1. SYNGAP1 encodes a component of the postsynaptic density.

Analyses of brain tissue from schizophrenia cases that included study of gene numbers and structure indicated that C4 levels were likely increased in schizophrenia brain. Sekar et al. also noted that in brain, components of the complement cascade have been implicated in synaptic pruning. They noted that in brain complement receptors for C4 are expressed primarily on microglia. They postulated that schizophrenia manifestations developing in adolescence could be intensified by increased synaptic pruning.

They also noted that rare mutations in genes that encoded synapse components had been identified in other schizophrenia cases (Fromer et al., 2014). They proposed that diverse synaptic abnormalities could interact with the complement components.

Sekar et al. also determined that human complement C4 localizes to neuronal synapses, dendrites, and neuronal cell bodies. Their studies in mice revealed that complement C4 plays roles in synapse elimination during development. Based on information on C4 activity, Sekar et al. noted that altered C4 after could lead to a reduction in synapses as is reported in schizophrenia.

Dhindsa and Goldstein (2016) stressed the important therapeutic relevance of the findings of Sekar et al. presented a review of the complement pathway in relation to schizophrenia. In addition, they presented an interesting discovery, the association of a specific schizophrenia risk allele in the CSMD1 locus, CUB, and Sushi multiple domains 1, on chromosome 8p23.3. CDMD1 is primarily expressed in brain and in testis. The CSMD1 protein is an inhibitor of complement C3 and C4.

All three complement pathways, classical, lectin, and alternative pathways, promote formation of C3 convertase. Woo et al. (2020) noted that C4 is a component of the classical complement pathways. The classical complement pathway is reported to play the key role in synaptic pruning.

Details have emerged indicating that different allelic forms of C4 occur including C4BL and C4BS. These forms vary with respect to their content of a retroviral gene sequence insertion. Also, Woo et al. noted that C4 and CSMD1 are being considered as potential drug target in the treatment of schizophrenia. The different structural forms differ with respect to their copy number. The Sekar study revealed that each of the C4 forms was associated with schizophrenia and led to higher levels of C4 expression. C4A expression was particularly associated with schizophrenia.

A specific variant, the A allele, in the CSMD1 gene, rs10503252, was found to associate with schizophrenia. Studies by Liu et al. (2019) reported that the A allele led to reduced expression of CSMD1, a C4 inhibitor. In patients with schizophrenia, antipsychotic treatment led to increased expression of the inhibitor.

Khandaker et al. (2015) reviewed evidence that immune brain interactions play roles in central nervous system disorders including psychiatric disorders. They noted evidence that indicated a role of the immune system in the pathogenesis of schizophrenia and depression.

Facets of the immune system include the innate immune response mediated by neutrophils and macrophages and the adaptive immune response that is involved in the initiation of immunologic memory of antigens through antibodies and activity of T lymphocytes.

Early immunological studies in schizophrenia drew attention to possible increases in the risk of schizophrenia following specific infections during childhood and also the increased risk of schizophrenia in individuals with certain autoimmune disorders.

Khandaker et al. also drew attention to genetic studies that revealed association of schizophrenia with certain genetic markers that mapped to the HLA region on chromosome 6p21.

Studies on specific immune components in blood were the focus of a number of studies. Specific studies revealed increased levels of interleukin 6 and other cytokines that have immune regulatory functions. Decreased levels of the antiinflammatory cytokine interleukin 10 were reported in cases with active schizophrenia manifestations. Furthermore, an increase in proinflammatory cytokines was reported during episodes of acute psychosis.

In considering possible deleterious effects of increased levels of cytokines, Khandaker et al. noted that cytokines increase oxidative stress and can increase levels of nitrous oxide. In addition, increased cytokines potentially activate the hypothalamic pituitary adrenal axis and increase cortisol release.

Cytokines also impact microglial function. Khandaker et al. drew attention to the abundance of microglia in the nervous system and noted that these cells derive from myeloid precursor cells that migrate to the nervous system during early development. Brain injury or inflammation can activate microglia. This activation includes activation of surface receptors and increased levels of cytokines.

Khandaker et al. noted that specific studies had revealed the presence of autoantibodies to specific brain region components in certain cases of schizophrenia. These included antibodies against NMDA glutamate neurotransmitter receptors and antibodies against voltage-gated potassium channels in patients with schizophrenia, including patients with first episode psychosis. Elevated levels of NMDA antibodies have been reported in patients with schizophrenia and in some patients with depression. These antibodies may also occur in cases with infections in the brain.

With respect to components of the cellular immune system, there were reports of increased levels of T and B cells reported in postmortem samples of the hippocampus in schizophrenia patients.

Khandaker et al. considered possible mechanisms through which immune dysregulation could lead to schizophrenia. A possible mechanism included impacts on neurotransmitters. One proposed mechanism involves effects of proinflammatory cytokines on activity of the cyclooxygenase (COX) pathway components, and this interference can impact tryptophan metabolism leading to increased levels of kynurenic acid that acts as an antagonist of the NMDA receptor. Support for importance of this mechanism derives from evidence that administration of COX2 inhibitors were reported to improve clinical symptoms.

Khandaker et al. emphasized that evidence for neurodegenerative processes in schizophrenia included evidence for neuronal atrophy, reductions in neuronal synapses, and dendrites. The question arises as to whether microglial activation contributes to neurodegeneration.

The human microbiome and evidence for bidirectional communication between gut and brain are being investigated. Questions arise as to whether specific metabolic components generated by the microbiome can be transported through the gut brain axis and influence neural function. There is evidence to increased incidence of intestinal infections in patients with schizophrenia.

Khandaker et al. emphasized that increased insights into inflammation and altered immune system function in schizophrenia could lead to development of new therapies in schizophrenia.

Immunoneuropsychiatry

In 2019, Pape et al. (2019) reviewed immunoneuropsychiatry and novel perspectives in brain diseases. They emphasized that immune processes play key roles in central nervous system homeostasis and that novel discoveries revealed interplay between neurons and glial cells.

Motile processes of microglia were reported to play an active role in synaptic pruning in response to active complement C3, and microglia were noted to participate in phagocytosis of synapses targeted by complement. In adult brain, microglia were reported to play roles in synaptic circuit homeostasis. Other factors important in brain homeostasis included chemokines that bound to chemokine receptors.

Microglial processes that interact with neurons were reported to release specific factors that impact synaptic activity and synaptic pruning.

Pape et al. noted that microglia could be considered as the innate immune cells of the brain. Other immune cells including T cells were noted to be present in meninges, choroid plexus, and cerebrospinal fluid.

The blood–brain barrier was noted to be comprised of endothelial cells of blood vessels and capillaries and tight junctions between these endothelial cells. Astrocyte end feet were reported to project onto endothelial cells. Cells in the blood including T cells were prevented from entering the brain through integrity of the blood–brain barrier. However, cytokines released from T cells can enter the brain.

Proinflammatory cytokines were noted to interact with receptors on cerebral vascular endothelial cells, and cytokines also interact with microglia.

The G lymphatic system is a more recently described perivascular channel system that promotes removal of solutes and small molecules, including antigens, from the brain. The G lymphatic system interacts with the meningeal lymphatic system and facilitates drainage from the brain to the cervical lymph nodes.

Links between the central nervous systems and the peripheral immune systems have been determined to lead to the psychiatric manifestation in autoimmune diseases.

Pape et al. noted that more recently, abnormal peripheral immune profiles and abnormal cytokine profiles have been identified in psychiatric diseases including schizophrenia, depression, and posttraumatic stress disorder. Abnormal levels of C-reactive proteins, proinflammatory markers including interleukins IL6, IL1B, IL17A and IL2 receptor may be present.

In addition, Pape et al. noted that in cases of schizophrenia and depression, increased microglial activity was documented on PET scans. Bloomfield et al. (2016) reported use of a second-generation radioligand 11C PBR 28 and PET tomography to image microglial activity. This ligand binds to 8 kDa translocator protein (TSPO), a marker of microglia activation. Their study revealed elevated microglial activity in schizophrenia patients relative to controls.

Pape et al. emphasized that aberrant genetic regulation and polygenic risk factors underlie psychiatric diseases, and they specifically noted that complement C4 was impacted in schizophrenia.

A particularly important aspect of recent findings is that immune-modulatory therapies are being considered in psychiatric disorders.

There is some evidence that serotonin reuptake inhibitors, which are frequently used in treatment of depression, have antiinflammatory effects.

Pape et al. (2019) presented a flowchart for evaluation of patients with psychiatric disease. They noted importance of determining if inflammatory processes may be involved through determining levels of C-reactive proteins, levels of interleukins, and interleukin receptors. Specific antibodies to brain antigens may be present in some cases, e.g., antibodies to aquaporin 4 anti-MOG (antimyelin oligodendrocyte glycoprotein) antibodies.

Cross-talk between immune system and dopaminergic system in schizophrenia

Vidal and Pacheco (2020) reported cross-talk between the immune system and the dopaminergic system in schizophrenia. They noted evidence that schizophrenia is associated with inflammation and noted that dopamine may regulate activation, migration, and proliferation of immune cells.

They noted also evidence for altered dopamine activity in schizophrenia, including hyperactive dopaminergic activity in the striatum and hippocampus and hypoactive dopaminergic activity in the prefrontal cortex.

Dopaminergic pathway and schizophrenia

Weinstein et al. (2017) reported that dopaminergic dysregulation in different brain regions had been implicated in schizophrenia pathogenesis. They noted that detailed studies had revealed complexity in dopamine expression, with differences in different brain regions. Weinstein et al. presented evidence that dopamine release and activity are increased in the striatum in schizophrenia; however, there is subregion heterogeneity with respect to dopamine release in the striatum. In extrastriatal regions, including the prefrontal cortex, dopamine activity was reduced.

Blood–brain barrier and psychosis

Pollak et al., 2018 reviewed studies carried out to investigate the blood–brain barrier in schizophrenia and other psychoses. They noted that genetic and environmental risk factors for psychotic disorders have been identified; however, they proposed that it was important to study the blood–brain barrier in these disorders since it serves as the interface between systemic factors and the brain.

Pollak et al. defined the blood–brain barrier as formed by the endothelium of microvessels and influenced by adjacent cells with tight junction between the cells that were noted to restrict diffusion through the barrier. Tight junction resulted in part from intercellular connections with junction proteins claudins and occludins that also formed links with the cellular cytoskeleton. However, certain lipid soluble molecules were noted to bind to the endothelial cell lipid membranes and to cross the barrier. In addition, specific transporters facilitate passage of essential nutrients such as glucose and amino acids into the brain. P-glycoprotein was noted to be an efflux transporter. Receptor-mediated passage of certain larger molecules, including growth factors, also occurs.

In vascular endothelial cells in small vessels in the brain, the vascular endothelial cells and basement membrane also interact with perivascular astrocytes, pericytes, and microglial cells. Cerebrospinal fluid secreted by the choroid plexus flows into the subarachnoid space and is reported to also be low along the perivascular space.

Pollak noted that limited numbers of postmortem studies had been carried out on the cerebral microvascular structures in psychosis.

One protein of interest is S100B that is reported to be abundant in brain and is below detection limits in blood of healthy individuals. Pollak noted that a number of studies revealed increased concentration of S100B in serum of patients with schizophrenia.

Specific neuroimaging studies designed to assess brain arteriolar blood flow have revealed decreased blood flow in schizophrenia. Polak also drew attention to studies that revealed microglial activation in psychotic disorders and also presence of auto antibodies to certain neuronal surface antigens in serum of patients with psychosis.

There is also some evidence that increased expression of the efflux transporter P-glycoprotein may play roles in decreased responses to antipsychotic medications.

Pollak et al. noted that under inflammatory conditions, leukocyte adhesion molecular expression could be increased and facilitate passage of leukocytes across the blood–brain barrier.

White matter microstructure changes in schizophrenia

A number of studies have revealed oligodendrocyte and myelin changes in schizophrenia (Ellison Wright et al., 2009). It is interesting to note that Konrad and Winterer (2008) questioned whether the structural and connective changes in schizophrenia were primary factors or if they were epiphenomena.

Kelly et al. (2018) carried out a study of 1963 schizophrenia patients who had participated in international studies and studies on 2359 healthy controls. Their study was focused on analysis of DTI fractional anisotropy. Metaanalyses of data were carried out on data obtained in five different countries by the ENIGMA schizophrenia consortium. Comparison of data from patients and controls indicated the 20 of the 25 different brain regions analyzed had lower FA measures in patients than in controls. The largest effect size was obtained when whole brain white matter skeleton was analyzed and revealed that the average FA values were lower in patients than in controls. Other significant reductions in patients versus controls were found in the anterior corona radiata (carries axons and signals to and from the cerebral cortex) and the whole corpus callosum.

In analyzing sex-specific findings, Kelly et al. reported that changes in females were more significant in females than in males. Degree of change was not correlated with duration of illness. Kelly et al. also noted that their study did not reveal correlation of FA scores with medication dosage.

They noted that several different potential biological mechanisms have been proposed for the DTI changes, and these included white mater deterioration and neuroinflammation.

Pasternak (2016) proposed that altered FA on STI in schizophrenia could be a marker for neuroinflammation.

Schork et al. (2019) in a cross-disorder study of psychiatric diseases, identified risk loci were predicted to regulate genes expressed in radial glia and interneurons in the developing neocortex during midgestation. They postulated that dysregulation of genes that direct neurodevelopment may result in increased liability for many later psychiatric disorders.

4. Psychiatric disorders, indications for involvement of different pathways

Mäki-Marttunen et al. (2019) emphasized that genetic studies had revealed large numbers of loci with variants associated with psychiatric disorders and that the loci impacted different pathways. These findings ruled out a reductive approach to psychiatric diagnoses. They noted that in schizophrenia, 145 GWAS loci were defined. The associated loci showed a broad range of function, including immune functions, synaptic function, neurotransmission, and neuronal electrogenesis.

They emphasized the influence the interplay of internal and environmental factors in the causation of psychiatric disorders. They also emphasized the importance of transdisciplinary studies in psychiatry. Recent relevant advances included progress in structural and functional genomics, systems biology, statistical computation, and models to define neuronal functions and neuronal networks. Analyses of ion channels and signaling molecules have also advanced.

Mäki-Marttunen et al. emphasized that no definitive microscopic parameters were established for mental illness. They considered opportunities for biophysical psychiatry approached. These could include computer-based analyses of neuronal dysfunction and of neuronal excitability.

In considering neuronal electrogenesis studies, they noted that precise identification of genetic factors related to ion channel electrophysiology was ongoing. Many different neuronal cell types exist in the central nervous system, and the neuron excitability of these different cell types may vary.

Mäki-Marttunen et al. noted the correlation of specific disease–associated phenotypes with specific physiological functions and with genes involved in those functions would be useful. One example could be demonstration of specific brain wave patterns and altered neurotransmission in a specific psychiatric disease.

They emphasized importance of analyzing the effect of specific variants also on biophysical properties. They also emphasized the importance of considering not only neuronal but also glial functions.

5. Bipolar disorders

Carvalho et al. (2020) reviewed bipolar disorders and noted that in *DSM5* (*Diagnostic and Statistical Manual*, Fifth Edition), bipolar disorders include bipolar I, bipolar II, and cyclothymic disorders; in all of three disorders, there are striking mood alterations. Bipolar I is reported to be characterized by episodes of mania with elevated mood, overconfidence, disinhibition, talkativeness, and diminished need for sleep. Psychotic episodes occur with high frequency in bipolar I. These episodes may be followed by episodes of depression. Bipolar II disorder includes episodes of hypomania and episodes of depression. Cyclothymic disorder includes less severe manifestations of depression and mania. Onset of bipolar disorders was reported to be most frequently around 20 years of age though earlier onset occurred in some cases.

Carvalho et al. noted that bipolar disorders are reported worldwide to be the 17th most common cause of disability. Suicide rates are reported to be 20 to 30 times more common in cases of bipolar disorder.

Heritability of bipolar disorders was reported to be between 70% and 90%.Genetic factors were reported to intersect with environmental factors, but the latter have not been well defined. In Stahl and Bipolar Disorder Working Group of the Psychiatric Genomics Consortium, (2019), Stahl and the Bipolar Disorder Working Group of the Psychiatric Genomics Consortium reported results of a GWAS on 32 cohorts of bipolar affected individuals of European descent from 14 different countries. Cases included 20,352 diagnosed individuals, and 137,760 controls were studied. Analyses led to identification of 30 significant loci. Significantly associated loci included genes that encode the following products:

- Ion channels SCN2A, CACNA1C
  - Synapse-associated proteins including neurotransmitter receptor, GRIN2A glutamate ionotropic receptor NMDA-type subunit 2A
    - RIMS1, regulating synaptic membrane exocytosis 1
    - ANK1, ankyrin 1; ankyrins link the integral membrane proteins to the underlying spectrin-actin cytoskeleton

Biological bases of bipolar disorder

In 2019, Kato (2019) reviewed investigations into the biological basis of bipolar disorders. Kato distinguished between pragmatic sciences and mechanistic science in psychiatry. He considered pragmatic science to be represented by diagnostic criteria and randomized clinical trials and mechanistic science to include neuroscience research.

Kato noted that bipolar disorder patients are defined as having interepisodes free of disturbance. However, he suggested that some level of impairment could occur in interepisodes, which could include sleep disorders and some level of cognitive impairment.

Kato noted that stress has been considered as a triggering factor for manic-depressive episodes.

In reviewing imaging studies, Kato noted that the ENIGMA international consortium study of 1837 bipolar patients and 2582 controls revealed that bipolar patients manifested gray matter volume reduction in specific cortical areas including frontal temporal and parietal cortices. In addition, there were suggestions of white matter reductions on diffusion tensor imaging.

PET scan studies had revealed alterations in serotonin transporter levels with increased levels in specific areas including thalamus and decreased levels in the pons.

Studies carried out in the manic phase revealed decreased expression of dopamine D1 receptor in the cerebral cortex and decreased levels of serotonin 2 receptor. PET scans revealed activated microglia indicating evidence of inflammatory response.

Kato noted that P^{32} magnetic resonance spectroscopy provided evidence for alterations in brain energy and phospholipid metabolism alterations. He proposed that these findings could be indicative of mitochondrial dysfunction.

Functional neuroimaging and analyses of different brain areas including regions defined as emotion-related regions, amygdala, nucleus accumbens, and anterior cingulate were reported to indicate emotion cognition imbalance.

EEG abnormalities have been documented in patients with bipolar disorder and in patients with schizophrenia. Specific findings included reduced P50, P100, and P300 amplitudes.

Molecular studies in psychiatric disorders

Kato emphasized studies of gene expression in postmortem brain specimens and risk being impacted by effects of medication. However, there are reports of altered expression of mitochondrial and synapse-related genes, (Gandal et al., 2018).

Kato noted that questions arise as to whether somatic mutations occur in psychiatric disorders. He noted that there are reports indicating increased retrotransposon activity, specifically involving LINE1 elements in schizophrenia and bipolar disorder (Bundo, 2014).

Mendelian inherited disorders manifesting comorbidity with bipolar disorder

Kato documented three such disorders. Darier disease is an autosomal dominant disorder with prominent skin lesions and was reported to occur in patients with mood disorders (Craddock et al., 1994). The gene defective in Darier disease encodes a protein ATPA2 that functions as a calcium pump on endoplasmic reticulum.

Nanko et al. (1992) reported occurrence of bipolar disorders in some cases with Wolfram syndrome. The WFS1 protein is a transmembrane protein that interacts with the endoplasmic reticulum. Patients with this syndrome have diabetes mellitus optic atrophy.

Ophthalmoplegia and other neurological manifestations can occur in patients with multiple mitochondrial DNA deletions, and some of these patients were reported to also have mood-related psychiatric disorders (Mancuso et al., 2008).

Genetics and bipolar disease studies reported in 2020

In 2020, Gordovez and McMahon (2020) reported that the highly polygenic architecture of bipolar disease overlaps with that of schizophrenia, major depressive disorder, and other psychiatric disorders. They noted that functional studies on specific gene products have increased and that there is some evidence for risk genes based on that data. This includes evidence for roles of the following genes.

ANK3 is encoded by a gene on chromosome 10q21.3, and ANK3 protein serves to link membranes with the cellular cytoskeleton and is thought to play a role in maintaining specialized membrane domains.

CACNA1C, the gene encoding this voltage-gated calcium ion channel protein, maps to chromosome 12p13.3. CACNA1C protein forms the pore of the calcium channel.

TRANK1, encoded by a gene on chromosome 3p22.2, is described as a tetratricopeptide repeat and ankyrin repeat–containing protein.

DCLK3 encodes by a gene also in 3p22; the protein is sometimes known as DCAMKL3 and acts as a doublecortin kinase.

Gordovez and McMahon (2020) noted that of the different molecular pathways proposed in bipolar disorder, the calcium signaling pathways is most frequently implicated pathways. They noted further that lithium that is often useful in treatments of bipolar disorder is known to impact calcium signaling.

Some investigators have drawn attention to altered circadian rhythm and possible implications of the CLOCK gene in bipolar disorder.

6. Calcium ion channels and neuropsychiatric disorders

Several timely reviews have shed light on calcium ion channels and their roles in neuronal functions and in specific neuropsychiatric disorders. In a 2018 review, Nanou and Catterall (2018) reviewed roles of calcium channels in synaptic function. They reported that the Cav2 channels play important roles at presynapses where calcium entry into these channels promotes neurotransmitter release. They noted that calcium channels not only bind calcium but also bind other proteins. The Cav2.1 and Cav2.2 proteins were reported to bind SNARE proteins that facilitate exocytosis and neurotransmitter release.

Presynaptic calcium channels were shown to be regulated by G protein subunits and by phosphorylation through activity of protein kinase. Proteins reported to be important in appropriately anchoring Cav2 channels include RIM proteins. The RIMS1 and RIMS2 proteins are RAS superfamily members that regulate synaptic vesicle exocytosis. Nanou and Caterall noted that a dense assembly of proteins was required to regulate neurotransmitter release.

Lory et al. (2020) reported that the Cav2 channels are referred to as high voltage-activated channels. Genes and proteins in this category are listed in the following:

- CACNA1A encoded on 19p13.3 Cav2.1, calcium voltage-gated channel subunit alpha1 A
  - CACNA1B encoded on 9q34.3 Cav2.2, calcium voltage-gated channel subunit alpha1 B
    - CACNA1E encoded on 1q25.3 Cav2.3, calcium voltage-gated channel subunit alpha1 E

Defects in the Cav2 channels were reported to play roles in ataxias including spinocerebellar ataxias and episodic ataxia and have also been reported to be associated with developmental epileptic encephalopathy associated with motor and cognitive defects.

Lory et al. (2020) reported the following information on Cav3 channels:

- CACNA1G encoded on 17q21.33, Cav3.1, calcium voltage-gated channel subunit alpha1 G
  - CACNA1H encoded on16p13.3 Cav3.2, calcium voltage-gated channel subunit alpha1 H
    - CACNA1I encoded on 22q13.1 Cav3.3, calcium voltage-gated channel subunit alpha1 I

Lory et al. noted that intracellular calcium concentrations are directly influenced by influx through the Cav3 channels, and another important observation was that the Cav3 channels do not widely interact with other proteins.

The Cav 3.1 channels were reported to be abundantly expressed in the cerebellum, and defects in function of CACNA1G encoded protein were reported in hereditary cerebellar ataxia.

CACNA1G-encoded proteins are also abundantly expressed in the thalamus and cortex and modulate firing of T-type channels that are considered to be involved in idiopathic generalized epilepsy. Some forms of movement disorder defined as essential tremor were reported to be due to defects in CACNA1G.

Defects in the CACNA1H gene that encodes Cav3.2 were reported to play role in some forms of epilepsy including absence epilepsy.

The CACNA1I gene that encodes Cav3.3 was reported to harbor variants that constituted risk factors for schizophrenia and autism.

The Cav1 family of calcium channels protein complexes occurs on postsynaptic membranes. Cav1.2 and Cav1.3 proteins play important roles in the influx of calcium into the postsynapse. Cav1.2 and Cav1.2 are referred to as L-type calcium channels. Important components of the Cav1 channels include the following.

- CACNA1S encoded on 1q32.1 Cav1.1, calcium voltage-gated channel subunit alpha1 S
  - CACNA1C encoded on 12p13.3 Cav1.2, calcium voltage-gated channel subunit alpha1 C
    - CACNA1D encoded on 3p21.1 Cav1.3, calcium voltage-gated channel subunit alpha1 D
    - CACNA1F encoded on Xp11.23 Cav1.4, calcium voltage-gated channel subunit alpha1 F

Kabir et al. (2017) reviewed L-type calcium channels. They noted that CACNA1C is widely reported as a candidate gene for neuropsychiatric disorders including bipolar disorders, schizophrenia, autism, and major depressive disorder.

The CACNA1D gene has been considered to be a risk gene for autism disorder and bipolar disorder. The specific subunits encoded by these genes Cav1.2 and Cav1.3 were reported to be primarily expressed in brain.

Brain connectivity in psychiatric diseases

Griffa et al. (2019) carried out studies on brain connectivity as measured by fractional anisotropy diffusion methods, relevant to psychotic episode. They noted that in schizophrenia, abnormalities in white matter and structural brain connectivity are well documented.

Their study was designed to determine if brain connectivity worsened with increases in number of psychotic episodes. Specific data analyzed included fractional anisotropy and apparent diffusion coefficients.

Data generated in their study provided evidence of association of degree of alteration of connectivity, clinical stages, and numbers of episodes of psychosis. Griffa et al. reported that although changes were spatially diffuse, there was evidence for convergence on a subnetwork that was reported to span interhemispheric frontal, corticothalamic, and striatal circuits.

Default mode network and psychiatric disorders

Doucet et al. (2020) noted that alterations in the default mode network have been reported in mood and anxiety disorders and in schizophrenia. Brain regions that are more active in the resting sate are represented in the default-mode network. This network includes the medial prefrontal cortex, the cingulate cortex, regions of the medial and lateral temporal cortex anterior and posterior, the precuneus and posterior cingulate cortex, and angular gyrus.

Doucet et al. noted that the default-mode network is considered to be involved in self-referential mental activity that includes autobiographical thought and stimulus-independent thought.

From their review of the literature including approximately 3000 patients and approximately 3000 controls, Doucet et al. reported evidence of dysconnectivity of the default-mode network in several psychiatric disorders. Shared changes involved anterior and posterior hubs of the default-mode network. They also documented evidence of disease-specific changes in specific disorders. Importantly that determined that unmedicated patients showed more evidence of default mode functional alterations.

In 2019, van den Heuvel and Sporns (2019) reviewed the connectome and brain dysconnectivity. They noted that alterations in the default-mode network were reported in autism, schizophrenia, epilepsy, amyotrophic lateral sclerosis, and Alzheimer disease. They concluded the connectome alteration in brain disorders "shared a common framework."

References

Anttila, V., Bulik-Sullivan, B., Finucane, H.K., Walters, R.K., Bras, J., Duncan, L., Escott-Price, V., Brainstorm Consortium, 2018. Analysis of shared heritability in common disorders of the brain. Science 360 (6395), eaap8757. https://doi.org/10.1126/science.aap8757.

Bloomfield, P.S., et al., 2016. Microglial activity in people at ultra high risk of psychosis and in schizophrenia. Am. J. Psychiatry 173 (1), 44–52. https://doi.org/10.1176/appi.ajp.2015.14101358.

Bundo, M., 2014. Increased l1 retrotransposition in the neuronal genome in schizophrenia. Neuron 81. https://doi.org/10.1016/j.neuron.2013.10.053.

Cannon, T.D., Chung, Y., He, G., Sun, D., Jacobson, A., van Erp, T.G., McEwen, S., et al., 2015. Progressive reduction in cortical thickness as psychosis develops: a multisite longitudinal neuroimaging study of youth at elevated clinical risk. Biol. Psychiatry 77 (2), 147–157. https://doi.org/10.1016/j.biopsych.2014.05.023.

Carvalho, A.F., Firth, J., Vieta, E., 2020. Bipolar disorder. N. Engl. J. Med. 383 (1), 58–66. https://doi.org/10.1056/NEJMra1906193.

Craddock, N., McGuffin, P., Owen, M., 1994. Darier's disease cosegregating with affective disorder. Br. J. Psychiatry: J. Ment. Sci. 165 (2), 272. https://doi.org/10.1192/bjp.165.2.272a.

Dhindsa, R.S., Goldstein, D.B., 2016. Schizophrenia: from genetics to physiology at last. Nature 530 (7589), 162–163. https://doi.org/10.1038/nature16874.

Bipolar Disorder and Schizophrenia Working Group, & Psychiatric Genomics Consortium, 2018. Genomic dissection of bipolar disorder and schizophrenia, including 28 subphenotypes. Cell 173 (7), 1705–1715. https://doi.org/10.1016/j.cell.2018.05.046 e16.

Doucet, G.E., Janiri, D., Howard, R., O'Brien, M., Andrews-Hanna, J.R., Frangou, S., 2020. Transdiagnostic and disease-specific abnormalities in the default-mode network hubs in psychiatric disorders: a meta-analysis of resting-state functional imaging studies. Eur. Psychiatry 63 (1), e57. https://doi.org/10.1192/j.eurpsy.2020.57. The Journal of the Association of European Psychiatrists.

Ellison-Wright, B., Gandal, M., Zhang, H., Walker, L., Chen, et al., 2009. Transcriptome-wide isoform-level dysregulation in ASD, schizophrenia, and bipolar disorder. Schizophr. Res. 108 (1–3) https://doi.org/10.1016/j.schres.2008.11.021. PMID.

Fromer, M., et al., 2014. De novo mutations in schizophrenia implicate synaptic networks. Nature 506 (7487), 179–184. https://doi.org/10.1038/nature12929.

Gandal, J.M., et al., 2018. Transcriptome-wide isoform-level dysregulation in ASD, schizophrenia, and bipolar disorder. Science 362 (6420), eaat8127. https://doi.org/10.1126/science.aat8127.

Garey, L., 2010. When cortical development goes wrong: schizophrenia as a neurodevelopmental disease of microcircuits. J. Anat. 217 (4), 324–333. https://doi.org/10.1111/j.1469-7580.2010.01231, 20408906.

Gordovez, F.J.A., McMahon, F.J., 2020. The genetics of bipolar disorder. Mol. Psychiatry 25 (3), 544–559. https://doi.org/10.1038/s41380-019-0634-7.

Griffa, A., Baumann, P.S., Klauser, P., Mullier, E., Cleusix, M., Jenni, R., van den Heuvel, M.P., Do, K.Q., Conus, P., Hagmann, P., 2019. Brain connectivity alterations in early psychosis: from clinical to neuroimaging staging. Transl. Psychiatry 9 (1). https://doi.org/10.1038/s41398-019-0392-y.

Kabir, Z.D., Martínez-Rivera, A., Rajadhyaksha, A.M., 2017. From gene to behavior: L-type calcium channel mechanisms underlying neuropsychiatric symptoms. Neurotherapeutics 14 (3), 588–613. https://doi.org/10.1007/s13311-017-0532-0.

Kato, T., 2019. Current understanding of bipolar disorder: Toward integration of biological basis and treatment strategies. Psychiatry Clin. Neurosci. 73 (9), 526–540. https://doi.org/10.1111/pcn.12852, 31021488.

Kelly, S., et al., 2018. Widespread white matter microstructural differences in schizophrenia across 4322 individuals: results from the ENIGMA Schizophrenia DTI Working Group. Mol. Psychiatry 23 (5), 1261–1269. https://doi.org/10.1038/mp.2017.170.

Khandaker, G.M., Cousins, L., Deakin, J., Lennox, B.R., Yolken, R., Jones, P.B., 2015. Inflammation and immunity in schizophrenia: implications for pathophysiology and treatment. Lancet Psychiatry 2 (3), 258–270. https://doi.org/10.1016/S2215-0366(14)00122-9.

Konrad, A., Winterer, G., 2008. Disturbed structural connectivity in schizophrenia – primary factor in pathology or epiphenomenon? Schizophr. Bull. 34 (1), 72–92. https://doi.org/10.1093/schbul/sbm034.

Liu, Y., Fu, X., Tang, Z., Li, C., Xu, Y., Zhang, F., Zhou, D., Zhu, C., 2019. Altered expression of the CSMD1 gene in the peripheral blood of schizophrenia patients. BMC Psychiatry 19 (1). https://doi.org/10.1186/s12888-019-2089-4.

Lory, P., Nicole, S., Monteil, A., 2020. Neuronal Cav3 channelopathies: recent progress and perspectives. Pflueg. Arch. Eur. J. Physiol. 472 (7), 831–844. https://doi.org/10.1007/s00424-020-02429-7.

Mäki-Marttunen, T., Kaufmann, T., Elvsåshagen, T., Devor, A., Djurovic, S., et al., 2019. Biophysical Psychiatry-How Computational Neuroscience Can Help to Understand the Complex Mechanisms of Mental Disorders Front Psychiatry, vol. 10. https://doi.org/10.3389/fpsyt.2019.00534 eCollection 2019. (PMID).

Mancuso, M., Ricci, G., Choub, A., Filosto, M., DiMauro, S., Davidzon, G., Tessa, A., Santorelli, F.M., Murri, L., Siciliano, G., 2008. Autosomal dominant psychiatric disorders and mitochondrial DNA multiple deletions: report of a family. J. Affect. Disord. 106 (1–2), 173–177. https://doi.org/10.1016/j.jad.2007.05.016.

Marder, S.R., Cannon, T.D., 2019. Schizophrenia. N. Engl. J. Med. 381 (18), 1753–1761. https://doi.org/10.1056/NEJMra1808803.

Nanko, S., Yokoyama, H., Hoshino, Y., Kumashiro, H., Mikuni, M., 1992. Organic mood syndrome in two siblings with Wolfram syndrome. Br. J. Psychiatry 161, 282. https://doi.org/10.1192/bjp.161.2.282.

Nanou, E., Catterall, W.A., 2018. Calcium channels, synaptic plasticity, and neuropsychiatric disease. Neuron 98 (3), 466–481. https://doi.org/10.1016/j.neuron.2018.03.017.

Pape, K., Tamouza, R., Leboyer, M., Zipp, F., 2019. Immunoneuropsychiatry – novel perspectives on brain disorders. Nat. Rev. Neurol. 15 (6) https://doi.org/10.1038/s41582-019-0174-4 (PMID).

Pardinas, A.F., et al., 2018. Common schizophrenia alleles are enriched in mutation-intolerant genes and in regions under strong background selection. Nat. Genet. 50 (3), 381–389. https://doi.org/10.1038/s41588-018-0059-2.

Pasternak, O., 2016. The extent of diffusion MRI markers of neuroinflammation and white matter deterioration in chronic schizophrenia. Schizophr. Res. 173 (3), 200–212. https://doi.org/10.1016/j.schres.2015.05.034.

Plana-Ripoll, O., Pedersen, H., Benros, M., Dalsgaard, J., et al., 2018. Exploring comorbidity within mental disorders among a Danish national population. JAMA Psychiatry 76 (3), 79–92. https://doi.org/10.1001/jamapsychiatry.2018.3658. PMID: 30649197.

Pollak, T.A., Drndarski, S., Stone, J.M., David, A.S., McGuire, P., Abbott, N.J., et al., 2018. The blood-brain barrier in psychosis. Lancet Psychiatry 5 (1), 79–92. https://doi.org/10.1016/S2215-0366(17)30293-6, 28781208.

Pouget, J., et al., 2019. Cross-disorder analysis of schizophrenia and 19 immune-mediated diseases identifies shared genetic risk. Hum. Mol. Genet. 28 (20), 3498–3513. https://doi.org/10.1093/hmg/ddz145.

Satterstrom, F.K., et al., 2020. Large-scale exome sequencing study implicates both developmental and functional changes in the neurobiology of autism. Cell. https://doi.org/10.1016/j.cell.2019.12.036.

Schork, A.J., et al., 2019. A genome-wide association study of shared risk across psychiatric disorders implicates gene regulation during fetal neurodevelopment. Nat Neurosci. 22 (3), 353–361. https://doi.org/10.1038/s41593-018-0320-0.

Sekar, A., et al., 2016. Schizophrenia risk from complex variation of complement component 4. Nature 530. https://doi.org/10.1038/nature16549.

Stahl, E.A., Bipolar Disorder Working Group of the Psychiatric Genomics Consortium, et al., 2019. Genome-wide association study identifies 30 loci associated with bipolar disorder. Nat. Genet. 51 (5), 793–803. https://doi.org/10.1038/s41588-019-0397-8 (PMID).

Sullivan, P.F., Geschwind, D.H., 2019. Defining the genetic, genomic, cellular, and diagnostic architectures of psychiatric disorders. Cell 177 (1), 162–183. https://doi.org/10.1016/j.cell.2019.01.015.

Van den Heuvel, M., Sporns, O., 2019. A cross-disorder connectome landscape of brain dysconnectivity. Nat. Rev. Neurosci. 20 (7), 435–446. https://doi.org/10.1038/s41583-019-0177-6.

Vidal, P., Pacheco, R., 2020. The Cross Talk between the Dopaminergic and the Immune System Involved in Schizophrenia Front Pharmacol, vol. 11. https://doi.org/10.3389/fphar.2020.00394 eCollection 2020. (PMID).

Weinstein, J.J., Chohan, M.O., Slifstein, M., Kegeles, L.S., Moore, H., Abi-Dargham, A., 2017. Pathway-specific dopamine abnormalities in schizophrenia. Biol. Psychiatry 81 (1), 31–42. https://doi.org/10.1016/j.biopsych.2016.03.2104.

Working Group of the Psychiatric Genomics Consortium, 2018. Genomic Dissection of Bipolar Disorder and Schizophrenia, Including 28 Subphenotypes. Cell 173 (7), 1705–1715.e16. https://doi.org/10.1016/j.cell.2018.05.046.

Woo, J.J., Pouget, J.G., Zai, C.C., Kennedy, J.L., et al., 2020. The complement system in schizophrenia: where are we now and what's next? The complement system in schizophrenia: where are we now and what's next? Mol. Psychiatry 25 (1), 114–130. https://doi.org/10.1038/s41380-019-0479-0, 31439935.

Wu, Y., Cao, H., Baranova, A., Huang, H., Li, S., Cai, L., et al., 2020. Multi-trait analysis for genome-wide association study of five psychiatric disorders. Transl. Psychiatry 10 (1), 209. https://doi.org/10.1038/s41398-020-00902-6, 32606422.

CHAPTER

12

Neurodevelopmental disorders, diagnosis, mechanism discovery, and paths to clinical management

1. Patient evaluation

Important information to gather includes family history, information on pregnancy, birth and perinatal history, history of early development, developmental milestone achievements, and history of illnesses.

Detailed physical examination and phenotype evaluation is important. Unusual phenotypic findings can be researched using standardized terms in phenotype database resources, e.g., Human Phenotype Ontology (HPO) https://bioportal.bioontology.org/ontologies/HP and https://hpo.jax.org/app/.

Reports of prior cases with similar phenotypic abnormalities, similar histories, and neurological or radiological findings may be useful in determining possible genetic associations.

Laboratory evaluations

First-pass metabolic testing frequently includes analyses of plasma and urine amino acids, urine organic acids, plasma urea and creatine analysis, plasma ammonia analysis, plasma carnitine and acyl carnitine analysis, and in some cases plasma lactate and pyruvate analyses. If lysosomal storage diseases are suspected, specific enzyme tests are ordered depending on the disease suspected.

First-pass genomic studies often include microarray analyses. However, approaches to first-pass genomic studies are changing.

Chromosomal abnormalities and neurodevelopmental defects

Microarrays contain short segments of DNA from defined chromosome regions mapped to specific genomic locations (oligonucleotide arrays) and/or single nucleotides known to map at specific genomic locations and known to show defined variations (SNP arrays), immobilized to a solid matrix. This matrix can be reacted with fluorescence-tagged segments of DNA derived from patient DNA. In recent decades, microarray analyses had largely replaced cytogenetics, chromosome, and karyotype analyses. It is important to note that microarray studies are particularly valuable in detecting chromosome number abnormalities and also in detecting deletions or duplications of chromosome segments.

Waggoner et al. (2018) reviewed genomic abnormalities that may not be detected on

Mechanisms and Genetics of Neurodevelopmental Cognitive Disorders
https://doi.org/10.1016/B978-0-12-821913-3.00003-2

microarray analyses. These include balanced chromosome translocations, balanced insertions of chromosome segments into aberrant locations, inversions of chromosome segments, or mosaicism where different population cells with different chromosome changes occur. Uniparental disomy cannot be detected on oligonucleotide arrays but can be detected on SNP arrays. In uniparental disomy, a particular chromosomal region is not biparental in origin but is derived from a single parent.

Complex chromosome rearrangements and structural chromosome abnormalities including ring chromosomes that do not change dosage of a chromosome region are usually not detected on microarray analyses. SNP microarrays are useful in demonstrating regions of homozygosity, i.e., chromosomal regions that are identical on both chromosomes indicating potential genome relationship of two parents.

Copy number variants of short chromosome segments are, however, more common, and determining the pathologic significance of these variants requires consultation with database information (Zarrei et al., 2019).

In 2019, (De Coster and Van Broeckhoven, 2019) reviewed new methods for detecting structural variants in chromosomes. They noted that several new DNA sequencing technologies had been developed that enable detections of structural genomic changes. These techniques include long-range sequencing and short-range sequencing. They noted that the majority of structural chromosome variants are novel and rare variants. However, there is some evidence for occurrence of population specific structural variants.

Structural chromosome variants can arise during meiosis due to nonhomologous recombination between chromosomes. Insertion of mobile DNA elements, e.g., Alu or Line elements can also lead to structural chromosomal variants.

Questions arise as to how chromosome variants, including chromosome variants, lead to pathology. Gene dosage changes induced by deletions or duplications may lead to pathology. Structural chromosome changes that disrupt continuity of a gene or changes that remove a gene from its cis-regulatory elements may lead to pathology.

When chromosome abnormalities are identified in a specific case, it is important to explore the literature and databases to determine if other cases have been reported with the same genomic changes and to compare phenotypes of cases with the same genomic findings. It is also important to carry out genome analyses in parents of a child who is found to have a chromosome and genomic abnormality. This is important to determine pathogenicity of the variant, possible mechanism of origin of the abnormality, and likely relevance for future reproduction.

Whole-genome sequencing in neurodevelopmental disorders

Lindstrand et al. (2019) reviewed the advantages of whole-genome sequencing as a first-line test in individuals with intellectual disability. In their study, whole-genome sequencing was carried out in three different cohorts with intellectual disability.

The first cohort included cases that had previously been identified with chromosome changes, including trisomies, deletions, or duplications.

The second cohort included individuals previously studied with multigene panels to determine disease causation.

The third cohort included individuals referred to their center for diagnosis.

In all three cohorts, the results of whole-genome sequencing were evaluated with at least five different bioinformatic tools.

Short-read whole-genome sequencing was carried out. Sequence data were analyzed to search for large and small genomic segmental copy number variants and for sequence changes that included single nucleotide variants, insertions or deletions of nucleotides, and expansions of nucleotide repeats.

Additional testing applied included analysis of gene panels. Specific genes to be analyzed took into account details of the patients' phenotypic information and entry of this into the HPO database that links phenotype abnormalities with potentially impacted genes.

Single nucleotide variants and nucleotide repeat expansions found on whole-genome sequencing that were considered to be potentially relevant to disease were reanalyzed in Sanger sequencing.

Results of whole-genome sequencing data in cohort one detected deletion characteristics of specific contiguous gene syndromes. Trisomies and mosaic trisomies were also detected. Results previously identified by microarray analyses and by comprehensive cytogenetic studies were confirmed. In addition, three complex chromosome rearrangements were detected through whole-genome sequencing. These included reciprocal translocations, and the breakpoint junctions of the translocations were defined.

In cohort 2, abnormalities previously detected in gene panel studies were confirmed, and in addition, diagnostic yield was increased by 8%.

In cohort 3 that comprised newly referred individuals, complex chromosome abnormalities were detected in three cases; in one of these cases, a ring chromosome was identified. Nucleotide analyses revealed that in seven patients, compound heterozygous or homozygous single nucleotide variants occurred indicating the presence of autosomal recessive disorders.

Lindstrand et al. demonstrated that in terms of genome analyses, diagnostic rates of detection using whole-genome sequencing more than doubled the diagnostic rates achieved on microarray testing.

In addition, whole-genome sequencing and analyses enabled detection of single nucleotide variants, detection of uniparental disomy, and detection of short tandem repeat sequence elements.

2. Clinical value of a genetic diagnosis

In a 2019 review, (Horton and Lucassen, 2019) noted that genomic testing brings greater opportunities for diagnosis. They noted that in the Deciphering Developmental Disease Project that had recruited patients with undiagnosed developmental disorders, exome sequencing was carried out and a diagnosis was achieved in 40% of patients.

Horton and Lucassen emphasized that through correlating genetic changes and phenotype information and through individual professionals sharing information, the understanding of impact of information increases. Details of the spectrum of abnormalities due to disruption of function of a specific gene grow, and furthermore insights on interactions between genes also expand.

Horton and Lucassen emphasized that genetic information may not always seem clinically useful but that in many cases it has the potential to make a difference in patient care and has implications for families.

Insights into the impact of gene changes on specific functions can frequently impact treatment through implementation of measures to bypass a functional block. This is particularly the cases in specific metabolic conditions.

Efforts to correct defective genes are beginning to emerge. One of the most striking has been the use of antisense oligonucleotides in treatment of specific disorders, such as spinal muscular atrophy.

Insights into pathophysiology of rare diseases

Sanders et al. (2019) noted that de novo and inherited rare genetic disorders are frequently associated with neurodevelopmental or neuropsychiatric manifestation. Specific single gene rare disorders often affect multiple neurologic and neurodevelopmental domains, leading to developmental delay, speech difficulties, impaired muscle tome, spasticity, altered movements, and sometimes seizures.

Sanders et al. emphasized the importance of progressing from definition of genetic etiology to understanding of disease mechanisms. They noted that a genetic defect can lead to specific biochemical changes, to alterations in specific cell types, to alterations in specific neuronal circuits, or to alterations in specific brain regions.

They noted the limited understanding we still have of effects of mutations on specific cell types. However, this information was noted to be expanding. Thus, defects in SCN1A that lead to epilepsy and Dravet syndrome were shown to impact fast-spiking neurons. Mutations in FOXG1 were shown to impact cortical glutamatergic neuron development. A number of gene defects impact gene expression through impairments of chromatin modifications. Also, genomic alterations may impact dosage of more than one gene, thus having broader effects.

Sanders et al. noted that studies on pluripotent stem cells that can potentially be differentiated into specific cell types allow investigation of effects of specific mutations. They noted further that studies on model organisms also continue to provide information on the direct impact of specific gene mutations.

A goal of genomic medicine is to apply genomic techniques to discover underlying disease-causing mechanisms and to use this information to guide patient management.

In 2019, (Wise et al., 2019) reviewed resources and techniques that can be used in the application of genomics to clinical care. An important aspect in this endeavor is the decreased cost of DNA sequencing. Another important factor is the development of phenotype databases and integration of phenotype information and sequence findings. Neurodevelopmental disorders frequently occur in the presence of abnormalities in other systems, and full clinical evaluation of the patient is important.

In addition to features of diagnosed diseases, neurodevelopmental impairments may occur as one of the manifestations in diseases defined as undiagnosed diseases. Wise et al. defined undiagnosed disease as follows: "a patient is considered to have an undiagnosed disease if the individual has received an appropriate extensive evaluation based on presenting symptoms and signs and yet remains without an etiologic diagnosis."

In the context of this book, we are considering undiagnosed diseases in individuals with developmental delay, cognitive impairment, epilepsy, or motor disorders as part of a complex phenotype that can include defects in other body systems. Many families may engage in long and costly "diagnostic odysseys" as they explore the cause of disease in a specific child.

The report of Wise et al. emphasized the utility of comprehensive DNA sequencing in parallel and utilization of standardized phenotype ontology information to search phenotype databases including HPO, Phenomizer, and Phenolyzer.

3. Diseases encompassed in the neurodevelopmental disorder category

In some recent reports, the disorders included in the neurodevelopmental disorder category seem extensive. Morris Rosendahl and Crocq (2020) included intellectual disability, autism spectrum disorder, and schizophrenia in the neurodevelopmental category. In a report, (Cardoso et al., 2019) also included bipolar disorder in the developmental disease category.

Patients and families

In a review of genetic studies in neurodevelopmental disorders, Vissers et al. (2016) emphasized the importance of determining whether pathogenic mutations were novel in the patients or whether they were inherited. If pathogenic mutations are found to be inherited, that finding is particularly relevant for genetic counseling. They noted that in families with one child with neurodevelopmental disorders, mutations were most often found to occur de novo. However, in consanguineous families, parents were more often found to be carriers of deleterious mutation, and the affected offspring were homozygous for the pathogenic mutation, or in some cases, the affected children were compound heterozygotes for deleterious mutations in a specific gene.

Vissers et al. (2016) noted that mutation detection could provide information on the particular biological pathway that was perturbed and led to intellectual disability. They emphasized the important role of signaling pathways, e.g., the RAS and MAPK signaling pathway in brain development. They also noted that disruption of chromatin remodeling and transcription were also frequently found in cases of intellectual disability. Vissers et al. concluded that there is some hope that following accurate diagnosis, therapeutic measures can be discovered that improve functions and capabilities.

Genomic approaches in undiagnosed developmental disorders

Wright et al. (2014) described implementation of exome sequencing and microarray analyses to achieve genetic diagnoses in a large cohort of patients with developmental disorders of unknown causation. Through efforts of their Deciphering Developmental Disorders study (DDS) on 1133 children in the United Kingdom, they reported achieving a diagnostic yield in 27% of cases.

The goal of the DDS was to facilitate translations of genomic technologies to the clinical services provided through the UK National Health Service. The specific workflow in this study initiated with recruitment of individuals with neurodevelopmental disorders and/or abnormal growth patterns, congenital anomalies, unusual behavior phenotypes, of undiagnosed causation. Sequencing analyses were carried out on DNA from saliva samples from children and parents. In addition, detailed clinical data, family history, and results of previous testing were gathered.

Potentially pathogenic variants were identified in DNA sequence using several different algorithms including SIFT, Polyphen and information in the Human Gene mutation Base (HMGD), and Leiden open variant database. Potentially pathogenic variants diagnosed on sequencing together with phenotype information were also analyzed in the Developmental Disorders Genotype to Phenotype Database DDG2P. The DDG2P database was reported to include more than 1000 genes that had previously been reported in specific neurodevelopmental diseases. They emphasized the utility of the DDG2P database in facilitating prioritization of variants.

(Wright, 2014). reported identification of likely diagnostic or contributory variants in 311 of 1133 children (27%). Importantly 82 of the reported genes were found to be associated with inborn errors of metabolism potentially amendable to therapy. These genes and conditions caused by variants are listed in the following.

IVD (isovaleryl-CoA dehydrogenase) (ACAD2): This is a mitochondrial matrix enzyme that catalyzes the third step in leucine catabolism. The genetic deficiency of IVD results in an accumulation of isovaleric acid, which is toxic to the central nervous system and leads to isovaleric acidemia.

DHCR7 (7-dehydrocholesterol reductase): mutations in this gene cause Smith–Lemli–Opitz syndrome (SLOS); characterized by reduced serum cholesterol levels and elevated serum 7-dehydrocholesterol levels and phenotypically characterized by cognitive disability, facial dysmorphism, syndactyly of second and third toes, and holoprosencephaly in severe cases to minimal physical abnormalities and near-normal intelligence in mild cases.

LMBRD1 (LMBR1 domain–containing 1): This gene encodes a lysosomal membrane protein that may be involved in the transport and metabolism of cobalamin. Mutations in this gene are associated with the vitamin B12 metabolism disorder homocystinuria–megaloblastic anemia complementation type F.

MTR (5-methyltetrahydrofolate-homocysteine methyltransferase): This enzyme, also known as cobalamin-dependent methionine synthase, catalyzes the final step in methionine biosynthesis. Mutations in MTR have been identified as the underlying cause of methylcobalamin deficiency complementation group G.

SLC2A1 (solute carrier family 2 member 1): This gene encodes a major glucose transporter in the mammalian blood–brain barrier. Glut 1 deficiency syndrome has been reported in cases of tonic–clonic seizures, myoclonic seizures, intellectual impairment, spastic diplegia, or ataxia.

Inborn errors of metabolism and neurodevelopmental disorders

Data from numerous reports indicate that conditions due to pathogenic genetic variants that lead to metabolic changes potentially have the greatest possibilities for disease-altering therapies.

Tarailo-Graovac et al. (2016) noted that in a number of metabolic disorders, the underlying disease mechanism had not been defined. However, use of exome sequencing was enabling identification of relevant pathogenic mutations that led to neurometabolic syndromes. They reported results of whole-exome sequencing on 41 patients with intellectual disability on whom deep phenotypic had been carried out and nondiagnostic metabolic phenotypes were encountered.

Results of exome sequencing led to identification of genetic diagnoses in 68% of patients. Importantly, gene discovery impacted clinical management in 44% of patients. In five patients, the altered management included supplementation with specific available biologic substances. Tarailo-Graovac et al. emphasized that the high diagnostic yield in their study was likely due to initial evidence of metabolic abnormalities and intense bioinformatic analyses of sequencing data.

With respect to family follow-up, it is important to note that 16 of the affected cases in the study of Tarailo-Graovac et al. were compound heterozygotes for autosomal recessive mutation. X-linked mutations were identified in five cases. Autosomal dominant de novo heterozygous mutations were found in nine different genes.

Graham et al. (2018) drew attention to the importance of early treatments in inborn errors of metabolism that can potentially lead to progressive central nervous system damage. They stressed the importance of integrating genomics and metabolomics for early accurate diagnoses. They noted the importance of improved gene and metabolite annotations.

Warmerdam et al. (2020) reviewed a category of diseases that cause progressive intellectual and neurological deterioration (PIND). They noted the importance of timely diagnoses in these disorders and the initiation of appropriate treatment to prevent irreversible damage.

Included in this category of diseases are inborn errors of metabolism that lead either to the accumulation of toxic metabolites, to energy deficiency, or to shortage of essential proteins that form structures.

Warmerdam et al. noted that progressive intellectual and neurological deterioration disorders (PIND) do include diseases other than those that fit into the inborn errors of metabolism category. The literature survey they conducted yielded information on 79 different inborn errors of metabolism that were in the PIND category.

Storage diseases due to lysosomal defects formed the largest category of disorders that could lead to progressive deterioration. Diagnoses included ceroid lipofuscinosis, Niemann–Pick disease, metachromatic leukodystrophy,

GM2 gangliosidosis, and Krabbe disease. Other frequent disorders encountered included X-linked adrenoleukodystrophy, Menkes disease, Leigh syndrome (subacute necrotizing encephalomyelopathy) that can be due to nuclear or mitochondrial genome defects, and mitochondrial disorders in including MELAS (mitochondrial encephalopathy, lactic acidosis, and stroke-like episodes).

Clinical manifestations in PIND disorders include global developmental delay, hypotonia, movement disorders, and intellectual disabilities. Warmerdam et al. emphasized the importance of two-tier testing that included biochemical and genetic testing.

They also emphasized the need for therapy, noting that treatment was available for 35 of the 85 (41%) of the conditions reported in the PIND categories. In some cases, the therapies included specific dietary restrictions, substrate reduction, and dietary supplements. In other cases, more costly and difficult interventions were necessary including hematopoietic stem cell therapy and enzyme or protein replacement therapies. Warmerdam et al. presented categories of pediatric disorders where there is evidence that therapy can prevent ongoing neurodegeneration.

Urea cycle disorders and ammonia removal

Disorders where protein restriction is important and where sodium benzoate therapy is often included include the following urea cycle disorders in which the removal of ammonia is impaired:

- Arginase deficiency due to ARG1 defects and carbamoyl phosphate synthase 1 defects in CPS1 encoding gene.
  - Hyperammonemia, hyperornithinemia, ornithine transcarbamylase deficiency OTC (X-linked)
  - Hyperornithinemia–hyperammo nemia–homocitrullinemia (HHH) syndrome
    - SLCA15 (ornithine transporter) deficiency

Disorders where specific dietary supplements are important in therapy

- Coenzyme Q deficiency can arise due to defects in COQ2, COQ4, COQ6, COQ8A, or COQB encoding genes involved in coenzyme Q synthesis; treatment includes administration of coenzyme Q.
  - Cerebral folate receptor deficiency (FOLR1 gene defects) can be treated by providing folinic acid.
    - Methylene tetrahydrofolate reductase deficiency may give rise to homocystinuria and can be treated with methionine, folate, and carnitine.
    - Methylmalonic aciduria can arise due to defects in methyl malonyl-CoA mutase (MUT), can give rise to acidosis, some cases respond to treatment with vitamin b12 hydroxy cobalamin.
    - Methylglutaconic aciduria, some cases respond to carnitine administration.
    - In mitochondrial disorders, arginine supplementation may be helpful in some forms.
    - Biotinidase deficiency results in inefficient recycling of biotin in the body and can lead to ataxia and neurodevelopmental impairments (Fig. 12.1). It can be successfully treated with biotin supplementation (Porta et al., 2017).

Treatments for storage diseases

- Hematopoietic stem cell transplants can be used for treatment, and in some cases, enzyme replacement therapies are available. These disorders include mucopolysaccharidosis, mucolipidoses, mannosidosis, aspartyglucosaminura.
  - Niemann–Pick disease can be treated with Miglustat.
    - Warmerdam et al. (2020) noted that enzyme replacement was available for treatment of neuronal ceroid lipofucinosis.

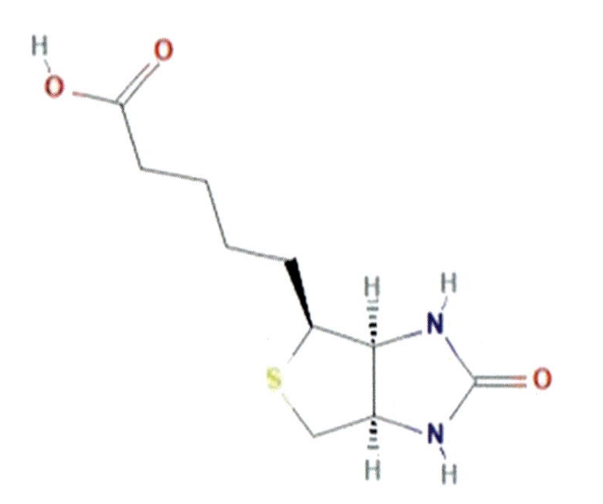

FIGURE 12.1 Biotin structure. Important in treatment of biotinidase deficiency. *From https://pubchem.ncbi.nlm.nih.gov/compound/biotin#section=2D-Structure.*

Translational research for rare diseases

(Hmeljak and Justice, 2019) reviewed the use of model organisms to explore functional pathways disrupted through causative gene defects in rare diseases. Specific model organisms utilized in translational research studies included yeast, zebra fish, Drosophila, and mice. Cultured cells including stem cells and organoids are also sometimes used. They defined organoids as multidimensional culture systems and noted that use of organoids solves problems that occur in monolayer cell cultures.

Pluripotent stem cells and methods to derive specific differentiated cell types from pluripotent stem cells are also increasingly being analyzed to explore cell type–specific and tissue type–specific functional changes that arise as a result of gene defects.

Discovery of the impact of gene mutations including regulatory gene mutation is facilitated through transcriptome analysis and RNA sequencing.

(Beneto, 2020) reported development of neuronal and astrocyte models of San Filippo syndrome, a lysosomal storage disorder with early onset neurodegenerative manifestations. They utilized transcription factor–based protocols to convert induced pluripotent stem cells developed from patients to astrocytes or neuronal cells. They determined that these cells exhibited the molecular phenotypes of San Filippo syndrome. Beneto et al. noted limitations of nonneural cells in analyzing disease mechanism and in investigating therapies for disorders that primarily impact the brain.

Disorders with impaired mitochondrial functions: therapies and potential therapies

In 2017, (El-Hattab et al., 2017) published a review on therapies and potential therapies of mitochondrial diseases, taking into account in-depth information on biochemistry and physiology of mitochondrial function.

In considering consequences of impaired mitochondrial function, they noted that inability to produce adequate energy particularly impacted organs with high energy demands including brain, both skeletal and cardiac muscle, liver, kidney, and endocrine systems. El Hattab et al. (2017) noted that other consequences of mitochondrial dysfunction include excess generation of reactive oxygen species and altered calcium homeostasis.

In addition to symptomatic treatment of the patient based on the specific systems impacted, they considered aspects of therapies based on the pathophysiology of specific mitochondrial disorders. General approaches to treatment of mitochondrial disorders included recommendations for exercise. Exercise was shown to increase levels of transcription factors that regulated mitochondrial biogenesis.

Another general approach noted was the importance of avoiding malnutrition but also avoiding high carbohydrate intake that was considered to be detrimental in cases of impaired oxidative phosphorylation.

Approaches to the treatment of electron transfer chain dysfunction

El Hattab et al. noted the important role of CoQ10 (ubiquinone) in ETC functions. They also noted that CoQ10 biosynthesis disorders lead to impaired mitochondrial function. Supplementation with CoQ10 (ubiquinone) or with reduced CoQ10 (ubiquinol)l or with the CoQ10 analog Idebenone has been reported to be useful in some cases with mitochondrial disorders.

The electron transport complexes I and II and ETFDH electron transfer flavoprotein dehydrogenase and other enzymes that impact mitochondrial function, e.g., acyl-coA dehydrogenases such as ACAD9, require the cofactor riboflavin (Fig. 12.2). El Hattab et al. noted reports that riboflavin supplementation proved useful in therapy of impaired mitochondrial complex 1 deficiency.

Increased levels of lactic acid constitute one of the prominent manifestations of impaired mitochondrial function. Specific methods are being implemented to increase pyruvate catabolism to acetyl-CoA and prevent pyruvate accumulation and generation of lactic acid. One such method involves administration dichloroacetate. Although beneficial in some mitochondrial disorders, there is evidence that it should not be used in MELAS (mitochondrial) syndrome patients. MELAS is an acronym for mitochondrial encephalopathy, lactic acidosis, stroke-like episodes. This syndrome is due to specific mitochondrial DNA mutations.

Thiamine (vitamin B1) is a cofactor for pyruvate dehydrogenase and was reported to also increase generation of the cofactor NADH and FADH. El Hattab et al. noted reports of clinical improvements in patients with Leigh syndrome when treated with thiamine, CoQ10, carnitine, and vitamins C and E.

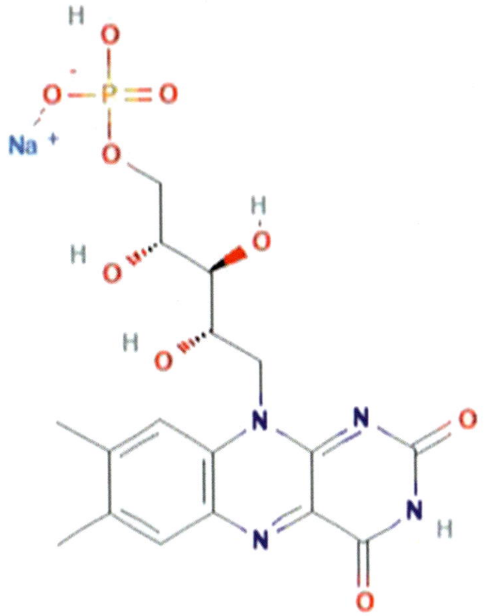

FIGURE 12.2 Riboflavin structure. Note importance of riboflavin in treatment of certain mitochondrial disorders. *From https://pubchem.ncbi.nlm.nih.gov/compound/Riboflavin-5_-phosphate-sodium.*

Leigh syndrome

This syndrome is characterized by encephalopathy with evidence of brain tissue loss. Leigh syndrome can arise as a result of nuclear or mitochondrial DNA mutations and is characterized by abnormally high levels of lactic acid and alanine. Mitochondrial complex 1 function is often aberrant in Leigh syndrome. It can also arise as a result of defective function of mitochondrial complexes II, III, IV, and V.

Creatine that generates phosphocreatine was reported to be at lower levels in brain and muscle in mitochondrial disorders. El Hattab et al. noted it has been prescribed for patients with mitochondrial disorders.

Treatment with antioxidants is carried out in patients with mitochondrial disorders based on the evidence for increased generation of reactive oxygen species in these disorders. Reactive oxygen species lead to damage in proteins, lipids, and DNA. El-Hattab et al. reported that specific antioxidants used in therapy include lipoic acid and cysteine donors. Glutamyl cysteine present in a whey supplement was reported to reduce oxidative stress in patients with mitochondrial disorders.

Hydrogen sulfide generated by gut microorganisms can normally be inactivated by a sulfur dioxygenase enzyme ETHE1. This enzyme is deficient in patients with ethylmalonic acidemia, a condition that can lead to brain atrophy. El Hattab et al. noted evidence that supplementation with cysteine or *N*-acetylcysteine is beneficial in patients with ETHE1 defects. Cystamine is converted to cystine in lysosomes.

Increased cystamine intake can potentially increase cellular glutathione levels, and this had been proposed as a useful therapy in mitochondrial disorders.

EPI743 is a pharmacologic substance that can potentially increase intracellular glutathione and protect against excess reactive oxygen species. EPI743 is reported to be a parazenzoquinine analog. El Hattab et al. noted that several clinical trials were in place to test efficacy of this compound in mitochondrial disorders.

There is evidence that nitric oxide deficiency occurs as a result of impaired mitochondrial function and may play a role in the increased incidence of stroke-like episodes in certain mitochondrial disorders, e.g., MELAS syndrome. Arginine supplementation and citrulline supplementation are being investigated in treatment, given that these substances can increase nitric oxide NOS production.

Cardiolipin is a phospholipid reported to be expressed in the mitochondrial inner membrane and to be important in anchoring the electron transfer complexes. El-Hattab et al. noted that oxidative damage and oxidized cardiolipin not only impairs inner membrane structure but also increases its permeability leading to loss of calcium and to release of cytochrome C.

They noted that a specific compound elamipretide had been developed that binds to and protects cardiolipin. In their 2017 report, El Hattab et al. noted that phase 3 clinical trials of this compound were in place in patients.

Enhanced mitochondrial biogenesis is reported to occur following treatment with agonist of PPAR (peroxisome proliferator–activated receptor). PPARs enhance expression of the PGC 1 alpha receptors that increase expression of mitochondrial genes. The PPAR agonist bezafibrate is under investigation for treatment of mitochondrial disorders.

El Hattab et al. reviewed mitochondrial DNA maintenance and its defects. They noted that mitochondrial DNA synthesis requires adequate nucleotides and that these are supplied through nucleotide recycling in the mitochondria and through import of nucleotides from the cytosol.

Inadequate nucleotide supply in mitochondria and impaired mitochondrial DNA synthesis leads to disorders characterized by multiple mitochondrial DNA deletions and leads to mitochondrial depletion syndromes.

They reported that pathogenic variants in 20 different nuclear genes could lead to inadequate mitochondrial DNA synthesis. These genes include POLG (POLG1) DNA polymerase gamma, catalytic subunit, POLG2 DNA polymerase gamma 2 accessory subunit, and TWINK mitochondrial DNA helicase.

Specific genes influence maintenance of nucleotide balance within mitochondria. Other genes that influence mitochondrial maintenance are genes involved in mitochondrial fission and fusion.

Emerging therapies in mitochondrial diseases

In a 2018 review, (Hirano et al., 2018) noted that recent successes in defining molecular causes and pathomolecular mechanisms and increasing awareness of clinical phenotypes were opening the way to new treatment modalities.

They noted that earlier and ongoing approaches to treatment of mitochondrial disorders involved the use of dietary supplements including vitamins defined as antioxidants and factors that act as cofactors for mitochondrial function. These included CoQ10 or idebenone, a synthetic analog of CoQ10, riboflavin, vitamin C, vitamin E, and carnitine.

Hirano et al. noted that the mitochondrial Medicine Society made specific recommendations that included offering CoQ10 (or related substances Ubiquinol), alpha lipoic acid, riboflavin, and in specific cases carnitine or folinic acid.

Hirano et al. noted that specific studies provided evidence of the value of exercise in patients with mitochondrial disorders. Exercise without resistance training was recommended.

Hirano et al. reviewed specific clinical trials in patients with mitochondrial disorders. He noted that a number of clinical trials had investigated utility of antioxidants. The latter includes CoQ20 and its analogs idebenone and Epi743. Other antioxidants investigated included vitamins C and E and the vitamin E analog trolox. They noted that definitive conclusions regarding efficacy of these compounds had not been reached. However, they noted that CoQ10, lipoic acid and folinic acid, and riboflavin are offered to patients with mitochondrial disorders. They noted that L-carnitine should be given when indicated when deficient and folinic acid should be given when deficient.

They noted that in Leber's hereditary optic atrophy (LHON) due to specific mutations in mtND1, mtND4, or mtND6 clinical trials of idebenone indicated possible beneficial effect.

Arginine, a precursor of nitric acid, was reported to be beneficial in patients at risk for stroke based on mitochondrial mutations, e.g., MTTL1 m3243 A > G that occurs in the disorder MELAS.

Hirano et al. reviewed emerging therapies in mitochondrial disorders and noted that these included 10 different strategies. These included the following:

1. Strategies to activate mitochondrial biogenesis
2. Discovery of factors that impact mitochondrial dynamics and mitophagy
3. Strategies to bypass oxidative phosphorylation defects
4. Mitochondrial replacement therapy as a preimplantation method
5. Application of chronic hypoxia
6. Use of compounds to scavenge toxic compounds
7. Nucleotide and nucleoside supplementation
8. Cell replacement therapy
9. Gene therapies
10. Impacting heteroplasmy shifting the frequency of mitochondrial with mutant DNA
11. Stabilizing mutant tRNAs.

Hirano et al. noted that bezafibrate was reported to activate PGC1alpha and to improve function in cases of mitochondrial respiratory chain defects. They noted that mitochondrial biogenesis is dependent upon PGC1alpha and that this acts as a coactivator of transcription factors involved in mitochondrial biogenesis. Other factors involved in mitochondrial biogenesis include NRF1 nuclear regulatory factor and PPAR alpha, beta, and gamma.

Activation of NRF1 was reported to lead to increased transcription of genes that encode oxidative phosphorylation–related products.

PGC1alpha activity was also reported to increase expression of SIRT and deacetylation activity. In addition, PGC1alpha increased activity of AMPK1, and to increase phosphorylation, PGC1alpha activity is known to be influenced by specific medications that include PPAR agonists.

Lysosomal storage diseases: pathophysiology and treatment

Parenti et al. (2015) reviewed pathophysiological defects and disease mechanisms in lysosomal storage diseases and directed therapies. More than 50 different disorders are included in the lysosomal storage degree category. They are characterized by accumulation of undegraded complex molecules leading to disease manifestations in a number of different tissues and organs, including liver, spleen, connective tissue bone, and eye. Parenti et al. emphasized that approximately two-thirds of patients with lysosomal storage diseases have progressive neurodegeneration leading to impaired cognition, epilepsy, and behavioral disorders.

They documented the pathogenic cascade from gene mutation that leads to the synthesis of mutant protein. The mutant protein may fail to undergo appropriate folding or may fail to undergo required posttranslational secondary modification and targeting for passage through endoplasmic reticulum and Golgi apparatus to lysosomes.

Failure of appropriate enzyme activity, e.g., hydrolase activity, leads to the accumulation of undegraded material in lysosomes.

Specific palliative therapies can be applied to compensate for impairments such as skeletal defects. However, during the past several decades, research has been carried out to design therapies based on underlying pathophysiology and in some cases to target molecular defects. This research has led to the initiation of clinical trials, and in some cases, these have led to approval for clinical use.

Parenti et al. discussed enzyme replacement therapies for specific lysosomal storage diseases. They noted that enzyme replacement therapies had been approved for Gaucher disease, mucopolysaccharidoses types I, II, and VI, Niemann–Pick disease, and Pompe disease.

Parenti et al. noted that in some patients, significant clinical benefits had been achieved with enzyme replacement therapy. Drawbacks include the fact that intravenously infused enzyme proteins do not readily cross the blood–brain barrier so that neurological problems are not successfully managed. Additional modifications of enzyme proteins are being carried out prior to their infusion in some of disorders to facilitate passage to the brain. Another important consideration includes the very high cost of enzyme replacement therapy.

In the case of disorders that are due to defects where the catalytic site of the enzyme is not impacted, therapies with small molecule ligands that act as chaperones for lysosomal enzymes are being investigated.

In some lysosomal disorders, substrate reduction therapies are being investigated. These therapies inhibit production of the substrate that a specific lysosomal enzyme must degrade.

A specific gene product TFEB, a transcription factor, was reported to enhance clearance of substrate accumulated in lysosomes through facilitation of exocytosis and autophagy and is being investigated as a possible therapeutic agent.

In Niemann–Pick disease, a cholesterol-sequestering agent cyclodextrin was reported to be useful in mobilizing stored cholesterol from lysosomes.

Parenti et al. noted that possibilities for gene therapy were being investigated. This requires delivery of a wild-type normal version gene to the patient through a viral vector.

Brain disease in mucopolysaccharide storage diseases and approaches to therapy

Scarpa et al. (2017) reviewed brain diseases in mucopolysaccharide storage diseases. Some of these disorders and the chromosomal location of the gene that encodes the relevant enzyme are described below.

MPS1 deficiency of alpha iduronidase normally breaks down glycosaminoglycans. Defects lead to Hurler syndrome and to Scheie syndrome less severe disorder IDUA, 4p16.3.

MPS2 deficiency of iduronidate sulfatase involved in glycosaminoglycan breakdown; leads to Hunter syndrome X-linked, IDS, Xq28.

MPS3 causes San Filippo syndrome particularly affecting the brain; there are types with somewhat different severity due to different defects in function of different enzymes involved in the breakdown of heparan sulfate a component of glycosaminoglycans. The relevant defective enzyme proteins in San Filippo syndrome include any one of the following:

GNS, glucosamine (N-acetyl)-6-sulfatase, 12q14.2; HGSNAT, heparan-alpha-glucosaminide N-acetyltransferase, 8p11.21-p11.1; NAGLU, *N*-acetyl-alpha-glucosaminidase, 17q21,2; SGSH, *N*-sulfoglucosamine sulfohydrolase, 17q25.3.

MPS4, also known as Morquio syndrome, mainly affects the skeleton and is due to defects in functions of enzymes involved in breakdown of glycosaminoglycans, GALNS galactosamine (*N*-acetyl)-6-sulfatase, and GLB1 galactosidase beta 1.

Scarpa et al. noted that MPS disorders are associated with accumulation of glycosaminoglycans (GAGs) in somatic tissue and brain. Although treatments such as hematopoietic stem cell transplants and enzyme replacement therapy have led to significant improvements and/or prevention of somatic cell defects in these disorders, central nervous system defects have responded poorly to these treatments due to the selective permeability of the blood–brain barrier.

Scarpa et al. noted that the blood–brain barrier is formed through cellular elements including endothelial cells and pericytes connected by tight junctions, also by basal lamina and perivascular glial processes. Small molecules can pass through the blood–brain barrier. There is evidence that certain lipid soluble molecules can also pass through. In addition, molecules that can bind to receptors or transporters on the blood–brain barrier may also cross the barrier.

Treatment approaches to circumvent the blood–brain barrier include administration of therapeutic entities to the fluids in the central nervous system, through intrathecal injection into the spinal fluid or injection into the cerebrospinal fluid in the brain ventricles.

Scarpa et al. reviewed aspects of hematopoietic stem cell transplants to treat disorders including storage diseases. In storage diseases, there is some evidence that monocytes and macrophages that are contained in bone marrow preparations can develop in such transplants and can penetrate the blood–brain barrier, but this penetration is very limited and has minimal impact on stored material in brain. Also, the processes required for successful transplant include significant immunosuppression in the host that can be quite hazardous.

New approaches include transferring genes that contain the normal version of the gene into host hematopoietic stem cells ex vivo and then transplanting the modified cells back into the host. Scarpa et al. noted that there are reports indicating modest successes of such treatment with some evidence that transplant at a young age, before the accumulation of significant quantities of stored material in the brain.

There is some evidence for benefits of enzyme replacement therapy for the brain on the administration of the therapeutic agent into the cerebrospinal fluid allowing direct access of the replaced enzyme to the brain tissue. This mode of administration is of course more complicated than intravenous infusion.

The Trojan Horse approach

This refers to fusing the enzyme molecules that need to be transfused, with antibody molecules that will bind to receptors on the blood–brain barrier cells and membranes to facilitate transcytosis of the enzyme into the brain. Scarpa et al. noted that receptors targeted in such approaches include the transferrin receptor, low-density lipoprotein receptor, and the insulin receptor.

They also reported investigation of coated nanoparticles, including nanoparticles of aminoglycopeptides to transport enzyme across the blood–brain barrier.

Also, under investigation, is chaperone use, where chaperones are small particles that facilitate appropriate folding of mutant enzymes proteins, thereby improving enzyme activity. Scarpa et al. noted that levels of enzyme, only 10% of that of normal levels, have been shown to prevent storage of glycosaminoglycans.

Substrate reduction therapy with the plant derived substance genistein is being investigated for treatments of MPS disorders at some sites.

Studies being carried out particularly in animal models include investigations of the value of gene therapy. A clinical trial of gene therapy using an adeno-associated viral vector and IV administration is recorded as being in progress in MPS VI, Maroteux Lamy syndrome arylsulfatase B deficiency. This condition primarily affects the skeleton. Neurological symptoms may result from thickening of the dura of the brain. Intelligence is usually normal in patients with this disorder.

4. Peroxisomal disorders

Wanders (2018) reviewed peroxisomal disorders and routes to treatment and noted that defects in at least 30 different genes could give rise to these disorders. Peroxisomes are involved in beta oxidation of fatty acids, in synthesis of ether phospholipids, in alpha oxidation of phytanic acid, in synthesis of bile acids, and in glyoxylate synthesis. Peroxisomal disorders can arise as a result of defects in peroxisome biosynthesis or as a result of defects in genes that encode specific proteins important in peroxisomal functions.

Defects in these genes that encode products involved in peroxisome biosynthesis lead to Zellweger syndrome. At least 13 different genes are involved in peroxisomal biogenesis. With respect to the nervous system manifestations, these can include seizures and hypotonia and intellectual disability in some cases. Zellweger syndrome is also associated with facial dysmorphology, hearing defects, and ocular abnormalities.

Wanders noted that degree of severity of manifestations is variable in Zellweger syndrome. Affected individuals may present in infancy, or later. Fibroblasts from affected patients manifest increased levels of very-long-chain fatty acids and decreased levels of short-chain fatty acids. Wanders noted that an increasing number or cases were being diagnosed on the basis of DNA sequencing methods.

Specific genes leading to altered metabolism in peroxisomes and neurological functional defects include defects in the ABCD1 (ATP-binding cassette subfamily D member 1). Defects in this gene lead to X-linked adrenoleukodystrophy. Males with this disorders may manifest impaired adrenal function. Significant cerebral pathology arises in males leading to leukodystrophy. Females who carry one copy of the damaged ABCD1 gene may develop peripheral neurological manifestation later in life.

D-bifunctional protein deficiency, also known as peroxisomal acyl CoA oxidase deficiency, is caused by a defect in the ACOX1 gene. Two different enzyme proteins are derived from this gene 2-enoyl-CoA hydratase and 3-hydroxyacyl-CoA dehydrogenase and three different disease subtypes are known, depending on which enzyme is primarily deficient or if both enzymes are deficient. The disorder is characterized by seizures and hypotonia. Brain abnormalities are present on MRI.

Nascent efforts in gene therapy in lysosomal and peroxisomal disorders

Ohashi (2019) reviewed gene therapy trials for lysosomal storage diseases and peroxisomal disorders and noted that advances in gene therapy had opened, due to developments in vector technology. He noted that current therapeutic strategies, including enzyme replacement therapy, were limited in cases with brain disease.

By 2019, three gene therapies had been approved for other genetic diseases, including therapies for RPE6 defects leading to Leber's congenital amaurosis, treatment for adenosine deaminase deficiency, and treatment for lipoprotein lipase deficiency.

Ohashi noted promising results in treatment of adrenoleukodystrophy due to ABCD1 functional defects. In the United States, a breakthrough therapy status was granted for therapy with ABCD1 inserted into a lentiviral vector.

Metachromatic leukodystrophy is a lysosomal storage disease due to defective function of the enzyme arylsulfatase A (ARSA). Deficiency of this enzyme leads to accumulation of sulfides and degeneration of oligodendrocytes responsible for the production of myelin. Encouraging results were reported in clinical trials of hematopoietic cells modified with lentiviral containing normal ARSA gene.

In the United States, clinical trials were noted to be in place for treatment of patients with mucopolysaccharidoses type IIIA due to defects in *N*-sulfoglucosamine sulfohydratase (SGSH) deficient in San Filippo syndrome type A and in patients with MPSIIIB due to defect in alpha-N acetylglucosaminidase San Filippo syndrome type B.

In Fabry disease, pathology primarily involves kidneys and heart; however, brain function may be impaired as a result of cerebrovascular events. Clinical trials of normal alpha galactosidase A gene cloned into hematopoietic stem cells were reported to be ongoing.

Along the path from mechanism discovery to therapeutic design

Tuberous sclerosis, due to defects in the functions of the product of the TSC1 gene, hamartin, or of theTSC2 gene product tuberin, can impact function of a number of different body systems (Fig. 12.3). However, epilepsy is one of the most common manifestations of the disorder.

Detailed analyses of the molecular of the TSC1 and TSC2 gene products and studies on the product of the more recently discovered gene TBC1D7 have revealed that they form a complex that inhibits the activity of RHEB a Ras GTPase. Active Rheb is required to activate the MTORC1 complex.

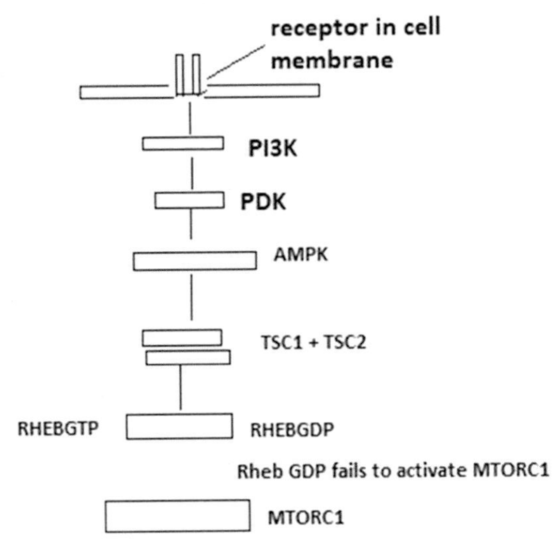

FIGURE 12.3 Activated TSC complex and RHEB GDP fail to activate MTORC1. TSC complex is important in control of MTORC1 activity.

Active MTORC1 promotes cell growth angiogenesis and translation of protein. Active mTORC1 increases the level of activity of S6 kinase. In the presence of pathogenic defects in TSC1 or TSC2, mTORC1 activity is increased leading to aberrant of abnormal foci in the brain that give rise to seizures (Laplante and Sabatini, 2012).

Detailed analyses of the mTOR complexes and of the mTOR signaling pathway have revealed that functional defects of several of the components in the mTOR pathway can lead to epilepsy (Fig. 12.4; Curatolo et al., 2018). These discoveries have stimulated design and investigation of pharmacological agents that impact mTOR activity. Several of these are related to rapamycin, a natural product that failed investigation as an antibiotic but was subsequently found to be an mTOR inhibitor.

Difficulties emerged in investigations of potential mTOR inhibitors as antiepilepsy medications. One significant difficulty was relatively poor passage of these compounds across the blood–brain barrier and side effects that emerged at doses required.

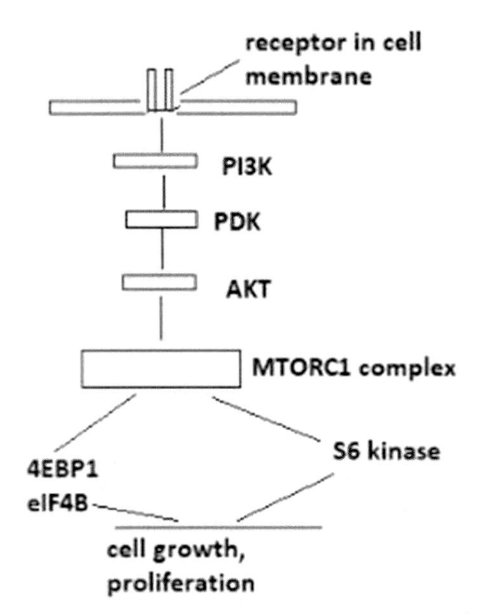

FIGURE 12.4 Activated mTORC1 complex. Stimulations of proteins synthesis and cell growth.

However, searches continue for novel mTOR inhibitors that will effectively cross the blood−brain barrier and have antiepilepsy effects and minimal side effects (Brandt et al., 2018).

In cell systems and in model organisms, the effectiveness of compounds in suppressing mTOR activity can be measured by determining if the compounds reduce expression of S6 kinase.

5. The mission of genomic medicine

Horowitz et al. (2019) emphasized that discovery, translation, and implementation of effective discoveries into clinical care are essential missions of genomic medicine. The mission of discovery research is to uncover influences of specifics factors in generating disease. Translational research seeks to evaluate the clinical impact of discoveries, while implementation research is required to investigate methods to integrate research into clinical care.

Comprehensive care

The focus of this chapter has been on design of care based on discoveries of underlying mechanisms leading to neurodevelopmental disorders. It is, however, important to emphasize comprehensive approaches to care of individuals with neurodevelopmental disorders. Their care can best be provided by a multidisciplinary team of healthcare providers and educators working with patients and their families.

In 2000, (Satcher, 2000) the US Surgeon General, proposed several measures to promote care of individuals with intellectual disabilities. In the United States, systems of Community Healthcare delivery of services to individuals with intellectual disabilities were established.

In 2014, (Ervin et al., 2014) emphasized the need for training of individuals to provide healthcare to individuals with intellectual and developmental disabilities.

Dr. Satcher's 2000 report was noted to end with a quote from John Gardner, Secretary of Health, Education, and Welfare in the 1960s, "Life is full of golden opportunities carefully disguised as irresolvable problems."

References

Beneto, N., et al., 2020. Neuronal and astrocytic differentiation from sanfilippo C syndrome iPSCs for disease modeling and drug development. J. Clin. Med. 9 (3), 644. https://doi.org/10.3390/jcm9030644.

Brandt, C., Hillmann, P., Noack, A., Römermann, K., Öhler, L.A., Rageot, D., Beaufils, F., Melone, A., Sele, A.M., Wymann, M.P., Fabbro, D., Löscher, W., 2018. The novel, catalytic mTORC1/2 inhibitor PQR620 and the PI3K/mTORC1/2 inhibitor PQR530 effectively cross the blood-brain barrier and increase seizure threshold in a mouse model of chronic epilepsy. Neuropharmacology 140, 107−120. https://doi.org/10.1016/j.neuropharm.2018.08.002.

Cardoso, A.R., Lopes-Marques, M., Silva, R., Serrano, C., Amorim, A., Prata, Azevedo, L., 2019. Essential genetic findings in neurodevelopmental disorders. Hum. Genom. 13 (1), 31. https://doi.org/10.1186/s40246-019-0216-4. PMID: 31288856.

Curatolo, P., Moavero, R., van Scheppingen, J., Aronica, E., 2018. mTOR dysregulation and tuberous sclerosis-related epilepsy. Expert Rev. Neurother. 18 (3), 185–201. https://doi.org/10.1080/14737175.2018.1428562.

De Coster, W., Van Broeckhoven, C., 2019. Newest methods for detecting structural variations. Trends Biotechnol. 37 (9), 973–982. https://doi.org/10.1016/j.tibtech.2019.02.003.

El-Hattab, A.W., Zarante, A.M., Almannai, M., Scaglia, F., 2017. Therapies for mitochondrial diseases and current clinical trials. Mol. Genet. Metabol. 122 (3), 1–9. https://doi.org/10.1016/j.ymgme.2017.09.009.

Ervin, D.A., Hennen, B., Merrick, J., Morad, M., 2014. Healthcare for persons with intellectual and developmental disability in the community. Front. Public Health 2. https://doi.org/10.3389/fpubh.2014.00083.

Graham, E., et al., 2018. Integration of genomics and metabolomics for prioritization of rare disease variants. J. Inherit. Metab. Dis. 41 (3), 435–445. https://doi.org/10.1007/s10545-018-0139-6.

Hirano, M., Emmanuele, V., Quinzii, C.M., 2018. Emerging therapies for mitochondrial diseases. Essays Biochem. 62 (3), 467–481. https://doi.org/10.1042/EBC20170114.

Hmeljak, J., Justice, M.J., 2019. From gene to treatment: supporting rare disease translational research through model systems. Dis. Model Mech. 12 (2), dmm039271 https://doi.org/10.1242/dmm.039271.

Horowitz, C., Orlando, L., Slavotinek, A., Peterson, A.,F., Biesecker, B., et al., 2019. The genomic medicine integrative research framework: a conceptual framework for conducting genomic medicine research. Am. J. Hum. Genet. 104 (6) https://doi.org/10.1016/j.ajhg.2019.04.006. Epub.

Horton, R.H., Lucassen, A.M., 2019. Recent developments in genetic/genomic medicine. Clin. Sci. 133 (5), 697–708. https://doi.org/10.1042/CS20180436.

Laplante, M., Sabatini, D.M., 2012. MTOR signaling in growth control and disease. Cell 149 (2), 274–293. https://doi.org/10.1016/j.cell.2012.03.017.

Lindstrand, A., et al., 2019. From cytogenetics to cytogenomics: whole-genome sequencing. Genome Med. https://doi.org/10.1186/s13073-019-0675-1.

Morris-Rosendahl, D.J., Crocq, M.A., 2020. Neurodevelopmental disorders-the history and future of a diagnostic concept. Dialogues Clin. Neurosci. 22 (1), 65–72. https://doi.org/10.31887/DCNS.2020.22.1/macrocq.

N.d. Retrieved October 25, 2020, from: https://pubchem.ncbi.nlm.nih.gov/compound/biotin#section=2D-Structure.

Ohashi, T., 2019. Gene therapy for lysosomal storage diseases and peroxisomal diseases. J. Hum. Genet. 64 (2), 139–143. https://doi.org/10.1038/s10038-018-0537-5.

Parenti, G., Andria, G., Ballabio, A., 2015. Lysosomal storage diseases: from pathophysiology to therapy. Annu. Rev. Med. 66, 471–486. https://doi.org/10.1146/annurev-med-122313-085916.

N.d. Retrieved October 26, 2020, from: https://pubchem.ncbi.nlm.nih.gov/compound/Riboflavin-5_-phosphate-sodium.

Sanders, S.J., Sahin, M., Hostyk, J., Thurm, A., Jacquemont, S., Avillach, P., Douard, E., Martin, C.L., Modi, M.E., Moreno-De-Luca, A., Raznahan, A., Anticevic, A., Dolmetsch, R., Feng, G., Geschwind, D.H., Glahn, D.C., Goldstein, D.B., Ledbetter, D.H., Mulle, J.G., et al., 2019. A framework for the investigation of rare genetic disorders in neuropsychiatry. Nat. Med. 25 (10), 1477–1487. https://doi.org/10.1038/s41591-019-0581-5.

Satcher, D., 2000. Mental health: a report of the surgeon general– executive summary. Prof. Psychol. Res. Pract. 31 (1), 5–13. https://doi.org/10.1037/0735-7028.31.1.5.

Scarpa, M., Orchard, P.J., Schulz, A., Dickson, P.I., Haskins, M.E., Escolar, M.L., Giugliani, R., 2017. Treatment of brain disease in the mucopolysaccharidoses. Mol. Genet. Metabol. 122, 25–34. https://doi.org/10.1016/j.ymgme.2017.10.007.

Tarailo-Graovac, M., Shyr, C., Ross, C.J., Horvath, G.A., Salvarinova, R., Ye, X.C., Zhang, L.H., Bhavsar, A.P., Lee, J.J.Y., Drögemöller, B.I., Abdelsayed, M., Alfadhel, M., Armstrong, L., Baumgartner, M.R., Burda, P., Connolly, M.B., Cameron, J., Demos, M., Dewan, T., et al., 2016. Exome sequencing and the management of neurometabolic disorders. N. Engl. J. Med. 374 (23), 2246–2255. https://doi.org/10.1056/NEJMoa1515792.

Vissers, L.E.L.M., Gilissen, C., Veltman, J.A., 2016. Genetic studies in intellectual disability and related disorders. Nat. Rev. Genet. 17 (1), 9–18. https://doi.org/10.1038/nrg3999.

Waggoner, D., Wain, K.E., Dubuc, A.M., Conlin, L., Hickey, S.E., Lamb, A.N., Martin, C.L., Morton, C.C., Rasmussen, K., Schuette, J.L., Schwartz, S., Miller, D.T., 2018. Yield of additional genetic testing after chromosomal microarray for diagnosis of neurodevelopmental disability and congenital anomalies: a clinical practice resource of the American College of Medical Genetics and Genomics (ACMG). Genet. Med. 20 (10), 1105–1113. https://doi.org/10.1038/s41436-018-0040-6.

Wanders, R.J.A., 2018. Peroxisomal disorders: improved laboratory diagnosis, new defects and the complicated route to treatment. Mol. Cell. Probes 40, 60–69. https://doi.org/10.1016/j.mcp.2018.02.001.

Warmerdam, H., Termeulen-Ferreira, e J., Tseng, E., Ferreira, C., Karnebeek, 2020. A scoping review of inborn errors of metabolism causing progressive intellectual and neurologic deterioration (PIND). Front. Neurol. 10 https://doi.org/10.3389/fneur.2019.01369. eCollection 2019. PMID: 32132962.

Wise, A.L., Manolio, T.A., Mensah, G.A., Peterson, J.F., Roden, D.M., Tamburro, C., Williams, M.S., Green, E.D., 2019. Genomic medicine for undiagnosed diseases. Lancet 394 (10197), 533–540. https://doi.org/10.1016/S0140-6736(19)31274-7.

Wright, C.F., et al., 2014. Genetic diagnosis of developmental disorders in the DDD study: a scalable analysis of genome-wide research data. Lancet. https://doi.org/10.1016/S0140-6736(14)61705-0.

Zarrei, M., Burton, C., Engchuan, W., Young, E., Higginbotham, E., MacDonald, J., et al., 2019. A large data resource of genomic copy number variation across neurodevelopmental disorders. NPJ Genom Med 4. https://doi.org/10.1038/s41525-019-0098-3 eCollection 2019. (PMID).

Index

Note: 'Page numbers followed by "f" indicate figures and "t" indicate tables.'

T

U

V

W

Z

Printed in the United States
by Baker & Taylor Publisher Services